HANDBOOK OF
NOISE
ASSESSMENT

HANDBOOK OF NOISE ASSESSMENT

EDITED BY
DARYL N. MAY, Ph.D.

Van Nostrand Reinhold Environmental Engineering Series

 VAN NOSTRAND REINHOLD COMPANY
NEW YORK   CINCINNATI   ATLANTA   DALLAS   SAN FRANCISCO
LONDON   TORONTO   MELBOURNE

Van Nostrand Reinhold Company Regional Offices:
New York   Cincinnati   Atlanta   Dallas   San Francisco

Van Nostrand Reinhold Company International Offices:
London   Toronto   Melbourne

Manufactured in the United States of America

Published by Van Nostrand Reinhold Company
135 West 50th Street, New York, N.Y. 10020
Published simultaneously in Canada by Van Nostrand Reinhold Ltd.

15 14 13 12 11 10 9 8 7 6 5 4 3 2 1

Library of Congress Cataloging in Publication Data

Main entry under title:

Handbook of noise assessment.

   (Van Nostrand Reinhold environmental engineering series)
   Includes bibliographical references and index.
   1. Noise pollution.   2. Noise—Measurement.
3. Noise—Psychological aspects.   4. Noise—
Physiological effect.   I. May, Daryl N.
TD892.H355   363.6   77-27657
ISBN 0-442-25197-1

Van Nostrand Reinhold Environmental Engineering Series

THE VAN NOSTRAND REINHOLD ENVIRONMENTAL ENGINEERING SERIES is dedicated to the presentation of current and vital information relative to the engineering aspects of controlling man's physical environment. Systems and subsystems available to exercise control of both the indoor and outdoor environment continue to become more sophisticated and to involve a number of engineering disciplines. The aim of the series is to provide books which, though often concerned with the life cycle—design, installation, and operation and maintenance—of a specific system or subsystem, are complementary when viewed in their relationship to the total environment.

Books in the Van Nostrand Reinhold Environmental Engineering Series include ones concerned with the engineering of mechanical systems designed (1) to control the environment within structures, including those in which manufacturing processes are carried out, (2) to control the exterior environment through control of waste products expelled by inhabitants of structures and from manufacturing processes. The series will include books on heating, air conditioning and ventilation, control of air and water pollution, control of the acoustic environment, sanitary engineering and waste disposal, illumination, and piping systems for transporting media of all kinds.

v

Van Nostrand Reinhold Environmental Engineering Series

Preface

How much noise is too much noise? Our knowledge of noise effects, being imperfect, makes an answer to this question more than the single decibel figure that most people would like to find. There are, of course, several criteria for assessing noise, as well as a plethora of limits, so that the answer to this question, far from being a single magic number, is a volume like this.

In this book, authors from the U.S., Britain, Canada and Australia have been invited to say how much noise is too much. The result, it is hoped, will be of assistance to acousticians, environmentalists, legislators, urban and transportation planners, equipment manufacturers and operators, physicians, audiologists, psychologists and sociologists, civil and criminal lawyers, and architects, as well as teachers, students, and researchers in these fields.

Besides the Introduction which follows, this book has two Parts and five Appendices.

Part I, "Psychological Effects Assessment," covers the assessment of noise for the annoyance it causes. Sounds that cause us physical damage are few in comparison with those that do no more than annoy us. The great majority of people suffering from traffic and aircraft noise, or construction noise, or their neighbor's air conditioner, do so without threat to hearing or of other damage to their health. The chapters of Part I deal in turn with various situations that annoy because annoyance is so governed by our attitudes that it differs from situation to situation. Only Chapter 1 in Part II is common to all noise annoyance situations, and is recommended as a unifying, tutorial chapter on our basic subjective responses to noise.

Part II, "Physical Effects Assessment," covers the assessment of noise for its physical effects. The effects in question are hearing damage, sleep interference, non-auditory system effects on health, and work disturbance. These effects differ from annoyance in two ways: they are more objectively assessed, and they are essentially independent of situation. Part II is therefore shorter than Part I, yet brevity is not the same as conclusiveness. Our knowledge of these effects is limited, and only Chapter 11 on hearing damage is able to be really explicit about the sound exposure that is excessive.

Appendix 1 contains some basic acoustics as a primer for those approaching this book without any. Appendices 2 and 3 provide reference information about acoustical standards and the agencies that regulate noise.

I am indebted to the authors not only for their hard work, but for their faith in entrusting their efforts to me. Less obvious but equal gratitude goes to the Institute of Sound and Vibration Research, University of Southampton, and the Ontario Ministry of Transportation and Communications. My associations with these organizations have contributed much to this book, though the responsibility for it is mine.

D. N. MAY

Introduction

Noise assessment procedure, like many other assessment procedures, consists of comparing an actual or proposed noise exposure with a noise exposure that is on the borderline between acceptable and unacceptable. Yet the task is never as simple as comparing, say, a car's speed with the posted speed limit. In noise assessment, neither the "speed" nor the "speed limit" is easily defined.

In the first place, noise is not always easily measured. Even when adequate instrumentation is available, a measurement procedure must be defined to take into account the place or places that the microphone should be sited in relation to the noise source, and the manner in which the source must be operated to display a representative noise output. Extraneous factors that might affect the results must be within defined bounds—weather that affects the source or its measurement, noise from sources other than those under study, sound reflecting or absorbing surfaces, or solid obstructions.

When a satisfactorily controlled sound signal is received, it still has to be processed to describe the magnitude of the sound. Here there is considerable disagreement. In our analogy with speeding cars we have, therefore, a considerable task establishing the "speed" in the first place.

The second step is even more difficult. There are few noise limits as well established as speed limits, so the comparison of a noise level with what is acceptable is a matter of considerable judgment. Often no limits at all are specified—and when they are, they are invariably disputed and under threat of alteration. The principle involved is therefore to compare the noise level with *all* relevant criteria. If there is a limit, does the noise exceed it? Does the noise have a harmful effect, interfering for example with work or sleep, or causing hearing damage? Is it accepted as too annoying?

Proper noise assessment is, therefore, not a task for those who deal in "instant answers." Instant answers in noise assessment probably neglect a large part of the truth.

How then does the truth emerge? Here are some suggestions on how to use this book to get at it.

1. Obtain at least a basic knowledge of noise generation, measurement, propagation, and control. Appendix 1 provides a crash course on these subjects for those approaching this book without prior acoustics training.

2. Ensure that rigorous standards are applied to the description of the noise under evaluation. If the source is already operating (as distinct from being planned or under contruction), it will need accurate measurement at the points where it is received and perhaps also at points defined in relevant measurement procedures. Appendix 2 contains a list of these procedures, and some are described in the text. If the source is only planned or under construction, its assessment depends on accurate prediction of its noise output. This is achieved by measuring similar operations elsewhere or by appropriately adding the estimated contributions of the individual sources that make up the whole operation. In many instances manufacturers are able to supply this information.

3. Ask the following questions about the noise.

a. Does it infringe defined statutory requirements, such as sound levels spelt out in regulations?

Such requirements may apply both to the time of sale of a product and to its operation, and different levels may be in force at different times into the future. Many of the more important statutory limits are described in the text, but readers are advised to check from time to time that this information is up to date; a list of rule-making agencies is given in Appendix 3.

If there are no limits defined by statute, the statutes may nevertheless contain catchall paragraphs about excessive noise, for which precedent has established certain limits. Such precedents are invariably those based on the considerations in b and c below.

b. Does the noise have harmful physical effects?

These effects include hearing damage, sleep interference and other health impairment, and work disturbance. The chapters in Part II deal with these effects in turn, and the criteria and limits that exist or may be recommended are included there. Of these effects hearing damage can be assessed reasonably meaningfully. For the others the conclusion that is frequently most applicable is that no levels can be authoritatively identified as the borderline of acceptability. Readers are cautioned not to lose patience with such a conclusion; it may be extremely valuable in refuting another party's charges that noise of certain levels is "well known" to have damaging consequences.

c. Is the noise excessive because of the annoyance it causes to individuals, or the individual or collective complaints it provokes?

Being an "attitudinal" matter, this is best assessed in the many individual situations that apply. The chapter in Part I most relevant to the circumstances under evaluation should therefore be referred to. As well, Chapter 1 of Part I is a general one on the basics of our subjective responses to noise and may be valuable as an introduction to the rest of Part I and, indeed, to Part II. Included in Part I are speech interference (Chapter 1) and noise in hospitals (Chapter 7) though it is arguable whether the material involved should not fall in Part II, where work disturbance (Chapter 14) and health effects (Chapter 12) are covered.

Assessing whether a noise will annoy is sometimes also a matter of the most basic common sense. Two obvious precepts are these. If a sound much exceeds the sound level that would occur without its presence, it will often annoy. And if a sound source is noisier than other sources that do the same job, it may also annoy; thus as discussed in Chapter 10, one way of assessing the noise of home appliances is to consider whether a particular appliance is significantly noisier than the average.

A final perspective on noise assessment is that a good part of it is founded on quicksand. Much of the job is no more than trying to predict what the community will find acceptable. The acoustician may try to determine this scientifically, and the legislator politically, but in either case public concern for noise is— like other environmental concerns—a matter of local opinion at a given time. The use of noisy snowmobiles and firearms in northern Canada may cause deafness, yet few of the many people who depend on them for their livelihoods find the situation unacceptable. In contrast, a quiet but new rail route through an upper-class rural community may be "too noisy" because an active consumer campaigner has successfully mobilized community reaction. Such local exceptions are not the rule, but they occur often enough to impress on us the fallibility of any oversimple approach to noise assessment.

Contents

HANDBOOK
OF
NOISE
ASSESSMENT

PART 1

Psychological effects assessment

I

Basic subjective responses to noise

D. N. MAY*

1-1. INTRODUCTION

This chapter is an introduction to the many specific noise situations independently considered in Chapters 2–10. We cover here the basic subjective response to noise, and in so doing introduce many of the noise descriptors that are referred to throughout the volume. The chapter is therefore a tutorial foundation for much of what is covered elsewhere, which is the reason we take a simple and generalized approach to a subject of considerable complexity. The purpose is to give newcomers to the subject an opportunity to get up to speed with the rest of the book; for this reason too, we have listed at the end a short selected bibliography rather than a great many state-of-the-art references.

We will cover here the loudness, noisiness, annoyance and speech interference induced by sound. These are all "re-

*Research and Development Division, Ministry of Transportation and Communications, Downsview, Ontario M3M 1J8, Canada.

sponses" to sound in the same way that laughter is a response to a joke and pain is a response to a burn. They differ from the other responses to sound described in Part II of this book, in that we deal here with subjective responses, i.e., responses which are primarily measured by direct human opinions rather than objective physical evidence.

When we come to discuss human responses, we are faced with the problem that it is difficult to ask people their opinions every time we have to judge the acceptability of a sound. Since people are different, we can never rely on a few opinions of a noise in order to judge its acceptability. People may, for example, have unduly acute hearing or simply be biased. And while we could conceivably remove the effect of these things by conducting a proper social survey, this would hardly be a practical way to evaluate the acceptability of each passing truck, or a neighbor's party, or a fleeting exhaust backfire.

The most practicable way to evaluate sound is to measure it physically, e.g., with sound level meters, and judge it with this objective evidence, because physical measures are more straightforwardly and less contentiously performable than subjective ones.

This still leaves the question unanswered, is each physically measured sound exposure acceptable or not? To try to answer this question, psychoacoustical experiments and social surveys have been performed. The first take place gen-

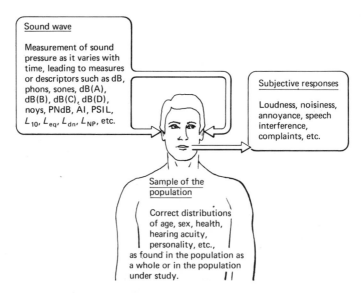

Sound wave

Measurement of sound pressure as it varies with time, leading to measures or descriptors such as dB, phons, sones, dB(A), dB(B), dB(C), dB(D), noys, PNdB, AI, PSIL, L_{10}, L_{eq}, L_{dn}, L_{NP}, etc.

Subjective responses

Loudness, noisiness, annoyance, speech interference, complaints, etc.

Sample of the population

Correct distributions of age, sex, health, hearing acuity, personality, etc., as found in the population as a whole or in the population under study.

Fig. 1-1   The structure of a psychophysical experiment or a social survey is to measure what people feel (their response) in relation to what they experience (in this case a sound wave). The aim of the experiment is to find objective measures of the sound wave that correlate best with the response.

erally in the laboratory, and the second in real life. The purpose of both of them has been to find out what characteristics of sounds promote the various subjective responses, and how much of these characteristics leads to "how much" response. In effect they seek relationships between physical stimuli and our subjective responses to them. They do this with carefully chosen groups of people so that the relationships they report are representative of the way the public at large or a particular portion of the public reacts. The fundamentals of a psychoacoustical experiment or a social survey are summarized in Fig. 1-1. (The meanings of the terms used there will become apparent as we progress further.)

1-2. THE RESPONSES DEFINED

1-2.1 Loudness

Loudness is defined as the subjective magnitude of a sound. The "unwantedness" of the sound (as it is sometimes known) should not be considered.

1-2.2 Noisiness

Noisiness is defined as the degree of unwantedness of a sound considered by itself. The important point is that we are talking of the sound in isolation. A laboratory is usually the only suitable place for trying to consider the noise by itself, so that noisiness, like loudness, is something that is best measured by psychoacoustical experiments only. It is sometimes called perceived noisiness.

1-2.3 Annoyance

Annoyance is defined as the overall unwantedness of a sound heard in a natural situation. A person's judgment of a sound's annoyance will include his assessment of the unwantedness of the sound itself (its noisiness), plus many other variables which depend on the source of the sound and the context in which it is experienced, and which can make a sound of a given noisiness induce different levels of annoyance. For example, a sound heard at night may be more annoying than one heard by day, just as one that fluctuates may be more annoying than one that does not (effects of time); a sound which resembles another sound we already dislike and which perhaps threatens danger may be especially annoying (emotional content); a sound which we know is mindlessly inflicted and will not soon be removed may be more annoying than one which is temporarily and regretfully inflicted (misfeasance); a sound the source of which is visible may be more annoying than one with an invisible source (accompanying visual stimulation); a sound which is new may be less annoying (novelty); a sound which is locally a political issue may have a particularly high or low annoyance (sociological influences); and so on.

What we are saying when we evaluate annoyance is that we are evaluating not only the sound itself but all the accompanying attitudinal variables that govern its unwantedness. Because these variables are not very well understood or measurable, we have often, unfortunately, to evaluate noises in the various situations in which they are found and apply a different, often quite pragmatic, assessment procedure to each situation, reflecting (but not necessarily dealing directly with) our attitudes in those situations. This is done in Chapters 2–10. The only annoyance measures introduced in this chapter are the few that apply to noise in all situations.

Thus this chapter serves only as an introduction to annoyance, and mainly covers loudness and noisiness. A study of loudness is important because in many instances measures of loudness are also measures of noisiness. Close attention to this chapter is essential, because it introduces measures like the dB(A) that are referred to throughout the book.

1-2.4   Speech Interference

This chapter also covers the question of speech interference. Whether or not a sound will interfere with one's speech perception is obviously a facet of the sound's acceptability, and we will treat it in a general way here, again leaving its specific applications to other chapters.

1-3.   LOUDNESS

1-3.1   Pure-Tone Frequency Response of the Ear

Let us suppose we listen to a pure tone (a sound of a single frequency, e.g., a whine or squeal) of 1000 Hz and 40 dB sound pressure level (SPL), and that we carefully remember its loudness. If, now, we listen to another tone, this time of 100 Hz, i.e., lower in pitch, and adjust the level of the second sound until it is as loud as the first, we will find that the second sound has a higher SPL than the first—in fact it is about 50 dB.

This illustrates immediately that the loudness of a sound depends not only on its sound pressure level but also on its frequency: the ear has a *frequency response*.

Figure 1-2 arises from experiments of this type performed on a group of individuals with normal hearing. Here each curve is an *equal loudness contour*, which is to say that each point on the same curve represents a tone of a frequency and SPL which has been judged to have the same loudness as each other point on the curve. Our example above, of a 40 dB/1000 Hz tone having the same loudness as a 50 dB/100 Hz tone, should help to explain Fig. 1-2.

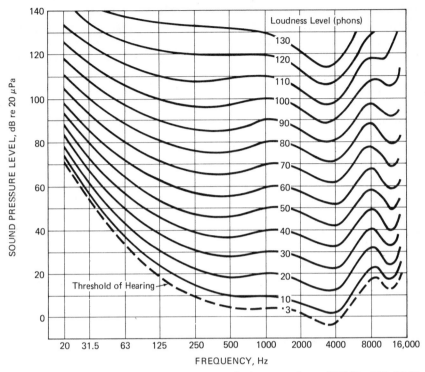

Fig. 1-2 Pure-tone frequency response of the ear, from ISO/R 226-1961. These curves are known as the equal-loudness contours, or phon curves.

1-3.2 Phons and Sones

The numbers written on the curves of Fig. 1-2 are seen to correspond with the SPL of each curve at the frequency of 1000 Hz. This is because the loudness of a tone of 1000 Hz is said to have a value in *phons* equal to its SPL: that is simply how phons are defined. Thus, to use our example again, the loudness of a tone of 40 dB at 1000 Hz is said to be 40 phons and, as each point on the equal loudness contour passing through the 40 dB/1000 Hz point has the same loudness, the loudness of a tone of 50 dB at 100 Hz is also 40 phons.

The bottom curve in Fig. 1-2 represents a loudness level of a just audible sound. It is helpful to remember that, approximately speaking, a sound of 1000 Hz is just audible at approximately 0 dB SPL and has the loudness of about 0 phons.

Now let us suppose that we listen to a sound whose loudness is, say, 50 phons, and that we then listen separately to a sound which we judge to be twice as loud as the first. How many phons is the sound?

Nothing mentioned above will in fact answer that question. However, it has been found from psychophysical experiments that that sound which is twice as loud will have a loudness which is approximately, but conveniently, 10 phons higher. In our example, that would be 60 phons. Likewise, a third sound twice as loud as the second would be 70 phons. This sound would also be judged to be four times as loud as the first.

So a useful fact in human response is that loudness doubles with every 10 phon increment in the stimulus. For pure tones at 1000 Hz, the equal loudness contours then demonstrate that loudness accordingly doubles with every 10 dB increase, a fact which is reasonably true across the frequency band generally, except at low frequencies, where the phon curves are closer and a lesser decibel increase corresponds to each 10 phon (loudness doubling) increment.

Loudness measured in phons makes it slightly tedious to deduce "how many times louder" a sound of, say 73 phons is than another of, say, 57 phons, because the 16 phon difference is not immediately meaningful. Accordingly, another loudness scale has been devised, known as the *sone scale*. The sone is a ratio scale: the number of sones doubles as loudness doubles (e.g., a sound of 3.2 sones is twice as loud as one of 1.6 sones). To match the phon and sone scales, 1 sone is arbitrarily defined as equal to 40 phons.

Thus if 1 sone = 40 phons, then 2 sones = 50 phons, 3 sones = 60 phons and so on. Converting between the two, when necessary, can then be effected with a graph (Fig. 1-3). Or it may be done using mathematical relationships, which are simply an expression of what we have just said:

$$\log_{10} S = 0.030(P - 40),$$

$$P = 40 + 33 \log_{10} S.$$

Here P and S are loudness in phons and sones, respectively.

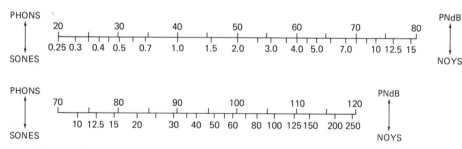

Fig. 1-3 These lines allow conversion between phons and sones (both loudness measures), and between PNdB and noys (both noisiness measures). It is not possible to convert between a loudness measure and a noisiness measure, except approximately. Examples: 40 phons = 1 sone; 40 PNdB = 1 noy; 40 PNdB ≠ 40 phons.

Loudness as expressed in phons, is sometimes called "loudness level;" expressed in sones it is simply called "loudness". We will not adhere to these terms here.

1-3.3 Masking

So far we have talked only about pure tones, though pure tones are about as common in real life as tuning forks are in orchestras. Most sounds are mixtures of many different frequencies, and we must know how to find their loudnesses from measurements of the sounds themselves. To know how to approach that, we must know a little about masking.

It is well known that some sounds seem louder at night. One can hear the crickets chirping, or the air-conditioner, or the fluorescent light ballast, or rustling leaves—all of which may have been present but unheard during the day. The reason for their new-found prominence is simply the fact that the sounds that had masked them are temporarily absent.

Masking of one sound by another can only occur when the sound that does the masking has a good amount of its energy around the frequencies of the sound to be masked. Thus a pure tone of 1000 Hz (possibly a cricket chirping) can be masked by a sound which is, say, in the range 900 to 1100 Hz, but would not be masked by another sound whose frequency content was in the 1900-2100 Hz range. Likewise, a sound which has its energy spread roughly evenly across the frequency band from, say, 500 to 4000 Hz could only be masked by another sound possessing at least as wide a frequency content.

If, now, we were to be presented with two sounds, one with a frequency content substantially at 990 Hz and the other at 1000 Hz, it is evident we could not calculate the overall loudness of the sound we hear by adding the loudnesses of the sounds when heard separately, because together the sounds would to some extent mask one another, and the overall loudness would be diminished.

Thus it is easy to see that the equal loudness contours are not sufficient to allow us to assess the loudness of composite sounds, unless we take account of masking.

Now when we listen to, say, a pure tone of 1000 Hz and at the same time to a sound with a range of frequencies (called a *bandwidth*) of, say, 980-1020 Hz (i.e., 40 Hz), we can find a level of the second sound for which the first sound is completely masked. We find, moreover that about the same amount of sound energy in the masking sound will still mask the first sound as we widen the bandwidth to 100 Hz, e.g., the range 950-1050 Hz. There comes a bandwidth, however—known as the *critical bandwidth*—when spreading the energy of the masking sound any wider (say, to the range 850-1150 Hz) ends the masking: the tone is again perceived. The critical bandwidth for the case we have described is, in fact, about 160 Hz.

We find also that this critical bandwidth is approximately proportional to the center frequency. That is to say, for a tone of 2000 Hz, the critical bandwidth is about 320 Hz, for 4000 Hz it is about 700 Hz, and for 8000 Hz, it is about 1600 Hz.

We see below how this can be used to determine the loudness of composite, real-life sounds.

1-3.4   Octave Bands and One-Third Octave Bands

The audible frequency range can be arbitrarily divided up into a series of adjacent frequency bands known as octave bands, for which the width of a given band is proportional to the center frequency. Each of these can be further divided into three one-third octave bands, also of widths proportional to their center frequencies. It is the similarity of these bandwidths to the ear's critical bandwidths that permit them to be used for calculating the loudness of composite sounds.

An octave band is defined as a band of frequencies for which the upper frequency (f_2, say) is twice as great as the lower frequency (f_1, say), i.e., $f_2/f_1 = 2$. The *center frequency* of that band is then arbitrarily defined by $f_c = \sqrt{f_1 f_2}$.

Although the octave bands could be arranged anywhere in the frequency spectrum, convention has established a preferred set of these bands, whose position is fixed once one band is fixed. There is a band whose center frequency is at 1000 Hz. It follows from the relationships above that the upper and lower limit frequencies ("cutoff" frequencies) are about 1420 and 710 Hz, respectively. The next highest band then has a lower cutoff frequency of 1420 and a higher one of 2840. Its center frequency is 2000 Hz, twice that of the previous octave band.

In a similar manner, one-third octave bands are defined by $f_2/f_1 = 2^{1/3}$, or by $10^{0.1}$, with $f_c = \sqrt{f_1 f_2}$ as before.

Table 1-1 sets out the octave and one-third octave bands, with the frequencies rounded off according to convention.

Frequency analyzers with sound signal inputs are able to measure the SPL in each octave or one-third octave band, thus presenting a spectrum of the noise, i.e., showing how much energy lies in the various frequency bands across the audible frequency range.

Such a frequency analysis is a first step in calculating the loudness of a composite sound.

1-3.5   The dB(A), dB(B), and dB(C) Scales

One approach to measuring sounds in a way reflecting on their loudness is to alter the measured frequency spectrum of the sound to take account of the fact that the ear responds less well to frequencies below about 500 Hz and above

TABLE 1-1. Center Frequencies f_c, and Lower and Higher Cutoff Frequencies f_1 and f_2, for the Octave and $\frac{1}{3}$ Octave Bands (Hz).

$\frac{1}{3}$ Octave Bands			Octave Bands		
f_c	f_1	f_2	f_c	f_1	f_2
12.5	11	14			
16	14	18	16	11	22
20	18	22			
25	22	28			
31.5	28	36	31.5	22	44
40	36	45			
50	45	56			
63	56	71	63	44	88
80	71	89			
100	89	112			
125	112	141	125	88	177
160	141	178			
200	178	224			
250	224	282	250	177	355
315	282	355			
400	355	447			
500	447	562	500	355	710
630	562	708			
800	708	891			
1000	891	1120	1000	710	1420
1250	1120	1410			
1600	1410	1780			
2000	1780	2240	2000	1420	2840
2500	2240	2820			
3150	2820	3550			
4000	3550	4470	4000	2840	5680
5000	4470	5620			
6300	5620	7080			
8000	7080	8910	8000	5680	11,360
10,000	8910	11,220			
12,500	11,220	14,130			
16,000	14,130	17,780	16,000	11,360	22,720
20,000	17,780	22,390			

about 8000 Hz than it does between those frequencies. Thus by subtracting some decibels from the lower and upper frequency bands while leaving them approximately the same elsewhere, and by then adding up the new levels in the bands, one can obtain a weighted sound level which to some extent correlates with the sound's loudness.

Three such weightings were originally proposed.

For sounds that were "not loud," about 40 phons, the A weighting curve was defined. For sounds that were "moderately loud," about 70 phons, the B weighting curve was defined. For "loud sounds," about 100 phons, the C curve was defined, which is, like the 100 phon loudness contour of Figure 1-2, fairly flat;

TABLE 1-2.   The A, B, C, and D Weightings for $\frac{1}{3}$ Octave Bands, and the A and D Weightings for Octave Bands.[a]

$\frac{1}{3}$ Octave Bands					Octave Bands		
f_c	A	B	C	D	f_c	A	D
25	-45	-20	-4	-19			
31.5	-39	-17	-3	-17	31.5	-39	-17
40	-35	-14	-2	-15			
50	-30	-12	-1	-13			
63	-26	-9	-1	-11	63	-26	-11
80	-23	-7	-1	-9			
100	-19	-6	0	-7			
125	-16	-4	0	-6	125	-16	-6
160	-13	-3	0	-5			
200	-11	-2	0	-3			
250	-9	-1	0	-2	250	-9	-2
315	-7	-1	0	-1			
400	-5	-1	0	-1			
500	-3	0	0	0	500	-3	0
630	-2	0	0	0			
800	-1	0	0	0			
1000	0	0	0	0	1000	0	0
1250	1	0	0	2			
1600	1	0	0	6			
2000	1	0	0	8	2000	1	8
2500	1	0	0	10			
3150	1	0	-1	11			
4000	1	-1	-1	11	4000	1	11
5000	1	-1	-1	10			
6300	0	-2	-2	9			
8000	-1	-3	-3	6	8000	-1	6
10,000	-3	-4	-4	3			
12,500	-4	-6	-6	0			
16,000	-7	-8	-9	-3	16,000	-7	-3
20,000	-9	-11	-11	-5			

[a]Weightings are to be added to SPL in dB re 20μPa; f_c is octave band center frequency in Hz.

this meant that the lower and higher frequencies were not much de-emphasized so that the C-weighted decibel is rather similar to the overall sound pressure level itself.

To calculate a weighted sound level, the sound is first analyzed into octave or one-third octave bands, the latter being best for all sounds except those which have flat frequency spectra, i.e., have their energy fairly evenly distributed across the frequency bands. (For such sounds, an octave analysis is adequate.)

From the one-third octave band sound pressure levels, certain decibel values are added or subtracted as listed in Table 1-2. For the A weighting (the most important, as we shall see), values are also given for octave bands, rounded off to reflect the lesser accuracy of a calculation performed in this way. Figure 1-4 shows the weighting curves graphically. Both Table 1-2 and Fig. 1-4 refer also to the D weighting to be described in Section 1-4.3.

Although derived for sounds of varying degrees of intensity, it has been found that the A-weighted sound level is very useful at all levels of intensity. It is widely used throughout this book, not only reflecting on loudness but also on noisiness and hearing damage. In contrast, the B weighting is in virtual disuse.

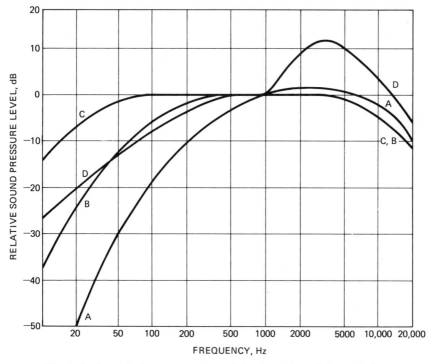

Fig. 1-4 Graphical representation of the A, B, C, and D weightings.

The C weighting is also not used to reflect human response, but its not quite flat weighting curve permits it to be a reasonable approximation of the overall SPL. It is simpler to build a sound level meter which gives a C weighting than the overall SPL, because the C weighting does not require as wide a frequency response.

Example

Problem: Calculate the A-weighted sound level of a sound which has been analyzed into the octave band sound pressure levels given below:

Octave band center frequencies, Hz	31.5	63	125	250	500	1000	2000	4000	8000
SPL, dB	65	67	73	75	75	78	80	73	67

Solution: From Table 1-2, write down the A weightings for each octave band, and add these to the SPLs to arrive at octave band weighted levels.

A weightings, dB	-39	-26	-16	-9	-3	0	1	1	-1
A-weighted octave band levels, dB	26	41	57	66	72	78	81	74	66

Finally we must summate these octave band weighted levels to arrive at the A-weighted sound level. This may be performed using a prepared chart for adding decibels (e.g., Fig. 1-5). One may also follow these steps: (a) Divide each weighted level by 10. (b) Take the antilogarithm (to the base 10) of each. (c) Add these together. (d) Take the logarithm (to the base 10) of the total. (e) Multiply by 10 to get the final result, in this case 84 dB (A).

A-weighted sound level is written as dB(A) or dBA. dB(A), dB(B), and dB(C) can be read directly on many sound level meters by switching in a set of filters to produce the appropriate weighting.

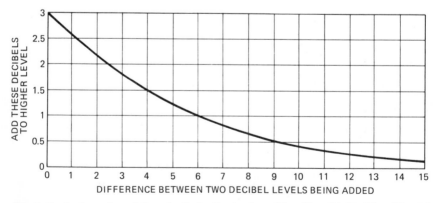

Fig. 1-5 A chart for adding decibels. Examples: 55 + 61 = 62 dB; 55 + 80 ≐ 80 dB; 60 + 60 = 63 dB; 55 + 55 + 59 = 58 + 59 = 61.5 dB.

How does dB(A) relate to loudness in phons or sones? In most instances we do not need to seek a direct relationship between them because it suffices that we relate the dB(A) level of a sound directly to the human response. What the mathematicians call a monotonic relationship linking dB(A) to phons or sones does not, in fact, exist. We *can* calculate the dB(A) and loudness levels of any one sound and compare these for that sound alone. Sacrificing accuracy, we can go a little further and do this for a given type of sound, making the assumption that the various frequency spectra measured in offices, for example, are rather similar. For many common noises, it has been found that the loudness in phons is greater than the sound level in dB(A) by an amount usually between 10 and 13. Such a finding can be helpful, even if it is like subtracting apples from oranges, because small differences in these values are not necessarily important in predicting human response with all its variabilities. We can often safely average the "10 to 13" at, say, 12. This is done in Section 1-4.7.

1-3.6 Loudness of Composite Sounds

We have seen that the ear is able to mask sounds close to each other in frequency. We have also noted that the critical bandwidths in this masking process are proportional to the frequencies involved, and some examples of actual critical bandwidths were given in Section 1-3.3. As we can see from Table 1-1, the octave and one-third octave bands resemble the critical bandwidths insofar as the bandwidths of the former are also proportional to frequency. Moreover, the one-third octave bandwidths are fairly similar in actual value to the critical bandwidths.

These facts are used in assessing the loudness of composite sounds. Experiments have been performed in which subjects judged the loudnesses, not of the pure tones which produced the equal loudness contours of Fig. 1-2, but of sounds composed of all the frequencies in a given octave or one-third octave band, so that the effect of masking is already accounted for in the loudness determination. Sets of equal loudness contours were thus derived for *bands* of noise instead of pure tones.

Two approaches to the calculation of loudness based on these considerations have emerged, one by Stevens and one by Zwicker. The Stevens approach has been progressively refined, and the version called Mark VI has been adopted by the American National Standards Institute as ANSI S3.4-1968. Particularly because it has similarities with the calculation of noys (see Section 1-4.2), we give this method here.

Figure 1-6 presents Stevens' equal loudness index contours. These are, as we have said, equivalent to the pure-tone equal loudness contours of Fig. 1-2, referring instead to sound in octave or one-third octave bands; the use of straight lines in Fig. 1-6 is an approximation found to yield acceptable accuracy.

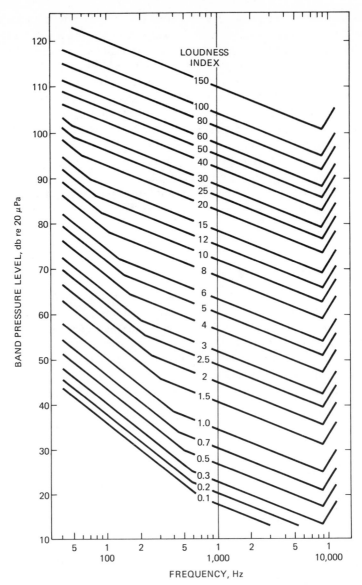

Fig. 1-6   Stevens' equal loudness index contours (From ANSI S3.4-1968)

The Stevens loudness calculation procedure uses either octave or one-third octave analysis, the latter being preferable. The approach is as follows:

Write down the SPL in dB for each band. With interpolation where necessary, write down for each such SPL the loudness index in sones found in Fig. 1-6.

Then perform the following calculation:

$$\text{Loudness (sones)} = S_m + f(\Sigma_i S_i - S_m),$$

where S_i is the sone value of a band i, S_m is the maximum sone value of all the bands, and f is a constant ($f = 0.3$ when octave bands are used; $f = 0.15$ when one-third octave bands are used). $\Sigma_i S_i$ indicates that one adds all the sone values. In other words, one writes down the maximum sone value and adds to that a quantity which is the constant value of f multiplied by the combined value of all the sone values except the maximum one. The loudness so obtained in sones may be converted to phons using Fig. 1-3 or the equations given in Section 1-3.2.

Further explanation is helpful with regard to this method. The prominence of the maximum band value (S_m) is because that band produces a partial masking of the other bands. The use of sone values for the loudness index allows us to add the loudness of each band arithmetically, which would not have been possible in phons. Finally, the phon or sone values given by the method correspond, as one would hope, with those for pure tones; e.g., a composite sound of 50 phons is judged to have the same loudness as a pure tone of 50 phons.

A revision of Stevens Mark VI method, called Mark VII, has been proposed.

Example

Problem: Calculate the loudness in phons, using the Stevens Mark VI method, of the sound whose octave band SPLs are given below.

Octave band center frequencies, Hz	31.5	63	125	250	500	1000	2000	4000	8000
SPL, dB	65	67	73	75	75	78	80	73	67

Solution: From Figure 1-6, write down the loudness indices for each frequency band as follows:

Loudness index, sones	1.8	2.5	5	8	10.2	15	20	16	12

The maximum loudness index value S_m is 20. The sum of all the values except this, $\Sigma_i S_i - S_m$, is 70.5. Since we are dealing with octave bands, the constant f is 0.3. Then

$$\text{loudness (sones)} = 20 + 0.3(70.5) \doteq 41.$$

Also, from Figure 1-3,

$$\text{loudness (phons)} \doteq 93.5.$$

This example uses the same data as the example used to calculate the dB(A) in Section 1-3.5. The sound level in dB(A) was 84 while the loudness in phons is 94, a difference of 10. This lies within the range of 10 to 13 suggested in Section 1-3.5 as being usually the difference between loudness in phons and sound level in dB(A).

1-3.7   Loudness of Transient Sounds

All that we have discussed so far refers to continuous sounds which, in a practical definition, may be regarded as those sounds having a duration of more than about half a second. The loudness of a given sound lasting more than this length of time does not depend on the duration of the sound, whereas a given sound lasting less than about half a second has a loudness which depends on the time we listen to it. Such sounds are known as transient or impulse or impulsive sounds.

The half second (500 msec) mentioned above is only a rough dividing line. Depending on the subject and the sound, it may be as little as 30 msec. One second may be taken as an extreme upper limit, and 30 msec as a lower.

The hearing phenomena which govern the loudness of transients are complex. Interested readers should refer to other texts for a discussion of aural integration time, masking of transient sounds, aural reflex (in the case of repeated transients) and impulse loudness generally. The philosophy of the loudness calculation is essentially the same as that for continuous sounds, the major difference being that the frequency analysis of a transient event is performed by finding its Fourier transform energy spectrum instead of the power spectrum performed for continuous sounds by bandpass filters. The Fourier transform of any real-life waveform generally requires a computer. Having obtained the spectrum, the energy is weighted in about the same way as it is with continuous sounds.

In the case of certain transient stimuli for which the waveform is geometrically simple (such as a sonic boom waveform which has a characteristic N-shape), it is sometimes possible to precalculate the energy spectrum for the various permutations the shape may take; this simplifies the procedure and obviates the need for a computer. For such shapes, it is also possible to find equations for loudness in terms of the waveform characteristics.

All these methods depend on accurately measuring the transient sound's waveform, something which generally requires a complex recording system. An impulse sound level meter is not usually adequate for obtaining anything but an approximation of the peak level of the wave.

1-4. NOISINESS

1-4.1   Frequency Response for Noisiness

The counterpart of the equal loudness contours (see Section 1-3.1 and Fig. 1-2) are the equal noisiness contours shown in Fig. 1-7. These were derived by Kryter and others and are the subject of the International Organization for Standardization Recommendation ISO/R 507. Unlike the equal loudness contours, they were derived for narrow bands of noise rather than pure tones; in this they resemble Fig. 1-6. Thus they take some account of masking.

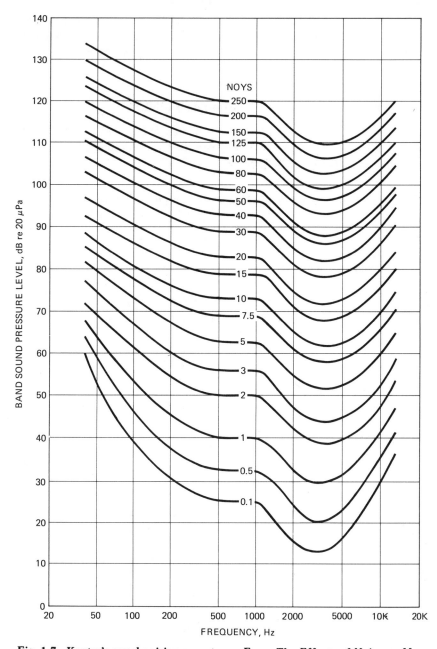

Fig. 1-7 Kryter's equal noisiness contours. From *The Effects of Noise on Man.*

Comparison of the equal noisiness contours and the equal loudness contours shows that there is a dip in the noisiness contours between about 2000 and 5000 Hz that is rather more marked than in the loudness contours. Sounds in this frequency range do not have to have very high SPLs to be noisy.

1-4.2   The Noy and Perceived Noise Level (PNL)

The noy is the counterpart in noisiness of the sone in loudness. Perceived Noise Level (PNL), which is given the unit of perceived noise decibel (PNdB), is the counterpart of loudness level in phons (see Table 1-3).

TABLE 1-3.   Similarities Between Measures of Loudness and Noisiness.

Loudness	Noisiness

40 phons = 1 sone
    judged loudness of a 40 dB tone at
    1000 Hz
	40 PNdB = 1 noy
	    judged noisiness of a 40 dB band of
	    sound centered at 1000 Hz

50 phons = 2 sones	50 PNdB = 2 noys
60 phons = 3 sones etc.	60 PNdB = 3 noys etc.

The exact nature of the band of noise used to compile the noy is, of course, defined. So is the duration of the sound, for, as we have noted, the length of a sound can affect our perception of its noisiness.

We can also convert between noys and PNdB using Fig. 1-3. Furthermore,

$$\log_{10} N = 0.030(\text{PNdB} - 40)$$

and

$$\text{PNdB} = 40 + 33 \log_{10} N$$

where N represents noys.

1-4.3   The dB (A) and dB (D)

Although the dB(A) was derived from the 40 phon loudness contour (see Section 1-3.5), we noted that it has been found to be a reasonable measure of loudness for all levels of SPL. We can now report further that it has also been found to be a good measure of noisiness. Despite its "loudness" origins, it is overwhelmingly accepted as the easiest commonsense way to measure a sound for its acceptability.

The dB(D) is derived from the 40 PNdB contour, and is intended to be a simple measure of noisiness. The decibel weightings for the dB(D) are given with the

A, B and C weightings in Fig. 1-4 and Table 1-2. The dB(D) has every reason to be as good a measure of noisiness as dB(A), but has not shown itself to be sufficiently superior to find its way into much real-life use.

The calculation of dB(D) is similar in approach to that of dB(A) given in Section 1-3.5. For the data of the example in Section 1-3.5, the dB(D) value is 90.

1-4.4 Noisiness of Composite Sounds

Calculating the noisiness of composite sounds closely resembles the loudness calculation procedure described in Section 1-3.6. We use the equal noisiness contours of Fig. 1-7. The approach is as follows.

Write down the SPL in dB in each octave or one-third octave band. With interpolation where necessary, write down for each such SPL the noisiness (in noys) found in Fig. 1-6. Then perform the following calculation:

$$\text{Noisiness (noys)} = N_m + f(\Sigma_i N_i - N_m),$$

where N_i is the noy value of a band i, N_m is the maximum noy value of all the bands, and f is a constant ($f = 0.3$ when octave bands are used; $f = 0.15$ when $\frac{1}{3}$ octave bands are used). $\Sigma_i N_i$ indicates that we add all the noy values. In other words, one writes down the maximum noy value and adds to that a quantity which is the constant value f multiplied by the combined values of all the noy values except the maximum one. The noisiness so obtained in noys may be converted to PNdB using Fig. 1-3 or the equations given in Section 1-4.2.

Various small amendments to this procedure are sometimes used, e.g., treating the lowest frequency octave bands together instead of separately.

Example

Problem: Calculate the noisiness in PNdB of the sound whose octave band SPLs are given below.

Octave band center
frequencies, Hz	31.5	63	125	250	500	1000	2000	4000	8000
SPL, dB	65	67	73	75	75	78	80	73	67

Solution: From Fig. 1-7 write down the noy values for each frequency band, as follows, though without a 31.5 Hz value, since the curves do not extend into this region of supposedly low noise sensitivity.

Noisiness index, noys –   2.0   5.9   10.0   11.3   14.0   28.0   21.0   10.0

The maximum noy value N_m is 28.0. The sum of all the values except this, $\Sigma_i N_i - N_m$, is 74.2. Since we are dealing with octave bands, the constant f is 0.3. Therefore,

$$\text{noisiness (noys)} = 28.0 + 0.3(74.2) = 50.3.$$

Also, from Fig. 1-3,

$$\text{noisiness (PNdB)} \doteq 96.5.$$

1-4.5   Tone Corrections

It has been found that where the sound spectrum has strong tonal components (e.g. in the fan noise one experiences when some jet aircraft land) that the Perceived Noise Level calculation underestimates the noisiness. To take account of this, a number of tone corrections are in use. A value is added to the PNL to give the *Tone Corrected Perceived Noise Level*, PNLT.

The calculation for this is lengthy, and reference should be made to, for example, ISO/R 507 "Procedure for describing aircraft noise around an airport." The principle involved is to consider the amount by which the highest $\frac{1}{3}$ octave band level (presumably containing the strongest tone) exceeds the level of the adjacent bands. The correction is equally applicable to A-weighted sound level.

For strong tonal components, the correction may amount to about 5 dB, while about 3 dB is more common.

Present-day tone corrections are seen as interim measures, since the need for them indicates imperfections in the procedures for calculating perceived noisiness and their accuracy is not well established. We will likely see them improved or, better, the calculations for perceived noisiness may be modified to make a tone correction unnecessary.

1-4.6   Noisiness of Transient Sounds

The state of knowledge of both the loudness and noisiness responses to transient sounds is relatively primitive; in fact, there are no real measures of the noisiness of transient sounds. Thus the measures for the loudness of transient sounds are used in default. Section 1-3.7 is an introduction to this subject.

The noisiness of transient sounds, when better understood, will probably take account of the startle that unexpected transient sounds can provoke; social surveys have revealed this as a component of annoyance, a component obviously seldom found with continuous sounds. Our knowledge of how to calculate startle from the sound itself is at present very limited.

1-4.7   Noisiness Measures Summarized. Which Measure is Best?

We have so far introduced the loudness and noisiness measures given in Table 1-4. Considering the loudness measures alongside the noisiness ones, in case they should also predict noisiness with reasonable accuracy, can we decide which of these measures best agrees with our sense of the sound's noisiness and should therefore be used?

TABLE 1-4. An Index to the Loudness and Noisiness Measures Introduced in Sections 1-3 and 1-4.

phon (and sone)[a]	see Sections 1-3.2 and 1-3.6
dB(A)	1-3.5 and 1-4.3
dB(B)	1-3.5
dB(C)	1-3.5
PNdB (and noy)[a]	1-4.2 and 1-4.4
dB(D)	1-4.3

[a]Grouped together because there is a one-to-one relationship between the two measures in each pair.

Answers to this question come generally from two opposing standpoints.

One standpoint is that the best measures are those found from psychophysical experiments to be the most highly correlated with human judgments of the noisiness of various different sounds. These experiments generally, but by no means always, rank the measures as follows, with the most accurate stated first:

PNLT
PNL
dB(D)
dB(A)
phon

The dB(B) and dB(C), along with the overall SPL in dB, are invariably much inferior, and they have been removed from the list. Supporters of this standpoint would maintain that for example, in the enormous expenditure involved in designing and certifying a passenger aircraft to conform with noise limits, those limits should be stated in the most accurate terms possible, even if a certain complexity is the result.

The other standpoint is that the difference in accuracy between the various measures in this list is not very large. One should note that these measures are derived from experiments performed in the laboratory on small samples of the population, and that in real life the sound of any given source and the response to that sound are subject to alteration by many important factors not considered, such as those discussed in Section 1-2.3. Supporters of this standpoint maintain that measures like the dB(A) are easily obtained with sound level meters, and should be universally accepted to allow us to get on with the business of reducing noise. They impatiently refer to research to improve these calculation procedures, which generally involves weighting the sound spectrum, as a "weighting game."

Anyone seriously interested in noise must acknowledge that the standpoints are compatible. Thus the dB(A) has been widely adopted as a measure of noisiness in all situations but those, as in aircraft noise, where more accuracy seems a reasonable requirement. As mentioned in Section 1-4.3, the dB(D) is as accurate but has not won much real-life usage.

If one refers at this point to the various example calculations performed in preceding sections, all for the same sound spectrum, one finds that we obtained (in round numbers) 94 phons, 84 dB(A), 90 dB(D) and 97 PNdB. These results are in accord with the useful finding that for many, but by no means all, sounds, the approximate conversions given in Table 1-5 may be used. These conversions are summaries of results from many sources, and the figures often vary one to two units to either side of those quoted.

Included in Table 1-5 for convenience are conversions for Noise Criteria (NC) and Preferred Speech Interference Level (PSIL). The NC curves are given in Chapter 6 and apply principally to noise in buildings. PSIL is introduced in Section 1-6.2 of the present chapter. Conversions to sones and noys cannot be effected with Table 1-5; Fig. 1-3 provides one way to obtain them.

The conversions of Table 1-5 are helpful only if one knows that they apply to the sound spectra under consideration. For new or unusual spectra, it would be necessary to calculate and compare these measures before assuming simple relationships between them. When quoting a measure derived through such a relationship, it is also essential to explain its origins and regard it as approximate.

An upper limit for the error introduced by these conversions is about 5 of whatever unit is involved. The error is usually less than this. One may place these errors in perspective by noting that a typical person who listens to the same sound on two different days will usually be unable to be any more consistent than this about his judgments of their noisiness.

TABLE 1-5. Conversion Table for Various Measures of Loudness, Noisiness, and Speech Intereference.[a]

To convert to one of these units . . .	. . . add to the units given here the amount shown below					
	phons	dB(A)	dB(D)	PNdB	NC	PSIL
phons	—	12	-18	-24	-2	-5
dB(A)	-12	—	-6	-12	10	7
dB(D)	18	6	—	-6	16	13
PNdB	24	12	6	—	22	19
NC	2	-10	-16	-22	—	-3
PSIL	5	-7	-13	-19	3	—

[a]Caution must be exercised in using this table; see the text of Section 1-4.7.

1-5. ANNOYANCE

In Section 1-2, it was explained that the annoyance caused by a sound differed from its noisiness because the latter was the unwantedness of the sound heard on its own, without regard to a list of additional factors that also govern the sound's acceptability.

Many of these additional factors cannot, in the state of present knowledge, be properly accounted for. There is, for example, no annoyance calculation procedure that properly deals with misfeasance or with the emotional associations that a sound may have for a person. It is partly because of these and other inadequacies in our understanding of human response that noise exposure limits are generally determined (by largely empirical methods) in the many different noise situations described in Chapters 2 to 10.

There is, however, one characteristic of the noise exposure that we are beginning to be able to take account of in a reasonably precise way, and that is the variation of the sound level with time. Obviously two sounds of the same noisiness will produce different amounts of annoyance if they last for different times, or rise and fall with different rapidities.

Thus the annoyance of an aircraft flyover is calculated by considering not only the Tone-Corrected Perceived Noise Level at any one instant in the flyover, but by deriving the Effective (Tone-Corrected) Perceived Noise Level from considering the way the PNLT rises and falls. Similarly, the annoyance of a whole day of aircraft flyovers is found from calculating the Noise and Number Index (NNI), or Noise Exposure Forecast (NEF), from a knowledge of the number of flyovers and the Effective PNLT of each. Fuller details of these procedures are contained in Chapter 3 on aircraft noise, just as descriptors particular to traffic noise are explained in Chapter 2. The purpose of mentioning them here is to explain that a measure of annoyance may sometimes be constructed from a noisiness measure and the time history of the sound.

There are, however, certain annoyance measures which are intended to be universal, that is to apply to most if not all sounds. These methods are described in Sections 1-5.1 – 1-5.7 though it should not be thought that their universality of application necessarily implies a universality of acceptance.

1-5.1 Levels Exceeded a Certain Proportion of the Time, L_n

A way of assessing the annoyance from sounds that vary with time is to describe the sound in terms of the level exceeded a certain proportion of the time. Thus L_n is the sound level, generally but not necessarily in dB(A), which is exceeded for $n\%$ of the time. Often used are:

L_0 or L_{max}: maximum level, never exceeded;
      L_1: a descriptor of the highest occurring sound levels, but neglecting

momentary peaks of combined duration less than 1% of the time surveyed;

L_{10}: a descriptor of the higher sound levels, i.e., those well above the ambient level, but neglecting levels occurring for less than 10% of the time;

L_{50}: the sound levels as often exceeded as not, a median level;

L_{90}, L_{95}: descriptors of the lower sound levels corresponding with what we normally think of as the ambient level.

These measures may be used to describe the noise climate for any period of interest, though typical sampling times are an hour, or a part of a day, or a 24 hour period. Of these descriptors, the one most used is L_{10}, since it describes intrusive events without paying overmuch regard to the unusual events; it correlates well with annoyance.

1-5.2   Equivalent Sound Level, L_{eq}

The equivalent sound level is the level of a steady sound which, in a given period of time, would contain the same noise energy as the time-varying sound level one is describing. In mathematical terms, equivalent sound level is defined as

$$L_{eq} = 10 \log_{10} \frac{1}{T} \int_0^T 10^{SL/10} \, dt$$

where SL is the sound level, generally expressed in dB(A), and T is the time period for which we are describing the sound.

The concept of equivalent sound level is that the same amount of noisiness occurs from a sound having a high level for a short period, as from a sound having a lower level but occurring for a long enough period that the same amount of energy is involved. For example, a sound of 70 dB(A) for 100 sec (1.67 min) has the same A-weighted sound energy as one of 50 dB(A) for 10,000 sec (2.78 hr). Most sounds vary irregularly in level, and the derivation of L_{eq} demands an integration of the sound intensity on a continuous basis.

The time for which L_{eq} describes the sound could be a single hour of interest, or for part of the day, or for 24 hours or longer.

Equivalent sound level has the advantage over L_n (see Section 1-5.1) that its value shows no abrupt change as the duration of a sound intrusion exceeds a particular proportion of the sampling period. It is a better descriptor of rare, short-duration events, like train pass-bys on low density routes, than a descriptor like L_{10}, which might not register these events at all. It has the further advantage that the L_{eq} from a number of different sources is the decibel addition of the L_{eq} from each individual source; such an addition is not necessarily accurate when L_n is involved.

L_{eq} is currently a noise descriptor favored by the U.S. Environmental Protection Agency, and a useful reference to its calculation and use is the EPA levels document (see bibliography).

A useful expansion of the above equation is

$$L_{eq} = 10 \log_{10} \frac{1}{T} [t_1 \cdot 10^{(L_1/10)} + t_2 \cdot 10^{(L_2/10)} + \cdots]$$

or

$$L_{eq} = 10 \log_{10} \frac{1}{T} [t_1 \text{ antilog } L_1/10 + t_2 \text{ antilog } L_2/10 + \cdots]$$

where t_1 is the time for which the sound level is L_1, and so on.

1-5.3   Day–Night Sound Level, L_{dn}

In order to represent the greater annoyance caused by a sound intrusion at night, the day-night sound level has been derived. It supposes that the equivalent sound level occurring between 10 pm and 7 am should be augmented by 10 dB(A) before being combined with the equivalent sound level for the period 7 am to 10 pm to give the day-night level. If L_D is the equivalent sound level for the 15 hour daytime period, and L_N is the equivalent sound level for the 9 hour nighttime period, the day-night sound level is

$$L_{dn} = 10 \log_{10} \frac{1}{24} [15 \cdot 10^{(L_D/10)} + 9 \cdot 10^{(L_N+10)/10}].$$

It has sometimes been suggested that the 24 hours should be divided into three segments rather than two—day, evening, and night—with an evening weighting of perhaps 5 dB(A). This has not been shown to be necessarily advantageous. It has also been suggested that the nighttime weighting is not worth incorporating, in which case we are left with just L_{eq}.

The relative merits of L_{dn} and L_{eq} are not fully resolved. However, the difference between them may be unimportant in practice. For a sound that is continuous over the 24 hours, L_{dn} exceeds L_{eq} by 6 dB(A); but for a sound occurring about one-quarter as frequently by night as by day, the difference is cut in half. Many noise intrusions, like those from transportation systems, occur less often at night than by day, and the difference between L_{dn} and L_{eq} may be small.

A noise descriptor similar to L_{dn} but featuring three rather than two time periods each 24 hours, is CNEL, the Community Noise Equivalent Level. CNEL is used mainly for aircraft noise (see Chapter 3).

1-5.4   Noise Pollution Level, L_{NP}

Robinson, in proposing this descriptor, suggested that annoyance is governed by two items—one an expression of the "average" sound level, the other an expression of the way the sound level varies.

He chooses to describe "average" sound level as the L_{eq} (see Section 1-5.2) and, for a Gaussian noise distribution, often a reasonable approximation to the actual distribution, we may take

$$L_{eq} = L_{50} + (L_{10} - L_{90})^2/56.$$

For the term describing the variation of the sound level, he proposes the constant 2.56 multiplied by the standard deviation of the sound level, σ, or alternatively the term $L_{10} - L_{90}$.

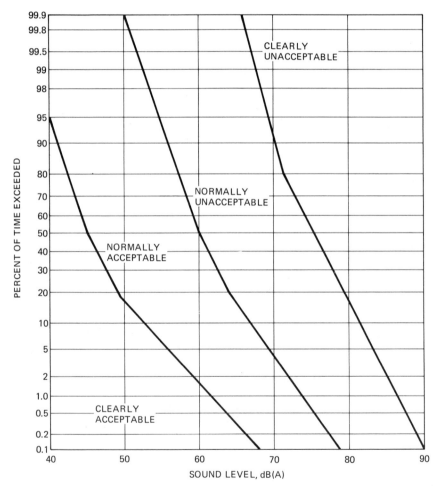

Fig. 1-8 HUD criterion for non-aircraft noise in residential areas outdoors. (From L. L. Beranek, *Noise and Vibration Control*)

TABLE 1-6. Tentative Conclusions to be Drawn from Various Values of L_{NP} in Outdoor Residential Locations from Noise Other than Aircraft Noise.

		L_{NP}		
	Less than 58 dB(NP)	Between 58 and 73 dB(NP)	Between 73 and 88 dB(NP)	Over 88 dB(NP)
Conclusion	Clearly acceptable	Normally acceptable	Normally unacceptable	Clearly unacceptable

Thus the Noise Pollution Level is given by

$$L_{NP} = L_{eq} + 2.56\sigma$$

or

$$L_{NP} = L_{50} + (L_{10} - L_{90})^2/56 + (L_{10} - L_{90}).$$

The unit of L_{NP} is sometimes termed dB(NP).

Noise Pollution Level has similarities to the Traffic Noise Index (TNI), from which it was derived (see Chapter 2).

Criteria in terms of L_{NP} are not yet established in their own right, but may be "translated" into L_{NP} from other units. In the case of the HUD criterion (see Section 1-5.5), the demarcation lines in Fig. 1-8 approximate to 58, 73, and 88 dB(NP) when going from left to right. One can thus make the tentative conclusion of Table 1-6.

The significance of L_{NP} is that it is a descriptor intended to be applied to all situations of noise nuisance, but further validation is necessary before this can be done.

1-5.5 HUD Criterion

A criterion for the acceptability of non-aircraft noise heard outdoors (Fig. 1-8) was developed for the U.S. Department of Housing and Urban Development (HUD). A cumulative distribution of actual noise exposure will invariably run more or less in line with the curves of Fig. 1-8 if it is approximately normal or Gaussian, thereby allowing us to decide on its acceptability from the portion of Fig. 1-8 that it occupies. It should, however, be noted that only a few sound levels extracted from Fig. 1-8 are included in HUD Circular 1390.2, which is the official document resulting from this work.

The HUD criterion can only be applied to the situation of residential exposure outdoors to non-aircraft noise, or to situations that are similar.

1-5.6   A Generalized Method for Predicting Community Response To Noise

A method for predicting community response to any sound intrusion has been developed in the U.S. from an analysis of 55 community noise situations. About half of these cases were of steady industrial and residential noises, and the other half were of intermittent transportation and industrial noise intrusions. The En-

TABLE 1-7.   Corrections to be Added to the Measured Day–Night Sound Level (L_{dn}) of Intruding Noise to Obtain Normalized L_{dn}.

Type of Correction	Description	Amount of Correction to be Added to Measured L_{dn} in dB(A)
Seasonal Correction	Summer (or year-round operation)	0
	Winter only (or windows always closed)	−5
Correction for outdoor noise level measured in absence of intruding noise	Quiet suburban or rural community (remote from large cities and from industrial activity and trucking)	+10
	Normal suburban community (not located near industrial activity)	+5
	Urban residential community (not immediately adjacent to heavily traveled roads and industrial areas)	0
	Noisy urban residential community (near relatively busy roads or industrial areas)	−5
	Very noisy urban residential community	−10
Correction for previous exposure and community attitudes	No prior experience with the intruding noise	+5
	Community has had some previous exposure to intruding noise but little effort is being made to control the noise. This correction may also be applied in a situation where the community has not been exposed to the noise previously, but the people are aware that bona fide efforts are being made to control the noise.	0
	Community has had considerable previous exposure to the intruding noise and the noise maker's relations with the community are good.	−5
	Community aware that operation causing noise is very necessary and it will not continue indefinitely. This correction can be applied for an operation of limited duration and under emergency circumstances.	−10
Pure tone or impulse	No pure tone or impulsive character	0
	Pure tone or impulsive character present	+5

Source: EPA Report 550/9-74-004.

vironmental Protection Agency has endorsed this method by including it in the levels document.

The method first involves calculating the normalized L_{dn} (see Section 1-5.3) of the intruding noise. This is the actual L_{dn} of the intruding noise normalized by adding or subtracting for the various factors which affect community reaction that are given in Table 1-7. The factors are based on the following reasoning: if the sound occurs only in winter, it will be heard behind closed windows; an already noisy community is less likely to react to a noise intrusion of a given sound level than is a quiet one; a community's previous experience with an intruding noise lessens the impact of further exposure to it; noise control consciousness on the part of the noise-makers (i.e., low misfeasance) lessens community reaction; pure tones and impulsive sounds are specially annoying.

Having normalized the L_{dn} with the corrections, reference is made to Fig. 1-9, which shows the different levels of community reaction observed for the various values of normalized L_{dn} in the 55 cases.

The community reactions may occur at a lower value of normalized L_{dn} than that predicted, should the absolute magnitude of the intruding noise be high enough to have some physical effect, e.g., to interfere with speech.

This method is described further in Chapter 8.

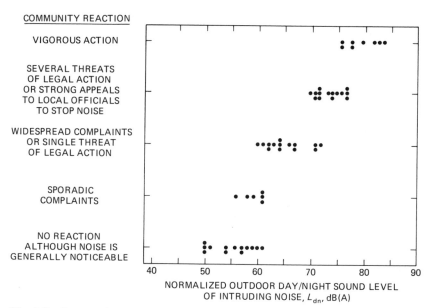

Fig. 1-9 Community reaction to intrusive noise measured in L_{dn} and normalized by the factors set out in Table 1-7. From EPA Report 550/9-74-004 (levels document).

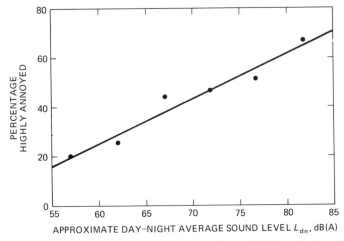

Fig. 1-10 Annoyance as a function of L_{dn}, from British and American surveys reviewed in the EPA levels document. An approximate equation for this relationship is Percentage Highly Annoyed = $2(L_{dn} - 50)$.

1-5.7 Community Annoyance and Complaints

The number of complaints, or proportion of complainants, is governed in any situation not only by the noise and its annoyance, but also by the presence of a body to whom to complain and the expectation of a successful result. The degree of complaint is also dependent on such political matters as the formation of citizens' groups to appeal to officials, seek publicity, and take legal action.

It is difficult to treat these matters quantitatively. We give here, however, an accepted general relationship between the noise, the annoyance, and the percentage of complainants. Figs. 1-10 and 1-11 are taken from the EPA levels document. Note that a generally held noise criterion, $L_{dn} = 55$ dB(A), lies on one extreme of these curves and corresponds to about 17% of the population highly annoyed, and 1% complaining.

1-6. SPEECH INTERFERENCE

One of the most annoying effects of noise is its interference with speech. The two basic methods for assessing whether (and to what extent) a sound will do so are to find the Articulation Index (AI) or the Speech Interference Level (SIL). Both these measures inherently reflect the degree of masking that the sound achieves over the speech. Affected is the intelligibility of the speech, where intelligibility is taken to be the proportion of words correctly perceived. SIL is the method most widely used, but it is based on the AI, which was derived earlier and is presented here first. Both methods are based on the fact that the frequen-

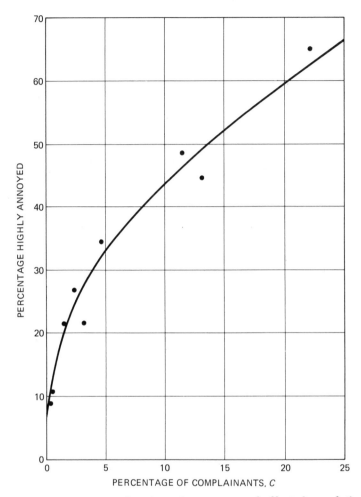

Fig. 1-11 Annoyance as a function of percentage of affected population who complain, from the EPA levels document. An equation for this curve is Percentage Highly Annoyed = $12.3\sqrt{C} + 4.3$.

cies of human speech lie in the range of *approximately* 200–5000 Hz, and that it is therefore this frequency range of the background noise that must be assessed for its masking properties.

1-6.1 Articulation Index (AI)

There are a number of forms of Articulation Index, one of which is the subject of ANSI S3.5-1969. Because, however, AI is no longer very much used, we present here the simplest method of calculation we know.

In Fig. 1-12 the area between the upper and lower curves is the area in which our normal speech most often falls. For example, the parts of our speech which a frequency analysis reveals to be in the $\frac{1}{3}$ octave band centered at 500 Hz is found to have SPLs in the range 43–73 dB for the great majority of people nearly all the time. (This is a 30 dB range.)

Furthermore, studies have shown that these frequencies are not all equally important in speech intelligibility: speech intelligibility depends on our accurately perceiving speech in approximately the 1500-3000 Hz range much more than in the frequency ranges to either side of this. The number of dots at each center frequency in Fig. 1-12 reflects the relative importances to speech intelligibility of the various frequencies.

To calculate AI, one plots on Fig. 1-12 the $\frac{1}{3}$ octave frequency spectrum of the background noise and simply counts the number of dots lying between the spectrum and the upper curve. Dividing this number by 100 gives AI.

Thus, an intense background sound having a spectrum lying on or above the upper curve will have an AI of zero—implying a large interference with speech. A sound of such low intensity as to have a spectrum lying on or below the lower

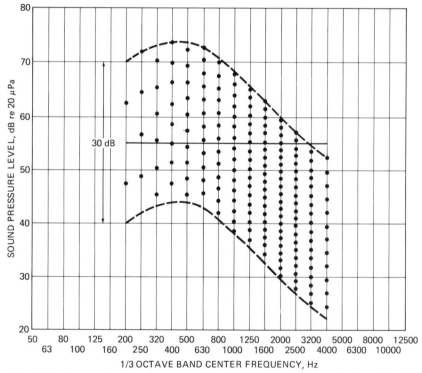

Fig. 1-12 Curves for the computation of the Articulation Index, AI. The straight line is the spectrum given in the example of Section 1-6.1.

curve will have an AI of at least unity—implying little or no interference with speech.

Generally an AI of less than 0.3 is taken to be unsatisfactory for good communication and greater than 0.7 is considered very good. Between 0.5 and 0.7 may be called good, while the range 0.3 to 0.5 is marginal but perhaps acceptable. Experiments have been performed to related these figures to the percentage of standard syllables, words or sentences uttered by various voices and perceived correctly by various subjects in various situations. These agree that an AI of 0 is associated with virtually zero intelligibility and an AI of 1 is associated with virtually complete intelligibility, but the various studies do not allow any single AI figure to be described as a firm borderline of acceptability. The figures given above have arisen empirically from everyday experience, and are summarized in Table 1-8. It should, however, be borne in mind that in special situations (such as aircraft communication, where trained individuals use oft-repeated, phonetically-efficient phrases) an acceptable AI may be less than 0.3.

TABLE 1-8. Conclusions to be Drawn from Various Levels of Articulation Index.

Articulation Index, AI	Above 0.7	Between 0.5 and 0.7	Between 0.3 and 0.5	Below 0.3
Conclusions	Very good communication	Good communication	Marginal but acceptable communication	Unsatisfactory communication

Example

Problem: A background noise in an office is expected to have a perfectly flat $\frac{1}{3}$ octave band spectrum of 55 dB between 200 and 4000 Hz. Calculate the AI and comment.

Solution: Plot the spectrum on Fig. 1-12 as shown there by the horizontal 55 dB line. Count the dots between this line and the upper curve; there are 59. Divide by 100 to get an AI of 0.59, say 0.6. From Table 1-8 we may expect good communication.

An example of how AI is used is contained in Chapter 6, which covers the matter of speech intelligibility and privacy in open-plan offices, and has helpful information on how AI is affected by speaker—receiver orientation and separation distance.

1-6.2 Speech Interference Level (PSIL)

A simplified derivative of AI are various speech interference calculations which demand an octave rather than a $\frac{1}{3}$ octave analysis of the background noise.

Rather than weight these levels according to the importance of their frequencies, the levels are simply averaged arithmetically and the result reported in decibels.

The original Speech Interference Level (SIL) was the average of the SPLs in octave bands that have now been supplanted by the preferred bands adopted universally and given in Table 1-1. The SIL has thus changed slightly, and should now be called Preferred Speech Interference Level or PSIL, though the word "preferred" is often dropped, while the "P" in PSIL is retained. PSIL is the arithmetic average of the SPL in dBs in the three octave bands centered on 500, 1000, and 2000 Hz. Thus it considers the frequency range from about 350 to 2850 Hz (the former being the lower cutoff frequency of the 500 Hz band, and the latter the upper cut-off frequency of the 2000 Hz band). Later modifications to PSIL may include the 4000 Hz octave band.

For reference, one may suppose that

$$PSIL \doteq SIL + 3$$

for many commonly found noises. (It is also so, as given in Table 1-5, Section 1-4.7, that an approximate relationship between PSIL and dB(A) is PSIL $\doteq$ dB(A) – 7 for most sounds other than speech—for which special case PSIL $\doteq$ dB(A) – 2.)

What values of PSIL imply excessive speech interference?

Experiments performed by Beranek for different voice levels and different distances between speakers have produced the data given in Table 1-9.

TABLE 1-9. Preferred Speech Interference Levels (PSIL) of Steady Continuous Noises at which Reliable Communication is Barely Possible. Table Applies to Male Vocal Effort and to Speaker and Listener Facing each other.

Separation Distance, ft (m)	Preferred Speech Interference Level, dB			
	Normal Effort	Raised	Very Loud	Shouting
0.5 (0.15)	74	80	86	92
1 (0.3)	68	74	80	86
2 (0.6)	62	68	74	80
4 (1.2)	56	62	68	74
6 (1.8)	52	58	64	70
12 (3.7)	46	52	58	64

Source: Beranek, *Noise and Vibration Control.*
Subtract 5 dB for female voices.
Add 7 to the PSILs shown for the table to refer approximately to dB(A).

This table applies to male voices, and one may subtract about 5 dB to apply it to females (and probably to children). As one can see, there is a 6 dB decrement in level for each doubling of distance—which derives from an assumption of open-air listening where sound decays at this rate according to the well-known inverse square law. For talkers indoors, where the speech may be reinforced by reflections, all the numbers should differ somewhat, though for simplicity one generally ignores this. One may note, further, that the different columns of the table increase in steps of 6 dB from left to right, making it possible to construct Table 1-9 from memory if one remembers, say, that for a man's normal voice effort at 4 ft, the maximum acceptable PSIL is 56.

Use of the table still requires one to make certain judgments regarding the distance between speakers and what constitutes reasonable vocal effort. How such judgments may be made is illustrated in the example.

Example

Problem: The results of an octave analysis of the sound at a passenger's head position on a long-distance bus are given below. Determine whether this spectrum is acceptable for speech.

Octave band center frequencies, Hz	125	250	500	1000	2000	4000
SPL, dB		90	87	80	75	70	67

Solution: PSIL = (80 + 75 + 70)/3 = 75 dB. We ignore the levels at frequencies other than 500, 1000, and 2000 Hz. To evaluate this PSIL value, one must judge that communication on any bus should be acceptable for females and children as well as males; and one may suggest that for a long-distance bus one should be able to talk with a normal voice to people one seat away, say 2 ft. The PSIL limit from Table 1-9 for a "normal" effort male voice at 2 ft is 62 dB, i.e., 57 dB for a female or child. Since the calculated PSIL of 75 dB is in excess of this, the sound is unacceptably loud from a speech standpoint.

This example of sound on board a transportation system illustrates the use of PSIL; Chapter 5 covers the subject of transportation noise to travelers in more detail.

1-6.3 Speech Interference from Time-Varying Sounds

Time varying sounds result in time-varying interference with speech, the acceptability of which is not easy to determine. However the EPA levels document describes L_{eq} = 60 dB(A) as being acceptable for speech outdoors, and L_{eq} = 45 dB(A) for indoors.

BIBLIOGRAPHY

On basic acoustics and acoustical measurements:

J. P. Broch. *Acoustic Noise Measurements*. Brüel and Kjaer, Naerum, Denmark, 1971.

A. P. G. Peterson and E. E. Gross, Jr. *Handbook of Noise Measurement*. General Radio (Gen Rad), Concord, Mass., 1972.

On noise descriptors, subjective responses and criteria:

"Information on Levels of Environmental Noise Requisite to Protect Public Health and Welfare with an Adequate Margin of Safety." Environmental Protection Agency Report 550/9-74-004, March 1974.

K. D. Kryter. *The Effects of Noise on Man*. Academic Press, New York, 1970.

L. L. Beranek (ed.), *Noise and Vibration Control*. McGraw-Hill, New York, 1971, Chapter 18.

2

Noise of surface transportation to nontravelers

MALCOLM J. CROCKER*

2-1. INTRODUCTION

Surface transportation noise is not a new problem. In the time
of Julius Caesar, because of their noise, chariots were banned
from the streets of Rome during darkness [1]. In the nineteenth century, horse-drawn carriages with iron-rimmed wheels
created such a noise on the cobblestoned streets of London
that they prevented conversation when they passed and created a continuous deep rumble of background noise throughout the city [2]. Outside some hospitals straw was spread on
the streets to reduce the noise. In the late nineteenth and
early twentieth centuries, New York City was reputed to be
the noisiest city in the world because of its elevated railways.
In 1929 a Noise Abatement Commission was appointed "to
study noise in New York City and develop means of abating
it" [1, 3, 4].

*Ray W. Herrick Laboratories, School of Mechanical Engineering,
Purdue University, West Lafayette, Indiana 47907.

In Great Britain a Committee on Noise in the Operation of Mechanically Propelled Vehicles was appointed in 1934, and recommended legislation to provide sound level limits for vehicles [1, 5]. The passage of this legislation was delayed by the Second World War and in fact took place only in 1968, and was enforced only from 1970 [6, 7, 8]. Even so, recent enforcement in Britain has been neither strict nor effective: in 1970, out of approximately 14,000 vehicle noise convictions, about 13,300 concerned faulty silencers, and only two were for exceeding the maximum permitted sound level so painstakingly established [7]. In contrast, vehicle sound level limits had been enacted in Germany in 1935.

In the U.S., "quantitative" limits for vehicle noise were first introduced by individual states, beginning in 1965 with New York, which set a limit of 88 dB(A) measured at 50 ft from the center of the traffic lane for a vehicle passing at less than 35 mph; 90 dB(A) was the level at which violations were issued. However, in the three-year period after passage of the act, only 45 arrests were made after checking of 9569 vehicles. The chosen noise levels appear to have been too lenient and insufficiently enforced. In 1969, California amended its vehicle code to include noise limits which, for vehicles traveling at more than 35 mph, were set at 86 dB(A) for cars and 90 dB(A) for trucks, again at 50 ft. California found that the levels had initially been set too high, and modified them in August 1970. The California code has been somewhat better enforced, and in the first six months about 500 citations were issued. Other states such as Oregon, Pennsylvania, Connecticut and Indiana, and cities such as Chicago [9], have adopted similar vehicle codes.

New traffic and individual vehicle noise regulations and standards have been introduced recently in many countries, and some are still in the planning stage. We will discuss several of them later in this chapter.

2-2. TRAFFIC NOISE SOURCES, EXPOSURE AND EFFECTS

2-2.1 Sources

The main sources of surface transportation noise are (a) traffic, i.e, trucks, cars and motorcycles; (b) railroads; (c) off-road recreational vehicles; (d) ships; and (e) hovercraft. Traffic noise is the most important of these and will be discussed in some detail. Railroad noise, and particularly ship noise, are generally of less importance and are discussed in less detail. Hovercraft noise is a problem in only very few isolated instances, and has earned little legislative attention, so it is not covered. Off-road recreational vehicles, which may at times include motorcycles, are discussed in Chapter 4.

There are several reasons for the emergence of traffic noise as the main source of noise annoyance in most industrial countries. The power/weight ratio of trucks and cars has been constantly increased to permit higher payloads, and more speed and acceleration, the resulting higher power engines being usually more noisy than lower power ones. The number of vehicles has increased dramatically in most countries over the last twenty or thirty years. This, combined with the movement of people from country to city and the natural increase in urban population, has exposed more and more people to more and more traffic noise.

The statistics to prove these tendencies are not difficult to find. In the U.S., there are now half as many cars as there are people, a ratio which has doubled in

TABLE 2-1.   Estimated Noise Energy for Different Transportation Vehicles in 1970 [14].

Major Category		Noise Energy (Kilowatt–Hours/Day)
Aircraft		
	● Commercial–4-engine turbofan	3,800
	● Commercial–2- and 3-engine turbofan	730
	● General aviation aircraft	125
	Helicopters	25
Highway Vehicles		
	● Medium and heavy duty trucks	5,000
	● Sports, compact, and import cars	1,000
	● Passenger cars (standard)	800
	● Light trucks and pickups	500
	● Motorcycles (highway)	250
	City and school buses	20
	Highway buses	12
Recreational Vehicles		
	● Minicycles and off-road motorcycles	800
	Snowmobiles	120
	Outboard motorboats	100
	Inboard motorboats	40
Rail Vehicles		
	● Locomotives	1,200
	Freight trains	25
	High speed intercity trains	8
	Existing rapid transit	6.3
	Passenger trains	0.63
	Trolley cars (old)	0.50
	Trolley cars (new)	0.08
		Total ~ 15,000

● Top ten categories which each generate at least 125 kilowatt-hours per day.

25 years from a value only now being reached by Western European nations who have, however, had approximately a tenfold increase in this ratio in the same period. Continued growth is dependent on energy supply, but the approximately hundred million US vehicle registrations [10] is a formidable number already.

U.S. domestic intercity travel by common carriers has continued to grow [11], through airline traffic has done so more than surface traffic, and passenger rail traffic has decreased. However, common carriers only account for about 10% of the total intercity passenger-miles traveled. The remaining 90% is contributed by private cars and other modes. Per capita domestic intercity passenger-miles have increased at an annual rate of about 3%, about twice the rate of the population increase [11]. Intercity freight transportation has also increased, though at a slower rate.

Vehicle average speed, at least up to the time of the U.S. 55 mph speed limit, has grown continuously [12]. Various highway noise surveys [1, 8, 13–20] have illustrated the dependence of traffic noise on vehicle speed, as well as on the proportion of trucks, the overall traffic volume, and the degree of inclines. That truck noise is especially important is shown in Table 2-1, which contains an estimate of the average A-weighted noise energy generated during a 24-hour day in the U.S., though the figures bear only an approximate relationship to the relative amounts of noise from the various sources that are actually heard. Figure 2-1 is another illustration, from the California Highway Patrol, of the same point.

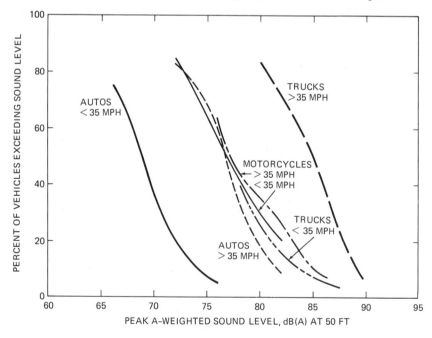

Fig. 2-1   Noise survey results of the California Highway Patrol [19].

2-2.2 Exposure

The available data on the sources of noise that annoy the most people underline the importance of traffic noise. Table 2-2 shows that the main sources of noise annoyance in typical U.S. urban communities are traffic noise, aircraft, and voices in that order, according to a 1971 survey of 1200 people [21]. Table 2-3, from Ref. 22, shows the number of people estimated to live in U.S. residential areas who are subjected to urban traffic noise and freeway traffic noise at or above outdoor L_{dn} levels of 60, 65, 70, 75, and 80 dB(A). An L_{dn} of 55 dB(A) (see Chapter 1) generally produces no public reaction, although the noise is noticeable; an L_{dn} of 60 dB(A) produces sporadic complaints; while an L_{dn} of 65 dB(A) normally produces widespread complaints and several threats of legal action [22]. Thus, from study of either Table 2-2 or Table 2-3, it seems about three or four times as many Americans are disturbed by road traffic noise as by aircraft noise.*

This statistic is confirmed in several other industrialized countries by social surveys. For example, in Britain 36% of people when at home complained of noise from road traffic, while only 9% complained of aircraft noise in a 1962 study in central London [24]. Also in Britain, a large number of people have accepted one of the house sound insulation grant schemes for protection against excessive aircraft noise but have then insulated rooms facing a busy road and have left unprotected the rooms and roofs facing the airport and flight routes [25]. In Norway, in a 1968 survey [26], 20% of people complained about road traffic noise, while only 5% about aircraft. These data tend to substantiate the hypothesis that about four times as many people are disturbed by road traffic noise as by aircraft noise in most industralized countries.

The well-known Wilson Committee Report on Noise [24] drew attention to

*See also Ref. 23.

TABLE 2-2. Percentage Contribution of Various Noise Sources Identified by Respondents Classifying Their Neighborhood as Noisy (72% of 1200 Respondents) [21].

Source	Percentage
Motor Vehicles	55
Aircraft	15
Voices	12
Radio and TV Sets	2
Home Maintenance Equipment	2
Construction	1
Industrial	1
Other Noises	6
Not Ascertained	8

TABLE 2-3.   Numbers of People in the U.S. (in Millions) Subjected to Various Levels of Noise [22].

Outdoor L_{dn} Exceeds	Urban Traffic	Freeway Traffic	Aircraft Operations	Total
60	59.0	3.1	16.0	78.1
65	24.3	2.5	7.5	34.3
70	6.9	1.9	3.4	12.2
75	1.3	0.9	1.5	3.7
80	0.1	0.3	0.2	0.6

the serious problem of road traffic noise in Britain in 1963. From 1963 to 1970 the number of vehicles increased by 40%, when it was estimated that between 19% and 46% of the UK urban population of 45 million lived on roads with traffic noise levels that were undesirably high for residential areas [8]. If the noise characteristics of individual vehicles in Britain do not change, and if the number of vehicles continues to grow at the present rate, by 1980 the percentage of the population undesirably exposed will increase to from 30% to 61% [8]. The percentage of the British rural population of 9 million undesirably exposed was estimated in 1970 to lie between 13% and 30% [8].

2-2.3 Effects

Though most people, when questioned, mention traffic noise as one of the *most* troublesome urban annoyances [24, 27] it is found that they then mention it *least* frequently in their complaints to authorities. Aircraft noise seems to get by far the most publicity in newspapers and on the radio and television, and also most of the U.S. government noise research money (by a factor of ten) [28]. What is the reason for this? The answers are not too hard to supply and are explained by the nature of traffic noise itself. Langdon [29] gives a good description of traffic noise.

In most cities, networks of roads are intimately interwoven with residential areas and there is a continuous fairly steady background of traffic noise. For houses close to roads with low traffic flow rates, fluctuations in noise may be more pronounced. However, for houses situated further away from roads the fluctuations decrease and the noise level is more steady [30]. In such locations it is difficult for people to identify individual sources. This steady quality is quite different from aircraft or railroad noise, where the noise levels fluctuate considerably with time and the individual noise sources can be easily identified.

Since most people drive cars they are also less likely to complain about traffic noise than aircraft noise. Still, a very small percentage of people pilot aircraft or travel as passengers on scheduled airline services. Thus, there is probably more acceptance of traffic noise than of aircraft or railway noise. There are other points to consider. With aircraft noise it is comparatively easy to complain to a responsible body or authority. In Sydney, Australia, for example, a number is listed in the telephone directory for aircraft noise complaints. Also, people living close to airports have often organized action or protest groups with considerable effect, sometimes winning court cases and damages in the U.S.A. With traffic noise the situation is different. There is rarely a central authority to which complaints can be directed, and then, as discussed already, it is hard to identify the noise culprit and more difficult to imagine that official action will be taken to rectify the situation. Noise action and protest groups on traffic noise are much less common than with aircraft noise.

In most cases, traffic noise varies from mildly annoying to very annoying. It is unlikely that hearing damage will be caused. However, traffic noise will interfere with conversation, listening to the radio and television, and with use of the telephone. It can interfere with sleep, although (except for some truck traffic) most traffic noise also decreases at night.

2-3. RATING TRAFFIC NOISE ANNOYANCE

Various attempts have been made to determine the relationship between traffic noise levels and annoyance. Other authors have discussed this in some detail [e.g., 29, 31]. Appendix D of Ref. 22 gives further examples of traffic noise annoyance studies which will not be discussed here.

As Chapter 1 indicates, annoyance due to noise is difficult to measure because of contaminating influences such as dissatisfaction with the environment or neighborhood, other forms of pollution, and many other public and personal factors. Annoyance is normally investigated with social surveys, and then considerable skill must be used to design the questionnaire so that the results are not unduly influenced by the phrasing of the questions. A choice of noise units must be made, and the annoyance must be measured on some scale.

2-3.1 Annoyance Scales

Two annoyance scales which have received frequent use [29] are shown in Table 2-4. The *verbal annoyance scale* is based on techniques originated by Guttman [32] and extended and refined by Borsky [33, 34], McKennell [35, 36] and others. A problem with this scale is the uncertainty as to whether equal increments of annoyance are represented by the intervals between the different verbal descriptors. The *semantic differential scale* has been used by Griffiths and Langdon [37] and others. This scale presents a sequence of numbers between opposites such as *good* and *bad* or *satisfactory* and *unsatisfactory*. A problem with this scale is that verbal definitions are not given to the numbers, and different

TABLE 2-4.   Scales of Annoyance.

a. *Verbal annoyance scale*

not at all	a little	moderately	very much

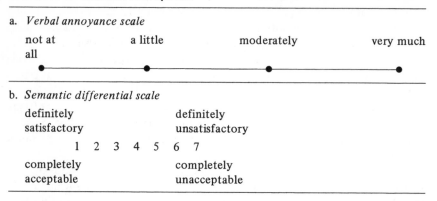

b. *Semantic differential scale*

definitely   definitely
satisfactory   unsatisfactory

        1   2   3   4   5   6   7

completely   completely
acceptable   unacceptable

people may associate different meanings with the same number. Combinations of such scales and other different scales have been devised and used. For example a Swedish study on road traffic noise used an eleven point scale [38].

2-3.2 London Noise Survey Results

The results of an early study, made in association with the London Noise Survey [39, 40], are given in Fig. 2-2. People were asked to judge the *loudness* of two noises on a four-point verbal scale. The first judgment was of an artificial noise made by a falling ball acoustic calibrator, producing a sound level of 72 dB(A) at

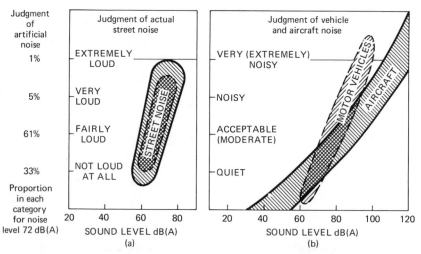

Fig. 2-2   Comparative judgments of different noises [39, 40].

a standard distance of one meter. The second judgment was of street noise existing at the time (normally just outside each person's house). These two *loudness* judgments are shown in Fig. 2-2(a). The inner shaded area at the right of the figure is drawn to include about 70% of the results for street noise. At the left of the figure is shown the proportion of people making judgments, on the same four-point scale, of the standard artificial noise at a level of 72 dB(A). Figure 2-2(b) shows, for comparison, the results obtained with groups of listeners under controlled conditions making judgments of the *noisiness* of motor vehicles and aircraft. There is some difference between people's judgment of loudness and noisiness. However, of most interest is the apparent difference between people's reaction to motor vehicle and aircraft noise. Such results could be used to set noise standards for vehicles. However, the considerable scatter in the results (different people judge the same traffic noise level very differently on the loudness or noisiness scales) and the problem of specifying proper noise test procedures for individual vehicles are problems which first must be overcome.

2-3.3   The Traffic Noise Index

In an attempt to develop acceptability criteria for traffic noise from roads in residential areas, Griffiths and Langdon [37] produced a new unit for rating traffic noise, the Traffic Noise Index (TNI):

$$\text{TNI} = 4\,(L_{10} - L_{90}) + L_{90} - 30$$

They measured A-weighted traffic noise at fourteen sites in the London area and interviewed 1200 people at these sites in the process. Griffiths and Langdon excluded sites with noise sources other than traffic (e.g., one site with 50 aircraft overflights per day was omitted). They chose roads which were straight and level, without intersections or traffic circles. Some sites were near two-lane roads and some near four-lane. Housing at the sites consisted of two-storey units of similar age and construction.

A seven-point noise dissatisfaction scale was used:

definitely satisfactory   1 2 3 4 5 6 7   definitely unsatisfactory

(see Table 2-4).

They then used regression analysis to fit curves to the data (see Fig. 2-3–2-6), which indicated that L_{10} was better at predicting dissatisfaction than L_{50} or L_{90}, and that TNI was also superior to L_{10}, L_{50}, and L_{90}. Certainly, the collapse of the data in Fig. 2-6 looks impressive. Each point in Figs. 2-3–2-6 represents the median dissatisfaction score at one site. Figure 2-7 shows that, at each site, there is a considerable variation in individual dissatisfaction scores of people. Individual dissatisfaction depends on a person's susceptibility to noise and, no doubt, on other factors.

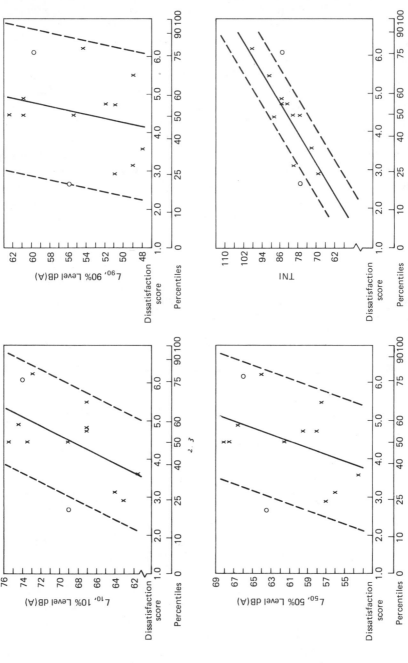

Figs. 2-3, 2-4, 2-5, and 2-6 Dissatisfaction scores compared with various traffic noise descriptors. Broken lines show confidence limits at ± 2 standard errors of estimate; open circles—pilot survey; crosses—main sites only [37].

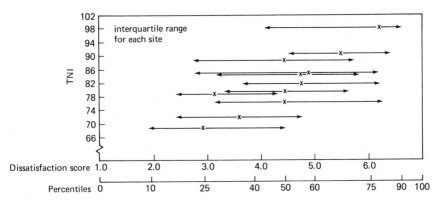

Fig. 2-7   Plot of median dissatisfaction vs. TNI, with interquartile ranges shown [37].

Use of the Traffic Noise Index has not been widespread. The index attempts to make an allowance for the noise variability (with the first term in the above equation), since fluctuating noise is commonly assumed to be more annoying. The second term, L_{90}, represents the background noise level, while the third term is simply a constant chosen to yield more convenient numbers.

Some doubt has been cast on the conclusions of Griffiths and Langdon by workers in France [41], the U.S.A. [42, 43, 44] and elsewhere [45]. Josse [41] and Schultz [42, 43] suggest that the very short sample times (100 seconds in each hour) used by Griffiths and Langdon may have resulted in underestimates of L_{10} and overestimates of L_{90}. If true, this would severely underestimate the difference $L_{10} - L_{90}$. This suggests that the coefficient 4 in the equation may be too large and, in fact, casts doubt on all the data on which the Traffic Noise Index is based. Schultz [31], however, although questioning the TNI data and results, feels that the TNI is a step forward over previous noise ratings because it includes an allowance for the annoyance due to noise level variations.

However the British Government has adopted L_{10}—averaged over 18 hours from 6 am to midnight—as the noise index to be used to implement planning and remedial measures to reduce the impact on people of road traffic noise [18]. TNI is not considered today to be significantly superior to either L_{10} or L_{eq} (see Section 2-3.5).

2-3.4   Noise Pollution Level

In a later survey, Robinson [46, 47] again concluded that, with fluctuating noise, L_{eq}, the equivalent continuous noise level on an energy basis, was an insufficient descriptor of the annoyance caused by fluctuating noise. He included another term in his Noise Pollution Level L_{NP} (see Chapter 1), which he defined as

$$L_{NP} = L_{eq} + k\sigma$$

where k is a constant and σ is the standard deviation of the sound level. He also offered an alternative expression:

$$L_{NP} = L_{eq} + a(L_{10} - L_{90}).$$

Robinson examined the available Griffiths and Langdon data [37] and concluded that $a = 1.0$ and $k = 2.56$ were good choices for the constants in these equations. He then examined the aircraft noise experiments of Pearsons [48] and found that L_{NP} predicted very well Pearsons' data points and the trade-off between duration and level for individual flyover events. A-weighted levels were used in L_{NP} with traffic noise, and perceived noise levels (PNdB) were used with aircraft noise.

The Noise Pollution Level is potentially attractive because, in principle, it allows annoyance from aircraft, traffic, and perhaps other sources such as industrial noise to be determined. If the Noise Pollution Level is in fact a good descriptor of fluctuating noise, this may explain the fact that in several studies [24, 49] the equivalent energy level alone has been shown to be a good descriptor of annoyance. The reason could be that for different noises where the level fluctuations are similar, the second term in the equations for L_{NP} is constant, and L_{eq} is a sufficient descriptor of annoyance.

Since L_{NP} is intended to describe *both* traffic and aircraft noise successfully, and since it describes traffic noise as well as or better than the Traffic Noise Index, the Noise Pollution Level has received more interest, attention, and use. However, the superiority of L_{NP} over all other forms of noise rating has not been proved in practice.

2-3.5   Equivalent Noise Level and Speech Interference

In 1974, a report in six volumes was prepared for the Transportation Research Board of the National Cooperative Highway Research Program of the National Academy of Sciences [49]. In Volume 6, Pearsons and co-workers give a literature review and report some of their own experimental work on the effects of traffic noise on sleep and the relationship between traffic noise and speech intelligibility. Figure 2-8 shows the annoyance results of Pearsons and co-workers, using six different noise ratings: $L_1, L_{10}, L_{50}, L_{90}, L_{eq}$, and L_{NP}. As expected for all of the noise measures, annoyance increases with level. The shapes of the curves, however, do vary considerably when the annoyance is less than very annoying. In particular, the L_{10} and L_{eq} measures exhibit a very steep rise in annoyance from the categories of slightly to moderately annoying for no increase in noise level—presumably one of their drawbacks. However, except for the case of L_{NP} in the extremely annoying category, the standard deviation of L_{eq} for a specified response category was in all cases less than or equal to the standard deviation of the other noise measures. This is an advantage in the use of L_{eq}, since

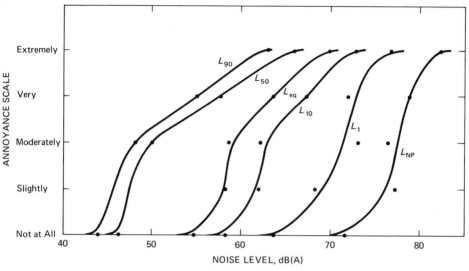

Fig. 2-8   Annoyance as a function of noise level [49].

there is more confidence in the annoyance scores predicted. There is one clear advantage of L_{eq} over L_{10}, however, in the case of noise containing short-duration, high-level single events. If the events do not occur for more than 10% of the time, then L_{10} will be insensitive to these high-level events and will tend to represent the "background" noise. In fact, for a noise measure L_n to be useful, the intruding noise events must be present for *more* than n% of the time. This suggests that L_{10} would be unsuitable as a measure of aircraft noise annoyance, and that it might be a possible source of error in Griffiths' and Langdon's results for low traffic flows.

One interesting and new result obtained by Pearsons and co-workers [49, Vol. 6] is that the annoyance caused by traffic noise is affected considerably if speech is being listened to (see Fig. 2-9). This is not an unexpected result, since it might be expected that the annoyance felt by a person would depend on the type of activity and degree to which it is interrupted. Three sets of data are shown in Fig. 2-9. When the speech-to-noise ratio is very low [-16.0 or -9.5 dB(A)] and the comprehension is only low or moderate, the annoyance caused is much higher at any level of traffic noise than if the speech-to-noise ratio is high [+7.5 dB(A)] and the comprehension is correspondingly higher. It is to be expected that if the traffic noise level is such that comprehension of speech becomes difficult, the annoyance will consequently increase, whatever the traffic noise level. This result serves to confirm intuition that annoyance is not independent of the activity in which an individual is engaged, nor probably of many other influencing factors.

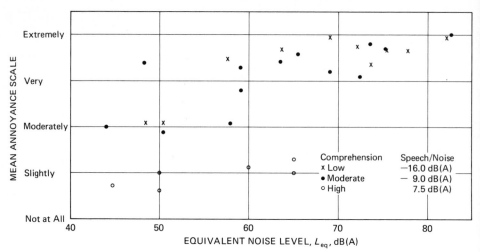

Fig. 2-9   Annoyance ratings vs. L_{eq} of traffic noise—with speech, for low, moderate, and high comprehension [49].

Pearsons and co-workers [49, Vol. 6] conclude that L_{eq} is the most stable and consistent and the least variable of the measures they used to predict traffic noise. They suggest that L_{eq} is superior to L_1, L_{10}, L_{50}, L_{90}, L_{NP}, and TNI in predicting annoyance from traffic noise. They also found that, for a constant L_{eq}, an increase in traffic noise variation tends to *decrease* annoyance whether speech is present or not. This result contradicts that of Griffiths and Langdon [37], and casts further doubt on their results and the basis for the units TNI and L_{NP} already discussed. Although it should be remembered that the data collected by Pearsons and co-workers [49] are limited there is a growing acknowledgment that L_{NP} and TNI in their present forms are not significantly superior to L_{eq}, which also has the virtue of being more simple.

2-3.6 Day–Night Sound Level

Another method of rating community noise annoyance has been developed by the U.S. Environmental Protection Agency [22, 50]. This is the day–night sound level, L_{dn} (see Chapter 1). It is based on the energy-equivalent, A-weighted sound level L_{eq}, with a 10 dB nighttime penalty applied. Eldred [51] has studied community reaction at 55 locations where there had been community noise problems and plotted these against a normalized day–night sound level (see Chapter 1, Fig. 1-9); the normalization is a modification of the type originally applied by Rosenblinth and Stevens [52] in rating aircraft noise in the community in the early 1950s.

Following the earlier approach, Eldred made corrections to L_{dn} in 5 dB steps to obtain a normalized L_{dn} (see Table 1-7). Eldred found that the normalization procedure reduced the standard deviation from 7.9 to 3.3 dB. The normalizing corrections have parallels in at least two community noise standards—British Standard 4142 [53] and the related ISO/R 1996 [54], as described in Chapter 8.

Obviously, for people to be annoyed by noise, they must first hear it. They cannot hear it if it is much below the background noise. However, once it is intense enough to be perceived, then the attitudinal factors in the normalizing corrections (see Table 1-7) become important.

Some authors [49 (Vol. 6), 22, 50] believe that speech interference; radio, television, and telephone interference; and sleep interference are equally important criteria on which to base vehicle noise regulations. For such interference, an absolute unit such as the uncorrected L_{dn} or L_{eq} must be used rather than an attitudinally corrected unit such as the normalized L_{dn}.

The speech–sleep interference and the community reaction criteria would at first sight appear to pose conflicting requirements. However, Eldred has pointed out that they need not [51]. He suggests that both absolute (uncorrected) L_{dn} and normalized L_{dn} should be considered in establishing community noise standards, regulations, and ordinances. The absolute value of L_{dn} should be used to compare with sleep and speech interference criteria, and its normalized value should be used with Fig. 1-9 to determine community reaction. No criterion should be exceeded.

2-4. MEASUREMENT OF NOISE EMISSION FROM INDIVIDUAL VEHICLES

The strategies by which traffic noise annoyance can be reduced include the following [7]:

a. quiet individual vehicles;
b. ensure that owners maintain and use their vehicles to minimize disturbance to others;
c. protect people from noise by house insulation schemes and by constructing roadside noise barriers;
d. reroute traffic away from residential areas and in particular from sensitive places such as hospitals and schools;
e. plan new roads and communities to reduce traffic noise effects on people by making use of the shielding effects of distance, of hills, cuttings, and valleys, and of industrial buildings in routing of new roads.

As with any noise control problem, it is useful to consider this in the *source-path-receiver* framework. The measures a–e above fall into each of the three categories. Although *source* control is often the best practice, sometimes it is necessary to treat the *path* and *receiver* also. Kihlman [55] has calculated that, in Sweden, about equal amounts of money should be spent on vehicle noise control and on building insulation (mostly installing double windows) to achieve optimum results economically. Traffic noise regulations in different countries may be aimed at achieving one or more of a–e above. In order to achieve noise control at the source (a, b), a standard, repeatable method must be devised of measuring the noise from individual vehicles.

2-4.1   Different Types of Individual Vehicle Noise Tests

Fifteen years ago many different ways of measuring vehicle noise were used in different countries. Mills [56] summarized the test procedures in use at that time as follows.

Tests with vehicle stationary

a. The engine is operated at constant speed in neutral gear; usually the engine speed corresponds to maximum brake horsepower.
b. The engine is operated at constant speed, as in a, but in gear, with the driving wheels supported by free-running rollers.
c. Test as b, but using braked rollers.

Tests with vehicle in motion

a. Constant speed (part throttle) tests in a specified gear and at a given road or engine speed.
b. Constant speed (full throttle) tests in a specified gear and at a given road or engine speed, the speed being controlled by some loading device, e.g., vehicle brakes (a towed dynamometer is also possible).
c. Acceleration tests in which the vehicle approaches the measuring position at a given engine or road speed, in a specified gear, accelerates past the microphone by opening the throttle fully at a prescribed point, and subsequently fully closing the throttle.

All the above types of tests had advantages and disadvantages: some were difficult to perform or gave results which were not representative of the noise produced by the vehicle in normal traffic conditions. However, a major cause of difficulty was that it was impossible to compare results from one country with those from another using a different test. This was a big disadvantage considering the fact that vehicles often cross state and national boundaries in use and are frequently imported or exported.

In 1958, an ISO Technical Committee started work on a vehicle noise measurement standard and in 1960 produced a draft which was finally adopted in 1964 as ISO Recommendation R362 [57]. All the above stationary and moving types of tests were considered. The stationary tests described are all difficult to perform, since it is difficult to maintain constant operating conditions with the engine in neutral gear or using free-running rollers. The braked roller test is easier to apply, and the noise closely resembles that which a vehicle would produce on the road, since the engine and transmission are loaded. However, some experience is still required to control the throttle and roller brake to obtain constant operating conditions. Such tests are still used by some manufacturers for diagnostic purposes, although sometimes an engine-mounted or engine-shaft loading dynamometer is used instead of rollers. Stationary tests are allowed in ISO/R 362 as an additional test to that recommended (an acceleration test at full throttle).

The ISO Committee felt that noise measurements should correspond to vehicle conditions which give the highest reproducible noise levels under normal driving conditions. Thus the constant-speed test described above was rejected because although the noise results are repeatable they tend to be low. Constant speed tests at full throttle give repeatable results, but with high-powered vehicles, very high brake temperatures can cause "brake fade" before completion of the tests. The acceleration test was recommended by ISO because it produces repeatable results which reasonably describe a vehicle's maximum noise potential provided the vehicle approaches at a specified speed with a high engine rpm.

2-4.2   The ISO Recommended Vehicle Noise Test [57]

ISO/R 362 calls for the acceleration test to be carried out in an extensive open space. A suitable test site would be a flat surface of at least 50 m radius, of which the central 20 m would be concrete, asphalt or a similar hard material. Deviations from ideal conditions can occur if: the ground surface is absorbing (snow, long grass, loose soil); reflections occur from buildings, trees, or people (the sum of the angles subtended at the test vehicle by buildings, etc., within the 50 m radius should be less than 90°); nonlevel ground; and wind noise (wind and instrumentation noise should be at least 10 dB less than vehicle noise).

The measuring positions are shown in Fig. 2-10. The microphone positions are 7.5 m from the vehicle path and at a height 1.2 m above the ground. At least two measurements should be made on each side of the vehicle as it passes. The vehicle approaches the line AA and on reaching it the throttle is fully opened as quickly as possible and held fully open until the rear of the vehicle reaches the line BB, when the throttle is closed as rapidly as possible. For a vehicle with a manually operated gearbox (transmission), if it is fitted with a two, three, or four speed gearbox, second gear should be used. If it is fitted with more than four gears, the third should be used. Overdrive and auxiliary gears are excluded. The vehicle should approach AA at a steady speed corresponding to either an engine speed of three quarters that at which the engine develops maximum power, or to three quarters of the maximum engine speed permitted by the governor (if fitted), or to 50 km/h, whichever is lowest. If the vehicle has an automatic gearbox, it should approach AA at a steady speed of 50 km/h or at three

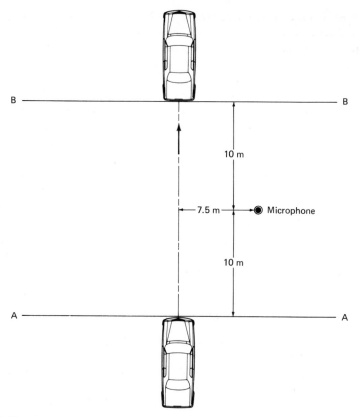

Fig. 2-10   Measuring positions in the ISO/R 362 vehicle noise measurement procedure [57].

quarters of its maximum speed, whichever is lower. Wherever alternative forward drive positions are available, the position which results in the highest mean acceleration of the vehicle between AA and BB should be selected.

A high quality sound-level meter should be used and A-weighting, fast response should be selected. The noise level recorded is the highest level measured during passage of the vehicle from AA to BB and the arithmetic average of the four sound levels recorded for each vehicle side should be reported.

2-4.3 Present Vehicle Noise Tests in Different Countries

Since the ISO/R 362 acceleration test was accepted by the ISO Council in 1964, many countries have adopted this test or devised tests similar to the ISO test, only making a few minor changes to suit local needs. A few countries such as Switzerland use stationary tests. Some other countries in Europe are considering the use of constant-speed drive-by tests. A survey [8] was made in 1970 of countries with vehicle noise legislation which is based on vehicle noise measurements and the result is given in Table 2-5. Several changes have since occurred. For example, Australia issued a draft standard [58] in 1975, in which seven different types of tests are discussed: (a) acceleration from standing start, (b) low-speed drive-by, (c) high-speed drive-by, (d) far stationary test (7.5 m), (e) close stationary (0.5 m), (f) acceleration, (g) moving body noise. It is proposed to adopt one or more of these tests in the final standard for use in new vehicle approval and vehicle-in-use regulations.

In the search for "ideal" vehicle noise tests, standards are continually under review or revision and any discussion such as this can become quickly outdated.

TABLE 2-5. Countries and States with Vehicle Noise Legislation Based on Measurements, from a 1970 Survey [8].

Country	Status		Type of Test				Vehicle Types		
	Current	Proposed	ISO Static	ISO Mobile	Special	Operating	Motor cycles	Cars	Commercial
Austria	X		X	X			X	X	X
Belgium	X			X			X	X	X
Bermuda	X				X		X		
California	X				X	X	X	X	X
Denmark	X				X		X		
Finland	X		X				X	X	X
France	X			X			X	X	X
Germany	X		X	X	X		X	X	X
Italy	X		X		X		X	X	X
Japan	X				X		X	X	X
Luxembourg	X			X			X	X	X
Netherlands	X			X			X	X	X
New York State	X				X		X	X	X
Norway	X			X		X·	X		
Portugal	X			X			X	X	X
South Africa		X		X			?	X	X
Spain	X				X		X	X	X
Sweden		X			X		X	X	X
Switzerland	X		X		X		X	X	X
U.K.	X			X		X	X	X	X

The ISO/R 362 test is presently under revision.* In the U.S.A., the Motor Vehicles Manufacturers' Association is presently funding a study at the Herrick Laboratories, Purdue University, to determine if indoor (all-weather) noise tests can be used to supplement drive-by ones.

The present vehicle noise tests in the United States and United Kingdom will now be briefly reviewed. Table 2-6 shows a summary of the major vehicle noise measurement standards used in the U.S.A. The Society of Automotive Engineers (SAE) has written three separate test procedures for cars and light trucks, for heavy trucks and buses, and for motorcycles. The three SAE recommended practices are all similar to ISO/R 362 although all slightly different from each

*See Revision of ISO/R 362 (Document 43/1 N 263E), May 1965. The revisions in this document are of a minor nature. However, the new draft "Acoustics–Survey method for the measurement of noise emitted by stationary motor vehicles," Document 43/1 N 262E, describes a method of measuring the noise produced by a stationary vehicle. Many countries have desired to use such a convenient method for checking the noise from vehicles. The noise is measured at 0.5 m from the exhaust or engine when the engine reaches half of the rpm for maximum power–or for diesel engines, at the governed speed.

TABLE 2-6.   Major Noise Measurement Standards in the U.S. for Aircraft (see Chapter 3) and for Vehicles, in Both Cases as Heard in the Community.

			Applicable Noise Measurement Standard				
Category	None	FAR Part 36	ISO R362	CHP[a] Article 10	SAE[b] J331 Proposed	SAE J366	SAE J986a
General Aviation Aircraft	X						
V/STOL	X						
Business Jets		X					
Subsonic Commercial Aircraft		X					
Trains	X						
Passenger Cars and Light Trucks GVW < 6000 pounds			X	X			X
Trucks and Buses GVW > 6000 pounds			X	X		X	
Motorcycles			X	X	X		

[a]California Highway Patrol.
[b]Society of Automotive Engineers.

other and from ISO/R 362. The major differences are: (a) microphone location is at a height of 4 ft (1.2 m), and a distance of 50 ft (15 m) from centerline of vehicle path instead of a distance of 7.5 m; (b) flat test site must be free of large objects within 100 ft (30 m) of either vehicle of microphone; (c) cars and light trucks must approach the microphone at 30 mph (47.5 km/hr) in a low gear; and 25 ft (7.5 m) ahead of the microphone they must accelerate at wide open throttle so that maximum rated rpm is achieved 25 ft (7.5 m) beyond the microphone; recommended maximum level being $86 + 2$ dB(A) at 50 ft (15.2 m); (d) heavy trucks and buses must approach the microphone in a gear ratio selected so that at a point 50 ft (15 m) ahead of the microphone, the engine rpm is no higher than $\frac{2}{3}$ the maximum rated or governed speed; and must then accelerate fully so that maximum rated engine speed is achieved between 10 and 100 ft beyond the microphone and without exceeding 35 mph (55.3 km/hr) at 100 ft (30 m) past the microphone, recommended maximum level being $88 + 2$ dB(A) at 50 ft; (e) motorcycles must approach the microphone at $\frac{2}{3}$ maximum rated engine rpm, then at a point at least 25 ft (7.5 m) ahead of the microphone they must accelerate fully to achieve maximum rated engine rpm at a point between 15 and 25 ft (4.5 and 7.5 m) past the microphone, recommended maximum levels varying according to engine capacity and date of manufacture.

British Standard BS 3425:1966 is very similar to ISO/R 362, differences being minimal.

2-4.4   Use of Drive-By Tests to Set Limits

Mills and Robinson [59] carried out subjective tests with vehicles undergoing three types of tests. These were (a) the ISO/R 362 drive-by test; (b) a constant-speed cruise-by test, and (c) a full-power, "wheelbraked," drive-by test. Groups of people were asked to rate the noise of individual vehicles on a six point subjective scale:

A	B	C	D	E	F
—	quiet	acceptable	noisy	excessively noisy	—

with the two end points A and F undefined. Diesel-engined vehicles, gasoline-engined vehicles, and motorcycles were examined separately. For convenience in expressing the results, a 10 point scale was later used with 2 assigned to quiet, 4 to acceptable, 6 to noisy and 8 to excessively noisy. It was found that the halfway point between acceptable and noisy corresponded to 80 dB(A) for the diesel or gasoline-engined vehicles and 82.5 dB(A) for the motorcycles. The result for trucks is shown in Fig. 2-11 (measured by Mills and Robinson in 1960) [24, 59]. Also shown are similar results for trucks measured in 1968, showing that they have become 10 or 12 dB(A) noisier.

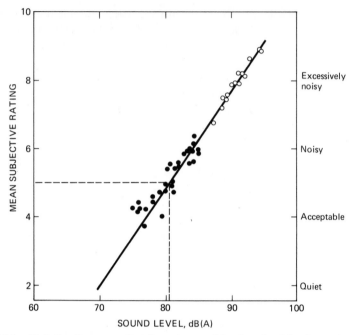

Fig. 2-11   Relation between diesel truck sound level and subjective rating; filled circles—values measured in 1960; open circles—values measured in 1968; dashed line—demarcation [24, 67].

There are problems in translating these acceptability judgments into regulations. With the exception of tire noise [60–65], drive-by tests give the *maximum* noise potentials of vehicles, even though most vehicle operations produce lower noise levels. Thus, drive-by test levels cannot be used directly to predict levels vehicles will make en masse in traffic, since some vehicles will be accelerating, some decelerating, and some traveling at constant speed. They may also be loaded or unloaded and traveling up or down inclines. Also, there are engineering and economic considerations which must be included before regulations can be written. The information on these subjects is extensive, and space cannot be found for it here. A discussion of some of the individual noise sources from the standpoint of inside-vehicle noise is contained in Chapter 5; interested readers are also referred to Refs. 14 and 60–81.

2-5. RAILROAD AND SHIP NOISE

2-5.1 Railroad Noise—Sources, Exposure, and Effects

In most countries, noise from railroads does not represent such a widespread source of community annoyance as traffic or aircraft noise [82]. This is particularly true in the U.S.A. and other countries where railroad operations have been declining for many years (as opposed to road traffic and aircraft operations, which have been on the increase). However, in some countries, notably Japan, where high-speed (250 km/h) trains were introduced in 1964, railway noise is posing a very severe community noise problem [83]. In Britain, West Germany, Canada, France, the U.S.A. and other countries, high-speed trains are either planned or already in service [84], and may increase in number as a result of energy problems. Such systems also pose similar community noise problems [85].

That railroad noise does not produce widespread public annoyance in most industrialized countries is claimed by Close and Atkinson [82] in the U.S.A. and by the Wilson Report [24] in Great Britain. One might also deduce that it is a lesser source of annoyance than road vehicle or aircraft noise by studying Table 2-1. However, there is evidence in the U.S.A. and Britain [24] that some people living near railroad tracks and marshalling (shunting) yards are greatly disturbed by noise. During 1960 the London Transport Executive received 98 noise complaints spread over 33 different causes [24]. Although this is not a large number, it does show that annoyance is present.

Data and literature on railroad noise are much scarcer than on vehicle or aircraft noise; however, they are not entirely lacking, and work on determining levels and mechanisms in railroad noise generation has proceeded for some years in several countries, including the U.S.A. [82, 85, 86, 87], Britain [84, 88, 89], France [90], Germany [91], Canada [92, 93] and Japan [83]. There has, however, been little evaluation of the disturbance it causes [83, 87,88].

Rail systems have many different applications and it is convenient to divide them into railroads, and rail transit systems, as shown in Fig. 2-12 [14].

In many countries, and particularly the U.S.A., long-distance passenger traffic has declined steadily in the last 25 years. Rail transit systems in cities have also declined, but this trend may be reversed if governments continue to encourage more use of public transportation in order to conserve energy. However, U.S. railroad freight traffic has shown not a decline but a small increase in ton-miles carried of about 25% over the last 25 years [14].

There are two basic sources of noise in railroads: the locomotive and the cars hauled. As seen from Fig. 2-12, in 1970 there were 27,000 locomotives in use in the U.S.A. Of these, 99% were diesel–electric powered and most of the rest were electric. The noise levels typically produced are shown in the shaded parts of Fig. 2-12.

2-5.2 Ship Noise—Sources, Exposure and Effects

Ships pose a smaller community noise problem than railroads, since only communities close to ports or on rivers or canals can be affected. Most of the main noise sources, the engine, gears and propellers, are below water level or are

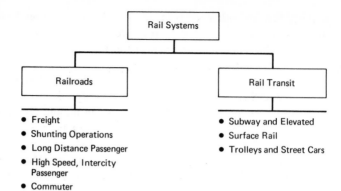

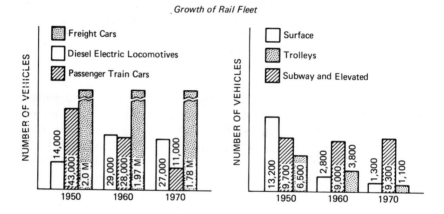

Growth of Rail Fleet

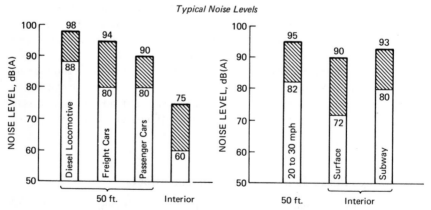

Typical Noise Levels

Fig. 2-12   Characteristics of rail systems [14].

enclosed by the ship structure, causing noise and vibration on board (see Chapter 5).

In Australia there are standards both for measurement of noise on board [94] and alongside [95] vessels. The second standard is for type, acceptance, and monitoring tests and requires the A-weighted sound level to be determined. The test site is required to be free of sound-reflecting objects like barriers, hills, rocks, bridges, and buildings within at least a 100 m radius of the microphone. The main propulsion machinery must be run at least at 95% of rated speed for the particular ship loading condition. Auxiliary equipment should be run at normal speed. The microphone should be placed on a boat, pier or bank not less than 3 m and not more than 6 m above the water surface. A distance of 25 m distance from the microphone to the vessel side is preferred, but it should in any case be not less than 20 m, nor more than 35 m. If the distance is different from 25 m, then the sound level should be corrected according to the following equation:

$$L_{A,25} = L_{A,d} + 20 \log (d/25),$$

where $L_{A,25}$ is the corrected, A-weighted sound level at 25 m, and $L_{A,d}$ is the sound level measured at d meters.

Most commercial, cargo, and passenger ships do not pose a severe community noise problem because maximum noise is normally developed at maximum power, which is not usually attained in a harbor. There are some exceptions, however. In some cases, ships in port require auxiliary power supplies for electricity. If these are run on deck they can be very noisy. Also, some ships run their main propulsion systems for an hour or two before leaving port, causing noise problems. Unfortunately, policing merchant ship noise regulations would be difficult because such vessels are normally from foreign countries and are only in port for a few days or a week at most. An international noise regulation would be desirable.

We know of no survey of ship noise effects on the community.

2-6. TECHNICAL BASIS FOR NOISE REGULATIONS

The preceding parts of this chapter constitute a technical background or basis for regulating the noise of road traffic, trains, and ships. The discussion has covered numbers of vehicles; numbers of people affected; effects of noise on people (such as annoyance and speech interference); different units for rating noise; and the measurement of noise from individual vehicles, trains, and ships. All these, as well as economic, practical, and other factors should be considered before designing noise regulations and limits.

Before proceeding further it is best to define the terminology used here. The terminology common in other environmental areas such as air and water quality, and also suggested by Franken [96] for noise control, will be adopted here. Note that this terminology has not been universally accepted, and other writers will use these same words with differing meanings. *Criteria* are statements of the effects produced by various levels of exposure to noise, including consideration of such effects as speech interference, sleep interference, and hearing damage. In this terminology, criteria do not amount to a statement about acceptability. *Standards* describe the noise environment considered acceptable in a given situation. Such standards represent long-term goals which a noise control program can be designed to reach. Lastly, *limits* are specific rules established for individual noise sources. Limits will require not only a knowledge of the criteria and standards, but also other factors such as numbers of noise sources and of people affected, technical and economic constraints, enforcement problems, and local, social; and other conditions. In order to approach noise standards in an orderly fashion, noise limits should be chosen with a schedule. In this way the standards may be thought of as the desired noise environment while the limits are means chosen to reach this environment.

It would be very convenient if there were a direct relationship known between the noise standards and the limits, so that the improvement in the noise environment could be directly predicted from an adoption of or change in noise limits. Unfortunately, the situation is usually too complex to predict such improvements accurately. The large numbers and variations in noise sources and in their use, and the complicated acoustic paths and local environmental conditions make such predictions very difficult. For this reason it is normally necessary to study the available criteria, to choose standards, and then to select noise limits for the most important or easily controlled sources and monitor the resulting environmental changes [96]. Such standards can also be used in national or local planning and community decisions on land use.

The noise criteria to be studied before legislation is created should extend beyond annoyance and the factors that influence it, such as speech interference, sleep interference, and interruption of work and thought. Also to be considered are hearing loss and the nonauditory effects of noise on health; Chapters 11 and 12 discuss these matters in detail. Here we will cover only speech interference and hearing loss, and those very briefly. It is the position of many U.S. acousticians including Kryter [97] that nonauditory diseases will not be caused by noise levels lower than those necessary to produce noise-induced hearing loss, a conclusion that is not universally accepted in other countries.

2-6.1 Community Reaction

Community reaction may be assessed according to the method of Eldred [51] introduced in Section 2-3.6.

2-6.2 Speech Interference

Speech interference due to transportation noise can be evaluated in three ways, all based on Beranek's findings [98] reviewed in Chapter 1.

a. A 6 dB(A) increase in sound level will require people to increase their vocal effort from normal to raised. If they are conversing in a normal voice, out of doors, at a maximum distance which ambient noise will allow, then a most conservative standard would be that traffic (or train or ship) noise must not be allowed to raise this ambient level by more than 6 dB(A), or that the noise source must not be more than 5 dB(A) above the ambient. For the purposes of this discussion, a constant relationship such as dB(A) = PSIL + 7 is assumed between dB(A) and PSIL (see Chapter 1).

b. A less conservative standard [93] would assume that people adjust their distance apart in the presence of traffic (or, again, train or ship) noise and that they accept that a raised voice will be necessary. Assuming a distance of 2 m apart, good speech intelligibility can be maintained until the sound level reaches 64 dB(A) for males and 59 dB(A) for females [98]. A standard then might be that the source noise level should not exceed 60 dB(A) out of doors. Indoor conversations should not be interrupted at such levels, since attenuations of about 15 dB(A) are usually found for houses with some open windows. If noise levels exceed this standard even for a short time then good speech intelligibility will be lost. This is still a conservative standard, and it is difficult to use with fluctuating sounds from road traffic or rail traffic.

c. The U.S. Environmental Protection Agency [22] has considered the intermittency of noise in identifying interfering noise levels. The U.S. EPA has suggested a goal of 100% speech intelligibility indoors, which requires a noise level not exceeding an L_{eq} of 45 dB(A). Assuming a house structure attenuates noise by 15 dB(A), then an outside noise level of L_{eq} = 60 dB(A) should not be exceeded. This outside level of 60 dB(A) allows normal conversation at distances up to 2 meters with 95% speech intelligibility. The U.S. EPA has applied a further 5 dB(A) margin of safety reducing the outside level recommended to 55 dB(A) and also applied a 10 dB nighttime penalty (to allow for other adverse effects of noise, e.g., sleep interference). Thus the EPA has recommended that an L_{dn} of 55 dB(A) should not be exceeded in residential areas.

2-6.3 Noise-Induced Hearing Loss

In Ref. 22, the U.S. EPA concluded that hearing will not be impaired if the noise level does not exceed an L_{eq} of 73 dB(A) when averaged over 8 hours. Making corrections for 24-hour exposure, 365 days/year instead of 250 days/year, and for intermittency of the various noise sources involved, a standard of L_{eq} = 70 dB(A) averaged over 24 hours has been selected [22].

To evaluate the effect of transportation noise, corrections are probably necessary for the percentage of time spent indoors shielded by a building structure, and by the fact that other noise must be added to the transportation noise when assessing exposure. The choice of corrections is necessarily arbitrary; interested readers are referred to May [93].

2-6.4 Sleep Interference

Experiments on sleep interference due to noise have often used transportation noise sources. However they have generally been carried out in the laboratory, using unrepresentative populations, and have not been prolonged enough to properly assess adaptation. Thus no useful standard can be confidently defined to prevent sleep interference due to surface transportation noise. Readers are referred to Chapter 13 for a general review of the subject.

2-6.5 Assessment of Overall Community Impact Due to Noise

As already discussed, it is very difficult to determine a relationship between changes in vehicle noise limits and the ambient noise in a community. However, some recent attempts have been made, and one by the U.S. EPA [20] is described briefly here. It should be realized that all such attempts are highly empirical and must be proved effective before we rely on them. The method defines the initial and final acoustic environment and also the relationship between any specified noise environment and the expected human impact.

The human impact in this study deals not with annoyance or community reaction, but with a single factor affecting those responses—speech interference. The effect on speech intelligibility is assessed for a mixture of known and unknown sentences as the level of the (presumably steady) noise environment is increased above 45 dB(A). The intelligibility for sentences (first presented to listeners) decreases to 90% when this level is increased by 19 dB(A) and to 50% when this level is increased by 24 dB(A). Intelligibility does not drop quite so rapidly if the sentences are known to the listener. Reference 20 suggests that for a mixture of known and unknown sentences, intelligibility decreases rapidly when the environmental noise level of 45 dB(A) is increased by more than 20 dB(A). For this reason an increased level of 20 dB(A) is chosen as the level for 100% impact on exposed people. The impact for levels between 0 and 20 dB above the chosen environmental level is assumed to vary linearly [20]. The impact is represented as a fractional index, FI, with an FI of 1 representing an impact of 100%:

$$FI = 0.05 (L - L_c), \quad L > L_c$$

where L is the L_{eq} for the environmental noise and L_c is the identified criterion level.* Note that FI can exceed one. The EPA has chosen $L_c = 45$ dB(A) for indoors and $L_c = 55$ dB(A) for outdoors. The number of people affected by environmental noise is introduced into the analysis and the total impact is assessed by multiplying the number of people P_i exposed at the ith level by the fractional impact FI_i associated with that level:

$$P_{eq} = \sum_i P_i FI_i.$$

The number P_{eq} is the magnitude of the impact on the population and is numerically equal to the equivalent number of people all of whom were 100% impacted.

The change in impact associated with some action leading to a noise reduction or increase, or a change in population subjected due to a change in land use, may be assessed by comparing the impacts before and after the change. The percentage change in impact Δ is:

$$\Delta = \frac{100 \, [P_{eq}(\text{before}) - P_{eq}(\text{after})]}{P_{eq}(\text{before})}.$$

The U.S. EPA has applied this impact concept to estimate the changes in number of people impacted by noise when state and city automobile noise regulations and federal motor carrier (truck) regulations become effective. Various empirical assumptions had to be made [20], e.g., vehicle mixture on urban streets of 1% heavy duty, 6% medium duty trucks, and 93% automobiles, and on freeways of 10% trucks and 90% automobiles. Other assumptions concerned the population density in the vicinity of urban roads and around freeways, the average number of lanes on a freeway, the average speed on roads and freeways, and the reduced use of crossrib tires on trucks. Assuming that new trucks of over 10,000 lb gross vehicle weight (GVWR) were required not to exceed the following sound level limits at 50 ft: 83 dB(A) after 1976, 80 dB(A) after 1980 and 75 dB(A) after 1982, then the total numbers of people in millions impacted were calculated to be: 37.3 in 1974, 33.6 in 1976, 29.8 in 1980, 27.2 in 1982, 17.0 in 1990 and 15.9 in 1992 [20]. Such predicted results, although empirical only, may give a useful indication of the trends to be expected when different noise limits are applied.

---

*Note that EPA has used *criterion* here with a different meaning from that defined at the beginning of Section 2-6.

2-7. NOISE LIMITS

2-7.1 U.S. Motor Carrier (Truck) Noise Regulations

The U.S. Noise Control Act of 1972 required the EPA to publish proposed noise emission regulations for motor carriers engaged in interstate commerce within 9 months after the date of enactment of the Act. The EPA proposed the following [99]:

a. 90 dB(A) at 50 ft in speed limits greater than 50 mph;
b. 86 dB(A) at 50 ft in speed limits equal to or less than 35 mph;
c. 80 dB(A) at 50 ft on level streets in speed limits equal to or less than 35 mph;
d. 88 dB(A) at 50 ft under stationary runup test; and
e. Visual exhaust and tire inspections.

The U.S. Noise Control Act of 1972 required the writing of "noise emission standards setting such limits . . . which reflect the degree of noise reduction achievable through application of the best available technology taking into account the cost of compliance." The proposed noise emission standards received considerable comment, some of which was reported in the final noise emission standards [100]. The final standards were essentially identical to those proposed earlier [99], and became effective on October 15, 1975. In February 1975 the U.S. Department of Transportation issued Proposed Compliance Procedures [101] mainly dealing with the measurement procedures to be used to determine whether vehicles are in compliance with the final noise standards [100].

The Environmental Protection Agency states [100] that "motor vehicles are the principal source of environmental noise in urban areas." The Environmental Protection Agency has also identified trucks as a major noise source both in urban situations and in high speed freeway situations [102].

The regulations so far discussed have been for trucks in-use. However, the EPA has also issued proposed regulations for new trucks [102]. The EPA initially proposed in these regulations to limit medium and heavy duty truck noise produced in a low speed acceleration test to the following: January 1977, 83 dB(A); January 1981, 80 dB(A); and January 1983, 75 dB(A). The measurement procedure on which these levels are based is an EPA test similar to SAE J366b. The EPA has estimated the approximate increased retail prices (shown in Table 2-7) to achieve these levels [108].

There was considerable comment from manufacturers that the costs shown in Table 2-7 were greatly underestimated. The revised new-truck noise regulation is less restrictive on truck noise and will limit the noise from new medium and heavy duty trucks to: 83 dB(A) after January 1978 and 80 dB(A) after January 1982. The 75 dB(A) limit specified in the earlier proposed standards [102] has

TABLE 2-7. Increase in Truck Prices Due to Noise Controls by Type
of Truck [108].

Type of Truck	83 dB(A)		80 dB(A)	
	Price increase	Percent increase	Price increase	Percent increase
Medium, gasoline	$ 35	0.6	$180	3.1
Heavy, gasoline	125	1.1	255	2.2
Medium, diesel	426	5.8	850	11.5
Heavy, diesel	356	1.4	589	2.3

temporarily been omitted because of the controversy it produced. However, the revised noise standard reserves a section for a more restrictive limit for new trucks manufactured after January 1985. This additional section would be added if data can be obtained to show that lower noise limits can be attained with available technology at an acceptable cost.

2-7.2   U.K. and Other European Vehicle Noise Regulations

In the last ten or fifteen years, many countries including the United Kingdom have produced noise limits which must be met by newly manufactured vehicles. In the U.K., regulations both for new vehicles and for those in use came into effect in April 1970 [6, 7]. The British limits for new vehicles, Regulation 23, follow fairly closely those recommended by the Economic Commission for Europe (ECE) and the more recent technical directive of the Commission of the European Economic Community (EEC). These new vehicle noise limits are shown in Table 2-8. Table 2-9 shows the U.K. in-use limits (Regulation 89), which as already mentioned have been only weakly enforced.

The limits given in Table 2-8 are seen to be above the levels found by Mills and Robinson [59] (see Section 2-4.4) to be acceptable in subjective tests. However, it may be argued that higher limits may be allowed, since vehicles are rarely operated at full power and on the average produce less noise. The lower limits in column 5 of Table 2-8 more nearly approach the Mills and Robinson levels with the exception of those from heavy vehicles.

2-7.3   Other Vehicle Noise Limits

In the U.S.A., various states (New York and California were the first) and cities (Chicago was the first) have issued noise limit regulations for new vehicles and those in use. In Tables 2-10 and 2-11 the Chicago noise limits are reproduced; note that they are to be measured at 50 ft (using the SAE J331, J184, J336,

TABLE 2-8. Noise Limits Which Motor Vehicles Must Be Constructed to Meet [6, 7]ᵃ.

	ECE Regulation, dB(A)	EEC Directive, dB(A)	U.K. Reg. 23 (Vehicles First Used After Apr. '70), dB(A)	Future U.K. Limits		
				Limit, dB(A)	Effective Date	
					Manufacture	First Use
Motorcycles over 125 cc	84–86 (depending on cylinder capacity)	(not covered)	86	(to be determined)	—	—
Cars	84	83	84	80	Apr. '73	Oct. '73
Diesel engined cars, cross-country vehicles	(not covered)	(not covered)	84	82	Apr. '74	Oct. '74
Light passenger vehicles	85	84	84			
Light goods vehicles	85	84	85			
Heavy vehicles up to 200 hp	89	90	89	86	Apr. '74	Oct. '74
over 200 hp	92	92		89		

ᵃAll measurements by ISO procedure at 7.5 m.

TABLE 2-9. In-Use Limits for the U.K. [6, 7].

Vehicle Type	Noise Level, dB(A) at 7.5 m Using ISO/R 362	
	Vehicles First Used Before Nov. 1, 1970	Vehicles First Used On or After Nov. 1, 1970
Cars	87	87
Motorcycles over 125 cc	90	89
Heavy vehicles	92	92

and J986 type practices). The European limits in Tables 2-8 and 2-9 are measured at 7.5 m (25 ft), and should thus be reduced between 5 and 6 dB(A) for comparison. When U.S. EPA limits become effective they pre-empt state or city regulations, which must then be identical with the federal limits. Thus state and city truck limits will soon be superseded.

2-7.4 Traffic Noise Standards

The limits described in Sections 2-7.1, 2-7.2, and 2-7.3 apply to individual vehicle noise. In addition to controlling the noise of each vehicle, however, stan-

TABLE 2-10. Chicago's New-Vehicle Noise Limits (at 50 ft) [9].

Type of Vehicle	Date of Manufacture	Noise Limit
1. Motorcycle	before 1 Jan. 1970	92 dB(A)
Same	after 1 Jan. 1970	88 dB(A)
Same	after 1 Jan. 1973	86 dB(A)
Same	after 1 Jan. 1975	84 dB(A)
Same	after 1 Jan. 1980	75 dB(A)
2. Any motor vehicle with a gross vehicle weight of 8,000 pounds or more	after 1 Jan. 1968	88 dB(A)
Same	after 1 Jan. 1973	86 dB(A)
Same	after 1 Jan. 1975	84 dB(A)
Same	after 1 Jan. 1980	75 dB(A)
3. Passenger cars, motor-driven cycle and any other motor vehicle	before 1 Jan. 1973	86 dB(A)
Same	after 1 Jan. 1973	84 dB(A)
Same	after 1 Jan. 1975	80 dB(A)
Same	after 1 Jan. 1980	75 dB(A)

TABLE 2-11.   Chicago's In-Use Vehicle Noise Limits (at 50 ft) [9].

Type of Vehicle	Noise Limit in Relation to Posted Speed Limit, dB(A)	
	35 mph or Less	Over 35 mph
1. Any motor vehicle with a manufacturer's GVW rating of 8,000 lbs. or more, and any combination of vehicles towed by such motor vehicle		
before 1 Jan. 1973	88	90
after 1 Jan. 1973	86	90
2. Any motorcycle other than a motor-driven cycle		
before 1 Jan. 1978	82	86
after 1 Jan. 1978	78	82
3. Any other motor vehicle and any combination of motor vehicles towed by such motor vehicle		
after 1 Jan. 1970	76	82
after 1 Jan. 1978	70	79

dards exist for traffic noise in general. Such standards are intended to ensure that highways are sited properly in relation to nearby development, and that nearby development (and especially residential and other noise-sensitive development) is properly located in relation to highways. Standards of this type also motivate highway engineers and architects to consider noise attenuation measures in their respective designs, whether these be to depress the highway, to erect noise barriers, or to insulate dwellings.

In the U.S., the Federal Highway Administration (FHWA) has issued regulations requiring highways constructed using federal funds to meet certain noise level standards on adjoining land [103]. These levels apply to the highway design year, perhaps 10 to 20 years from the start of construction, and refer to the noise in any hour, rather than over a longer period, on the reasoning that the worst-case noise level will thereby be considered.

The FHWA standards are set out in Table 2-12. Either L_{eq} or L_{10} may be used as the noise descriptor, though FHWA recognizes that L_{eq} may be superior where traffic volumes are light. The standard is deliberately drawn up to strike a balance between the desirable and the feasible. The levels are above those which result in no noise problem, but are below those of many present-day expressways.

The FHWA standards are, in general, in line with similar standards and guidelines in other countries. In the United Kingdom, where recent research [104]

TABLE 2-12.   Federal Highway Administration Noise Standards for
New Highway Construction.[a]

Activity Category	Design Noise Level, dB(A)		Description of Activity Category
	L_{eq}	L_{10}	
A	57 (exterior)	60 (exterior)	Tracts where serenity and quiet are especially important
B	67 (exterior)	70 (exterior)	Residences, motels, schools, churches, hospitals, etc.
C	72 (exterior)	75 (exterior)	Developed lands other than those above
E	52 (interior)	55 (interior)	Building interiors

[a]Either L_{eq} or L_{10} may be used—not both—and an hourly measure applies. The land use descriptions are further qualified in Ref. 103, and a category D is also reserved for undeveloped land. The interior noise levels may be established by subtracting from outdoor levels the attenuation expected of the particular wall and window constructions involved.

confirms the adequacy of the descriptors L_{10} and L_{eq} over the more complex L_{NP} or TNI, similar compromise values have been adopted. If one compares the recommended outdoor noise level in residential areas—often the situation of greatest concern—the FHWA hourly L_{10} (or L_{eq}) of 67 (or 70) dB(A) compares strikingly with the British L_{10} of 68 dB(A) [105]. The British figure also applies to a period well into the future, and though an 18-hour measure of L_{10} is used, the hours considered are the more heavily traveled hours, and the two standards are comparable.

In France, a proposed guideline [105] for outdoor residential land uses specifies an L_{eq} of 65 to 68 dB(A), which translates to an L_{10} of 68 to 71. The guideline would apply to the design hour volume when the expressway opens, and is also comparable to the U.S. and British standards.

The Swiss [105] have an existing guideline of $L_{50} = 60$ dB(A) for the period 6 am to 10 pm, which also translates to comparable values of L_{eq} and L_{10}. It is applied to Level of Service C traffic volumes when the highway is opened. This level of service describes a traffic situation for which noise levels are at about their maximum for any highway.

2-7.5 U.S. Railroad Noise Emission Standards

In 1974, the U.S. EPA issued proposed railroad noise emission limits [106], and in 1975 issued final noise emission limits [107]. The EPA was directed under the Noise Control Act of 1972 to produce "Federal regulations on interstate railroad equipment and facilities to include noise emission standards setting such limits on noise emissions which reflect the degree of noise reduction achievable [using] . . . the best available technology taking into account the cost of compliance."

The railroad noise emission standard applies to moving and stationary locomotives and to railroad cars. Horns, bells, whistles, and other warning devices—though to many people the most noticeable sources of railroad noise—are not regulated. The EPA believes that state and local authorities are better able to regulate noise from warning devices, and that national uniformity of treatment is not necessary. The regulation became effective on December 31, 1976. On that date and until December 31, 1979 the limit for stationary locomotives measured at 30 m (100 ft) is 93 dB(A) under any throttle setting and 73 dB(A) under idle; for moving locomotives it is 96 dB(A) under any condition of grade or acceleration or deceleration on track with a less than a two-degree curve (radius of curvature greater than 873 meters). After December 31, 1979 the limits for stationary locomotives are 87 dB(A) at any throttle setting and 70 dB(A) at idle, and for moving locomotives 90 dB(A) under any operating condition.

For rail cars the standard sets the following limits. Effective December 31, 1976, rail cars or combinations of such must not produce sound levels at 30 m (100 ft) in excess of 88 dB(A) for speeds less than 72 km/hr (45 mph) or 93 dB(A) at speeds greater than 72 km/hr. The track must be free of special track work or bridges and must again have less than a two-degree curve.

Measurement criteria are specified. These include restrictions on the acoustic environment including a clear test site within a radius of 30 m (100 ft) of the locomotive or rail car and within a radius of 30 m from the microphone. The microphone is situated 30 m from the track centerline. There must be no precipitation and there are statements on how level the ground must be and on visibility of the rails at the microphone location.

REFERENCES

1. M. E. Delaney, "Traffic Noise." R. W. B. Stephens and H. G. Leventhall, eds., *Acoustics and Vibration Progress Volume 1* Chapman and Hall, London, 1974, Chapter 1.

2. B. Winter, *Past Positive*. Chatto and Windus, London, 1971.

3. R. W. Young, "Scales for Expressing Noise Level—Summary." In J. D. Chalupnick, ed., *Transportation Noises*, University of Washington Press, Seattle and London, 1970, pp. 129–150.

4. H. Fletcher, "Apparatus Used in Noise Measurements." In *City Noise*, Noise Abatement Commission, New York, 1930, p. 120.

5. "Final Report on Noise in the Operation of Mechanically Propelled Vehicles." HMSO, London, 1937.

6. "Motor Vehicles." Construction and Use Regulations 1969, S.I. No. 321, London, 1969, Regulations 23 and 89.

7. The Noise Advisory Council, "Traffic Noise: The Vehicle Regulations and Their Enforcement." HMSO, SBN11 750478 5, London, 1972.

8. Working Group on Research Into Road Traffic Noise, "A Review of Road Traffic Noise," Road Research Laboratory, RRL Report LR357, Crowthorne, England, 1970.

9. City of Chicago, Department of Environmental Control, Noise Ordinance, July 1, 1971.

10. United States Department of Commerce, "The Noise Around Us," COM 71-00147, September, 1970.

11. D. O. Dickerson, ed., "Transportation Noise Pollution Control and Abatement," NASA Contract NGT 47-003-028, 1970.

12. W. H. Close, "Surface Transportation Noise," Paper presented at Inter-Noise 74 Seminar, Washington, D.C., September, 1974.

13. D. R. Johnson and E. G. Sanders, "The Evaluation of Noise From Freely-Flowing Traffic." *J. Sound Vib.*, 7, 287, 1968.

14. Wyle Laboratories, "Transportation Noise and Noise From Equipment Powered by Internal Combustion Engines." EPA Report No. NTID 300.13, 1971.

15. E. J. Rathé, "Ueber den Laerm des Strassenverkehrs."*Acustica*, 17, 268, 1966.

16. T. Priede, "Noise and Vibration Problems in Commercial Vehicles," *J. Sound Vib.*, 5, 129, 1967.

17. M. E. Delaney, "Prediction of Traffic Noise Levels." National Physical Laboratory Acoustics Report AC56, 1972.

18. Department of the Environment, "Motorway Noise and Dwellings," Digest 153, Building Research Establishment, Garston, Watford, England, May 1973.

19. California Highway Patrol, "Noise Survey of Vehicles Operating on California Highways." Sacramento, California, 1971.

20. U.S. Environmental Protection Agency, "Background Document for Proposed Medium and Heavy Truck Noise Regulation." EPA Report No. 550/9-74-018, October, 1974.

21. Bolt, Beranek and Newman, "Survey of Annoyance from Motor Vehicle Noise," BBN Report No. 2112, June, 1971, for Automobile Manufacturers Association, Inc.

22. U.S. Environmental Protection Agency, "Information on Levels of Environmental Noise Requisite to Protect Public Health and Welfare With An Adequate Margin of Safety." EPA Report No. 550/9-74-004, March, 1974.

23. C. Bartel, L. Godby, and L. C. Sutherland, "National Measure of Aircraft Noise Impact Through the Year 2000." Wyle Laboratory Research Report WCR-74-13, April, 1975.

24. Wilson Committee, "Noise—Final Report." HMSO, London, 1963.

25. P. J. Dickenson, Letter to the Editor, *Noise Control Engineering*, 4 (1), 4, 1975.

26. "Urban Traffic Noise." OECD, Paris, 1975.

27. A. Alexandre, "Traffic Noise Control in Europe." *Noise Control Engineering*, 2 (2), 69–73, 1974.

28. M. J. Crocker, "Training, Education and Research Problems in Noise Control Engineering." *Noise Control Engineering*, 15 (2), 50, 1975.

29. F. J. Langdon, "The Problem of Measuring the Effects of Traffic Noise." In A. Alexandre et al., *Road Traffic Noise*, Halstead Press, John Wiley & Sons, New York, 1975, Chapter 2.

30. G. J. Thiessen, "Community Noise due to Ground Transportation." *Noise Control Engineering*, 2 (2), 64–67, 1974.

31. T. J. Schultz, *Community Noise Ratings*. Applied Science Publishers, Ltd., London, 1972.

32. L. Guttman, "A Basis for Scaling Qualitative Data." *Amer. Social Review*, 9, 139–150, 1944.

33. P. N. Borsky, "Community Reactions to Air Force Noise," Parts I and II. WADD Technical Report 60–689, AF 33 and 41, 1961.

34. P. N. Borsky, "The Use of Social Surveys for Measuring Responses to Noise Environments." In J. D. Chalupnik, ed., *Transportation Noises*, University of Washington Press, Seattle and London, 1970, pp. 219–227.

35. A. C. McKennell, "Aircraft Noise Annoyance Around London (Heathrow) Airport." COI Report No. SS 337, HMSO, London, 1963.

36. A. C. McKennell, "Noise Complaints and Community Action." In J. D. Chalupnik, ed., *Transportation Noises*, University of Washington Press, Seattle and London, 1970, pp. 228–244.

37. I. D. Griffiths and F. J. Langdon, "Subjective Response to Road Traffic Noise." *J. Sound Vib.*, 8 (1), 16–32, 1968.

38. H. Fogg and E. Johnsson, "Traffic Noise in Residential Areas." Report 36E, National Building Research Institute, Stockholm, 1968.

39. P. H. Parkin et al., "London Noise Survey." S. O. Code No. 67-266, HMSO, London, 1968.

40. A. C. McKennell and E. A. Hunt, "Noise Annoyance in Central London." The Government Social Survey, COI Report No. SS.332, March, 1962.

41. R. Josse, "Comment Mesurer et Caractériser les Bruits." Revue d'Acoustique, 3 (10) 131–139, 1970.

42. T. J. Schultz, "Some Sources of Error in Community Noise Measurement." Sound and Vibration, 6 (2), 18–27, 1972.

43. T. J. Schultz, "Instrumentation for Community Noise Surveys." Inter-Noise 73 Proceedings, Copenhagen, 1973, pp. 559–568.

44. H. B. Safeer, J. E. Wesler and E. J. Rickley, "Errors due to Sampling in Community Noise Level Distributions." J. Sound Vib., 24, 365–376, 1972.

45. D. J. Fisk, "Statistical Sampling in Community Noise Measurement." J. Sound Vib., 30 (2), 221–236, 1973.

46. D. W. Robinson, "The Concept of Noise Pollution Level." NPL Aero Report Ac 38, National Physical Laboratory, Aerodynamics Division, March, 1969.

47. D. W. Robinson, "An Outline Guide to Criteria for the Limitation of Urban Noise." NPL Aero Report Ac 39, National Physical Laboratory, Aerodynamics Division, March, 1969.

48. K. S. Pearsons, "The Effects of Duration and Background Noise Level on Perceived Noisiness." Report FAA ADS-78, April 1966.

49. Bolt, Beranek and Newman, "Establishment of Standards for Highway Noise Levels (Final Report)," 6 volumes. Prepared for Transportation Research Board, National Cooperative Highway Research Program, National Academy of Sciences, NCHRP 3-7/3, November, 1974.

50. U.S. Environmental Protection Agency, "Impact Characterization of Noise Including Implications of Identifying and Achieving Levels of Cumulative Noise Exposure." EPA Report No. NTID 73.4, July, 1973.

51. K. M. Eldred "Assessment of Community Noise." Noise Control Engineering, 3 (2), 88–95, 1974.

52. W. A. Rosenblinth, K. N. Stevens, and Bolt, Beranek and Newman Inc., Handbook of Acoustic Noise Control. Noise and Man, Vol. 2, WADC TR-52-204. Wright Air Development Center, Wright Patterson Air Force Base, Ohio, 1953.

53. British Standards Institution, "Method of Rating Industrial Noise Affecting Mixed Residential and Industrial Areas." British Standard BS 4142: 1967, British Standards Institution, 2 Park Street, London W1.

54. "Assessment of Noise with Respect to Community Response." ISO Recommendation R 1996.

55. T. Kihlmann, "Traffic Noise Control in Sweden." *Noise Control Engineering*, 5 (3), 124–130, 1975.

56. C. H. G. Mills, "The Measurement of Traffic Noise." In *The Control of Noise*, National Physical Laboratory, Symposium No. 12, HMSO, London, 1962, pp. 345–357.

57. International Organization for Standardization, "Measurement of Noise Emitted by Vehicles." ISO Recommendation R 362, February, 1964.

58. Standards Association of Australia, "Draft Australian Standard Method for Measurement for the Determination of Motor Vehicle Noise Emission." DR 75075, June, 1975.

59. C. H. G. Mills and D. W. Robinson, "The Subjective Rating of Motor Vehicle Noise." *The Engineer*, 1070–1074, June 30, 1961.

60. W. A. Leasure, "Truck Tire Noise—Results of a Field Measurement Program." In M. J. Crocker, ed., *Noise and Vibration Control Engineering*, Purdue University, W. Lafayette, Indiana, 1972, pp. 38–45.

61. D. Tetlow, "Truck Tire Noise." *Sound and Vibration*, August, 1971.

62. F. M. Wiener, "Experimental Study of the Airborne Noise Generated by Passenger Automobile Tires." *Noise Control*, 6 (4), 13–16, 1960.

63. E. J. Rathé, "Ueber den Laerm des Strassenverkehrs." *Acustica*, 17, 268, 1966.

64. R. E. Hayden, "Roadside Noise from the Interaction of a Rolling Tire With the Road Surface." In M. J. Crocker, ed., *Noise and Vibration Control Engineering*, Purdue University, W. Lafayette, Indiana, 1972, pp. 59–64.

65. A. C. Eberhardt and W. F. Reiter, "Application of the Coherence Function in Truck Tire Noise Analysis." Proceedings of Symposium on University Research in Transportation Noise, Raleigh, North Carolina, 1974, pp. 520–532.

66. Department of Transportation, "Interstate Motor Carrier Noise Emission Standards—Proposed Compliance Procedures." *Federal Register*, Washington, D.C., 40 (41, Part II), 8658–8666, February 28, 1975.

67. T. Priede, "Origins of Automotive Vehicle Noise." *J. Sound Vib.*, 15 (1), 61–73, 1971.

68. C.-I. J. Young and M. J. Crocker, "Prediction of Transmission Loss in Mufflers by the Finite Element Method." *J. Acoust. Soc. Amer.*, **57**, 144–148, 1975.

69. C.-I. J. Young and M. J. Crocker, "Acoustical Analysis, Testing and Design of Flow-Reversing Muffler Chambers." *J. Acoust. Soc. Amer.*, **60**, 1111–1118, 1976.

70. M. J. Crocker and D. R. Tree, "Acoustic Enclosures for Diesel Engines in Trucks." In M. J. Crocker ed., *Reduction of Machinery Noise*, Revised Edition, Purdue University, W. Lafayette, Indiana, 1975, pp. 349–364.

71. D. T. Aspinall and J. West, "The Reduction of the External Noise of Commercial Vehicles by Engine Enclosure." Motor Industry Research Association Report No. 1966/17, Nuneaton, England, 1966.

72. F. Juhasz, "Noise Reduction in a Commercial Vehicle Through Complete Enclosure of the Engine." *Automobiltechnische Zeitschrift*, **67** (11), 380–385, 1965.

73. F. Juhasz, "Noise Reduction by Acoustic Enclosures for Internal Combustion Engines." *Nachrichtentechnische Zeitschrift*, **29** (1) 11-14, 1968.

74. R. S. Lane, "Sources and Reduction of Diesel Engine Noise." In M. J. Crocker, ed., *Reduction of Machinery Noise*, Revised Edition, Purdue University, W. Lafayette, Indiana, 1975, pp. 191–202.

75. T. Priede, "Origins of Automotive Engine and Vehicle Noise." Lecture delivered to the Engine Manufacturers Association, Chicago, Illinois, June 28, 1968.

76. W. C. Stevenson, "The Problem of Brake Squeal—A Literature Survey." *Motor Industry Res. Assoc. Bulletin*, **1968** (6), 3-6.

77. R. A. C. Fosberry and Z. Holoubecki, "Disc Brake Squeal, its Mechanism and Suppression." Motor Industry Research Association Report No. 1961/22.

78. W. Wagenfuehrer, "Noise From Brakes." *Automobiltechnische Zeitschrift*, **66**, (8), 217–22, 1964, (see Motor Industry Research Association Translation No. 15/65).

79. R. W. Lange, "Excavator Clutches and Brakes, Behavior of Drums and Linings." SAE Preprint 2240, September 1960.

80. Swiss Federal Police Department, Draft Construction Regulations, Bern, 1967.

81. R. L. Staadt, "Truck Noise Control." SAE Paper SP-386, 1974; reprinted in M. J. Crocker, ed., *Reduction of Machinery Noise*, Revised Edition, Purdue University, W. Lafayette, Indiana, 1975, pp. 158–190.

82. W. H. Close and T. Atkinson, "Technical Basis for Motor Carrier and Railroad Noise Regulations." *Sound and Vibration*, 7 (10), 28–33, 1973.

83. T. Nimura et al., "Noise Problems with High-Speed Railways in Japan." *Noise Control Engineering*, 5 (1), 4–11, 1975.

84. S. Peters, "Prediction of Rail–Wheel Noise from High-Speed Trains." *Acustica*, 28, 318–321, 1973.

85. G. P. Wilson, "Community Noise from Rapid Transit Systems." In M. J. Crocker, ed., *Noise and Vibration Control Engineering*, Purdue University, W. Lafayette, Indiana, 1972, pp. 46–53.

86. J. E. Manning and L. G. Kurzweil, "Prediction of Wayside Noise from Rail Transit Vehicles." Proceedings of INTER-NOISE 74, Washington, D. C., 1974, pp. 265–268.

87. T. J. Schultz, "Development of an Acoustic Rating Scale for Assessing Annoyance Caused by Wheel–Rail Noise in Urban Mass Transit." Dept. of Transportation Report No. UMTA-MA-66-0025-74-2, February, 1974.

88. J. G. Walker, "Noise From High Speed Railway Operations." Technical Report 75, ISVR, Southampton University, England, 1975.

89. G. Allen, "Railway Noise." MSc Dissertation, Dept. of Civil Engineering, Southampton University, England, 1974.

90. D. Aubrée, "Acoustical and Sociological Survey to Define a Scale of Annoyance Felt by People in Their Homes Due to the Noise of Railroad Trains." Centre Scientifique et Technique du Batiment, June, 1973. (Translation available as BBN Technical Information Report No. 88, August, 1973.)

91. C. Stuber, "Noise Generation of Rail Vehicles." German Federal Railway, Munich Testing Laboratory, Report P1, 1973.

92. T. F. W. Embleton and G. J. Thiessen, "Train Noises and Use of Adjacent Land." *Sound*, 1 (1), 10–16, 1962.

93. D. N. May, "Criteria and Limits for Wayside Noise from Trains." *J. Sound Vib.*, 46, 4, 537–550, 1976.

94. Standards Association of Australia, "Draft Australian Standard Method for Measurement of Noise on Board Vessels." DR 74073, June 1, 1974.

95. Standards Association of Australia, "Draft Australian Standard Method for Measurement of Noise Emitted by Vessels on Waterways and in Ports and Harbours." DR 74074, June 1, 1974.

96. P. A. Franken, "Criteria, Standards and Limits." *Noise Control Engineering*, 1 (2) 86–89, 1973.

97. K. D. Kryter, *Effects of Noise on Man*. Academic Press, New York, 1970.

98. L. L. Beranek, ed., *Noise and Vibration Control*, McGraw-Hill, New York, 1971, Chapter 18.

99. Environmental Protection Agency, "Noise Abatement: Interstate Motor Carrier Noise Emission Standards" (40 CFR Part 202). *Federal Register*, 38, (144), 20102–20107, July 27, 1973.

100. Environmental Protection Agency, "Motor Carriers Engaged in Interstate Commerce" (FRL281-8). *Federal Register*, 39 (209), 38208–38216, October 29, 1974.

101. Environmental Protection Agency, "Identification of Products as Major Sources of Noise." *Federal Register*, 39 (100) 11197–11199, June 21, 1974.

102. Environmental Protection Agency, "Transportation Equipment Noise Emission Controls—Proposed Standards for Medium and Heavy Duty Trucks." *Federal Register,* 39 (210), 38338–38362, October 30, 1974.

103. Federal Highway Administration, "Procedures for Abatement of Highway Traffic Noise and Construction Noise." Part 772 of Title 23, Chapter I, Subchapter H of the Code of Federal Regulations. *Federal Register* 41 (80), April 23, 1976.

104. F. J. Langdon, "Noise Nuisance Caused by Road Traffic in Residential Areas." *J. Sound Vib.*, 47 (2), 243–282, 1976.

105. F. A. Behrens and T. M. Barry "European Experiences in Highway Noise." Federal Highway Administration Report No. FHWA-RD-75-123, November 1975.

106. Environmental Protection Agency, "Railroad Noise—Proposed Emission Standards" (40 CFR Part 201). *Federal Register*, 39 (129), 24580–24586, July 3, 1974.

107. Environmental Protection Agency, "Railroad Noise Emission Standards" (FRL 469-3). *Federal Register*, 41 (9), 2184–2195, January 14, 1976.

108. Environmental Protection Agency, "Noise Emission Standards for Transportation Equipment: Medium and Heavy Trucks" (FRL 511-6). *Federal Register*, 41 (72), 15538–15558, April 13, 1976.

3

Noise of air transportation to nontravelers

MALCOLM J. CROCKER*

3-1. INTRODUCTION

Aircraft noise is a much more localized problem than surface transportation noise, since it is significant only around major airports. As seen from Table 2-1 in Chapter 2, most of the noise energy is produced by scheduled airliners, the contribution of the large numbers of general aviation aircraft being relatively small. Since scheduled intercity airline service has been a significant competitor to intercity train and bus service only from the 1950s, the aircraft noise problem is also more recent than that of surface transportation noise.

From 1960 to 1970, the decade when jet passenger travel really got into its stride, the advantages of increased speed and comfort, and reduced operating cost per seat-mile, resulted in a doubling of passenger-miles every six years. In 1976, world-wide passenger-miles approached 500×10^9, and nearly 600

*Ray W. Herrick Laboratories, School of Mechanical Engineering, Purdue University, West Lafayette, Indiana 47907.

million passengers were carried. The early passenger jet aircraft also brought much higher noise levels than piston-engine aircraft during both takeoff and landing (see Fig. 3-1). The majority of piston-engine airliners have now been phased out of service, although twin-engine prop-jet aircraft continue to be used on many low-density, short-range routes. However, in the early 1960s the first fanjet engines (also known as turbofans or bypass jets) entered service and, though developed mainly to improve fuel economy, were quieter than the first pure jet (or turbojet) engines of the same thrust. Current wide-body fanjets (747, DC-10, L-1011, and A-300B) have considerable noise control technology in their engines, and are much quieter than early pure jet and fanjet aircraft like the 707, 727, 737, DC-8, DC-9 and BAC-111. Many of these earlier aircraft remain in service, however, and until they are modified, equipped with quieter power plants, or completely phased out of service, the aircraft noise problem will remain. At the end of 1976 the U.S. Federal Aviation Administration (FAA) issued regulations (Amendment 136 to 14CFR 91, December 23, 1976) requiring all domestic commercial aircraft to meet the present federal noise standards (FAR Part 36) in accordance with a phased-in time schedule not to exceed eight years. The modification or replacement of these jet aircraft which generate noise levels exceeding those specified in FAR Part 36 should reduce the aircraft noise problem in the next few years even if air traffic continues to increase.

The effect of supersonic transports like the Concorde, which is considerably noisier than the latest fanjets, is largely dependent on the numbers that enter service, the significance of the airports they serve, and any improvements in noise emission levels they exhibit. It is too early to judge the severity of the SST noise problem.

3-2. AIRCRAFT NOISE EXPOSURE

3-2.1 The Air Transport Fleet

A general idea of the numbers and noise levels of the U.S. airliner fleet can be gained from Fig. 3-2 [1]. Since 1970, when these data were assembled, there have been increases in the numbers of wide-body turbofans and 2–3 engine turbofans, but the sound levels, passenger capacities, and airplane ranges have not substantially altered. Figure 3-3 indicates the dramatic rise in U.S. airline capacity over the last decade, and some projections for the future. The projections, however, will probably be adversely affected by increasing fuel costs and a declining U.S. share of world traffic.

How different aircraft types contribute to the noise problem is indicated in Table 3-1. The aircraft are categorized as noisy in the left-hand side of the table, and quiet in the right-hand side. The noisy aircraft are those which do not meet present FAR Part 36 noise certification standards for subsonic jet aircraft [3]. It is seen that the great majority of the U.S. airline fleet do not meet the standard.

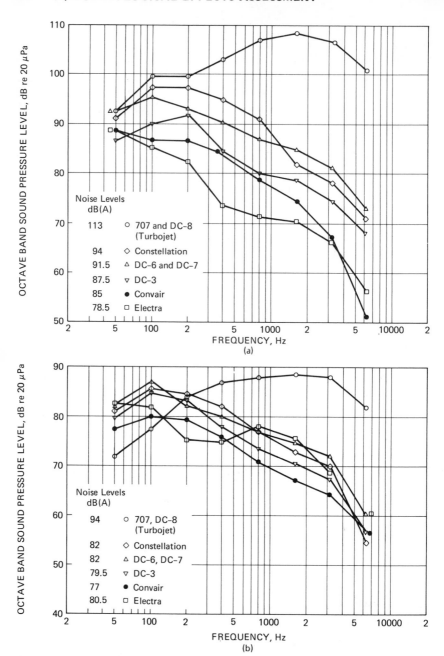

Fig. 3-1   Mean sound level spectra for various types of aircraft at approximately 1000 ft altitude: (a) during takeoff; (b) during landing [1].

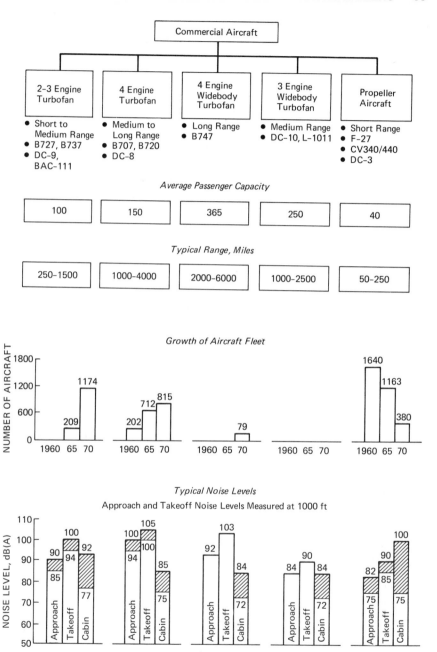

Fig. 3-2 Characteristics of the aircraft in U.S. airline service that are described in this chapter [1].

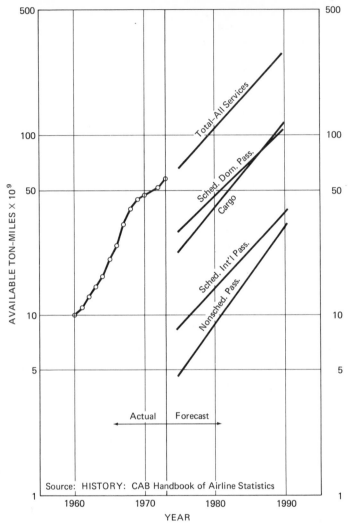

Fig. 3-3   Actual and forecast capacity in available ton-miles of the principal U.S. air carriers [2].

Several contracts awarded by NASA and the FAA to aircraft and engine manufacturers have resulted in the development of feasible methods to reduce the noise from the JT8D and JT3D engines which power most 707, 727, 737, DC-8, and DC-9 aircraft [4-8]. Various noise reduction alternatives are available, including fitting new front fans and installing sound-absorptive intake and exhaust linings. Modifications to thus retrofit existing aircraft have been a subject of continuing debate in the U.S. for several years as is witnessed by several Notices

TABLE 3-1. Aircraft Types in Service Among the Principal
U.S. Carriers, Grouped with Respect to FAR
Part 36 Noise Standards [2].[a]

	Noisy			Quiet	
Type	Number of Aircraft[b]		Type		Number of Aircraft[b]
Old Jets			*Quiet Fanjets*		
BAC-111	58		B727-200Q		4
B707-300	12		B747		106
B720	37		DC-10-10		60
CV-880	41		DC-10-10F		–
CV-990	8		DC-10-40		2
DC-8-20	40		DC-10-30		–
DC-8-30	24		DC-10-30F		–
	Total:	220	L-1011		18
				Total:	190
JT-8D (Retrofit			*Turboprops*		
Candidates)			CV-580/600		132
B727-100	299		F-27/227		59
B727-100C	122		L-188		46
B727-200	262		L-382		18
	Total:	683	YS-11		21
B737-200	150		Other		17
B737-200C	5			*Total:*	293
	Total:	155			
DC-9-10	72				
DC-9-10F	19				
DC-9-30	235				
DC-9-30F	7				
	Total:	333			
JT-3D (Retrofit			*Piston Aircraft*		140
Candidates)				*Total:*	140
B707-100B	96				
B707-300B	111				
B707-300C	119				
B720B	48				
	Total:	374			
DC-8-50	44				
DC-8-50F	28				
DC-8-61	55				
DC-8-61F	9				
DC-8-62	16				
DC-8-62F	1				
DC-8-63	4				
DC-8-63F	45				
	Total:	202			
Total–Noisy: 1,967			*Total–Quiet:* 623		
Total–All Types: 2,590					

[a]There has been an increase in the number of wide-body, quiet fanjets since these data were gathered.
[b]Number of aircraft in operation by principal U.S. carriers as of December 31, 1972.

of Proposed Rulemaking issued by the EPA and FAA [9-12]. One prognosis for the effect on the U.S. airline fleet of a retrofit program, had it been launched in 1974, is given in Fig. 3-4. The emergence of quieter aircraft from 1975 is evident in this study, while by 1980 nearly all aircraft in service would have met FAR Part 36 standards. Other countries have adopted a "wait and see" attitude toward retrofit programs, since they are not easily able to enact legislation that differs from the U.S. legislation in this area; in Britain, the feasibility of quieting

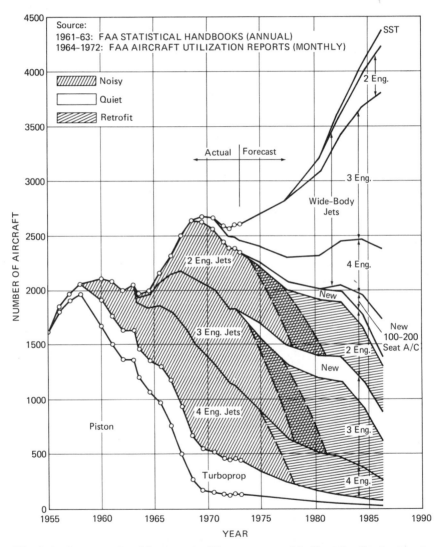

Fig. 3-4   An example of how a retrofit program would affect the fixed-wing aircraft in the U.S. airline fleet [2].

existing engines like the Spey and the Viper has been explored. Now the FAA has at last issued regulations (Amendment 136 to 14 CFR 91, December 23, 1976) requiring all commercial jet aircraft to comply with existing federal noise standards by January 1, 1985, and the situation will no doubt change. The FAA estimates that retrofitting the engines of the older four-engine aircraft (707, 720, DC-8 and Convair 990) would cut their noise levels by 13 or 14 EPNdB.* Retrofitting the engines of the two- and three-engine jets (727, 737, DC-9 and BAC-111) is estimated to reduce the noise of these aircraft by 5 to 7 EPNdB, while the 747-100 noise should be reduced by about 7 EPNdB.

The FAA suggests that 1600 of the 2100 jet aircraft in service in 1977 do not meet FAR Part 36 noise standards.* The FAA estimates that 1300 of these non-complying aircraft would remain in service in the 1980s and as many as 50 percent would still be flying by 1990 if no federal action had been taken. The FAA 1977 rule will not *force* the modification or retrofit of older airplanes but will encourage the adoption by operators of whatever means of achieving compliance is best suited to individual economic situations. Some operators may choose instead to replace older aircraft with new technology airplanes.

Initially the December 23, 1976 FAA rule does not apply to jets operated into U.S. airports by foreign airlines. However if discussions with the International Civil Aviation Organization (ICAO) do not establish worldwide aircraft noise standards comparable with FAR Part 36 by January 1, 1980, the FAA is expected to require all foreign jets and U.S. jets operated on international routes to comply with FAR Part 36 by January 1, 1985 also.

3-2.2 Numbers of People Disturbed

Table 2-2 (Chapter 2) shows that in a U.S. survey of 1200 people in 1971, motor vehicles, aircraft, and voices were judged to contribute most to the noise climate by 55 percent, 15 percent, and 12 percent of the people, respectively. In a similar survey of 1400 people in central London in 1962, approximately the same result was found. It is estimated (see Table 2-3) that there are about 7.5 million people in the U.S. who are seriously disturbed by aircraft noise [taking this to begin at $L_{dn} = 65$ dB(A)], while the number equally exposed to road traffic noise is 26.8 million [12]. Figure 3-5 gives a plot of the numbers experiencing various levels of aircraft noise exposure in 1972 [12].

3-3. SOURCES OF AIRCRAFT NOISE

Not only did the introduction of the jets bring increased aircraft noise levels as compared with propeller aircraft levels, but they brought two further noise problems: increased air traffic as a result of their popularity, and an increase in the numbers of people exposed to their longer approach paths.

*See Noise Control Engineering, Vol 8, 1977, p. 42.

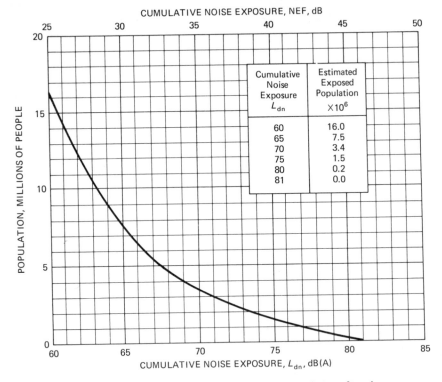

Cumulative Noise Exposure L_{dn}	Estimated Exposed Population $\times 10^6$
60	16.0
65	7.5
70	3.4
75	1.5
80	0.2
81	0.0

Fig. 3-5 Number of people impacted by various levels of aircraft noise exposure in the U.S. in 1972 [12]. The approximation L_{dn} = NEF + 35 is assumed.

Jet aircraft operate at a much higher cruising speed than do propeller ones, and the aerodynamic configuration necessary to do this results in higher takeoff and landing speeds. Required runway length is greater, partly as a consequence of these higher speeds, and partly because jet engine thrust is reduced when aircraft are stationary or nearly so. This brings the airport closer to the community. The generally large size and inertia of jet aircraft, and their slower throttle response times compared with those of propeller aircraft, require them to use a long approach path, resulting in low-level flight over surrounding neighborhoods. Moreover, they use considerable power on approach to counteract the drag of their high-lift devices.

The noise produced by powered aircraft is primarily from their engines. The aerodynamic noise produced by the passage of the aircraft through the air [13] is insignificant in comparison.

3.3.1 Jet Engine Noise

Figure 3-6 shows a cross section of a pure jet engine and Figure 3-7 shows a fanjet. The predominant noise source with the pure jet engine is the turbulent

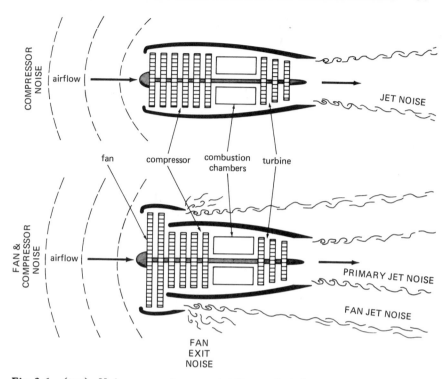

Fig. 3-6   (top). Noise sources in a pure jet (i.e., turbojet) engine.
Fig. 3-7   (bottom). Noise sources in a fanjet (i.e., turbofan, bypass jet) engine.

mixing that occurs along the boundary between the high velocity jet exhaust and the stationary atmosphere. (Secondary noise sources are the compressor and turbine, and unsteady combustion.) The exhaust noise is generally proportional to the eighth power of the exhaust velocity, so reducing this velocity is very beneficial. This is achieved in the fanjet engine by having a portion of the airflow bypass the main compressor and the combustion chambers; it is thus only slightly compressed and, as a consequence, has a low velocity. The mixing of this low velocity airflow with the higher velocity airstream that has passed through the core of the engine results overall in a lower velocity exhaust than occurs in a pure jet. Jet exhaust noise is thus much reduced in fanjet engines, particularly when a large proportion of the flow bypasses the engine core in this way.

The noise reduction achieved by the early fanjets was incidental to the prime motivation of increasing engine efficiency by avoiding the loss of kinetic energy in the exhaust, so it is not surprising that there remained other noise problems, such as the radiation of fan noise through the inlet. In modern fanjets, however, extensive use of sound-absorptive liners in the intake duct has been a deliberate and effective noise control measure.

3-3.2   Noise from Typical Jet Aircraft

Figures 3-8, 3-9, and 3-10 give octave band spectra and the relative contributions of the fan–compressor and jet exhaust during takeoff and approach for three different types of aircraft. The first two of these figures deal with early fanjets, and the third with the modern variety. For all three types of aircraft, the high frequency noise (above 800 or 1000 Hz) is caused mainly by the fan and compressor, while lower frequency noise is mostly from the exhaust.

3-3.3   Noise from Propeller Aircraft

As shown in Fig. 3-1, noise levels from propeller-driven passenger aircraft are generally much lower than from jet aircraft. With piston-engine planes, low frequency noise is mainly from the propeller and high frequency noise from the engine and exhaust. Prop-jet planes also have low frequency noise from the propeller, but their high frequency noise comes from the turbine.

Of these noise sources, the propeller is usually most important. The noise consists both of broad-band noise and of pure tones at the blade passage frequency and integer multiples of it. The blade passage frequency in Hz is given by $nN/60$, where n is the number of propeller blades and N is the propeller speed in rpm. Normally only this frequency and the first few multiples (harmonics) are significant, except in the rare case of supersonic propeller tip speeds, when higher harmonics can be important. The broad-band noise has its origins in the vortex shedding from the trailing edges of the blades.

Although there are about 130,000 general aviation propeller aircraft in the U.S., and about 1000 executive jet aircraft [1], their contribution to the noise energy produced by aircraft is quite small (see Table 2-1). It cannot be neglected, however, and the EPA is proposing to regulate some of it [14, 15]. Figure 3-11 shows the noise levels and spectra of general aviation propeller aircraft.

3-4.   RATING AIRCRAFT NOISE ANNOYANCE

3-4.1   Perceived Noise Level

Kryter has produced a method of evaluating the noisiness of aircraft noise [16, 17]. His method uses the contours shown in Fig. 1-7 and is described in Section 1-4.4 (Chapter 1). As an exercise, the reader may like to calculate the noisiness of some of the spectra given in Figs. 3-8 to 3-11. For example, the takeoff noise shown by the upper dashed line of Fig. 3-10 has a noisiness of 170 noys and 114 PNdB.

Perceived Noise Level has received wide acceptance as a measure of aircraft noisiness with and without tone corrections [3, 18]. As given in Table 1-5, there is a useful, approximate relationship between dB(A) and PNL. For aircraft noise spectra this is generally taken to be PNL = dB(A) + 13.

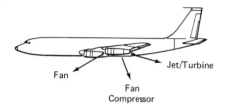

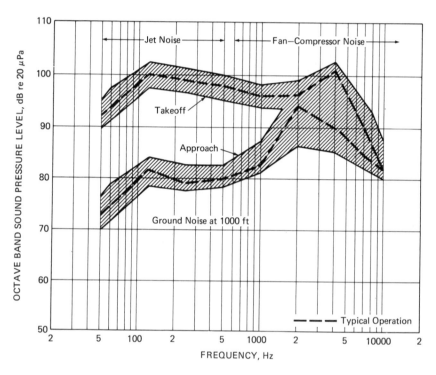

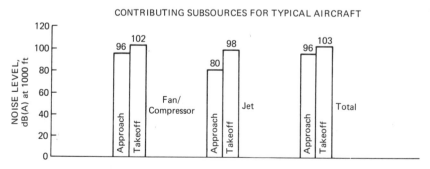

Fig. 3-8 Noise levels and spectra of 4-engine early fanjet aircraft (e.g., the Boeing 707-320B) [1].

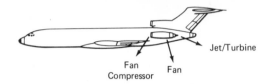

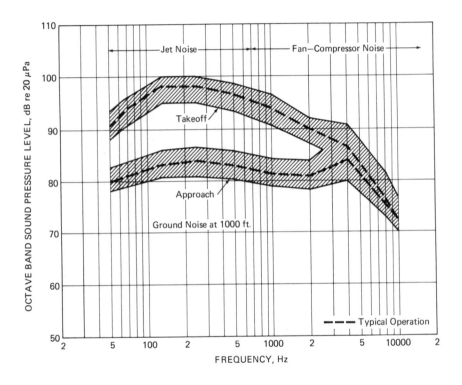

CONTRIBUTING SUBSOURCES FOR TYPICAL AIRCRAFT

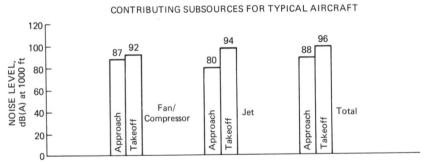

Fig. 3-9 Noise levels and spectra of 2-3 engine early fanjet aircraft (e.g., the Boeing 727) [1].

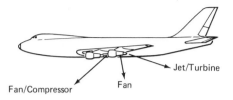

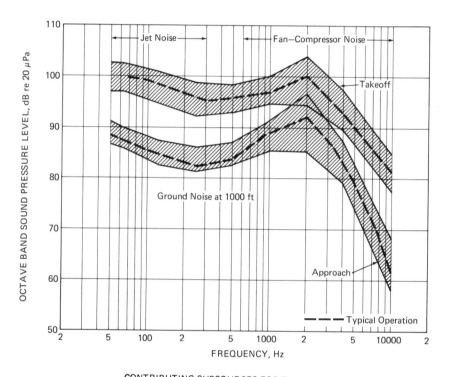

CONTRIBUTING SUBSOURCES FOR TYPICAL AIRCRAFT

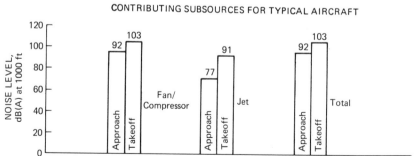

Fig. 3-10   Noise levels and spectra of wide-body modern fanjet aircraft (e.g., the Boeing 747) [1].

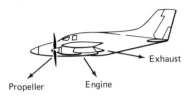

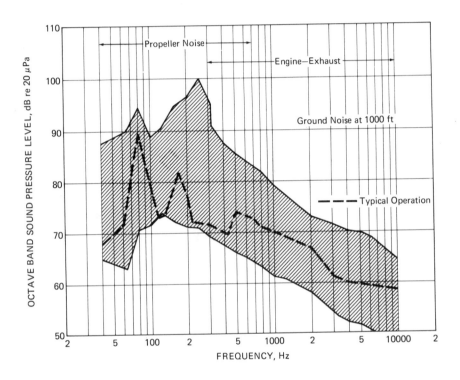

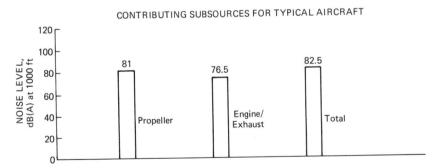

Fig. 3-11   Noise levels and spectra of general aviation propeller aircraft [1].

3-4.2 Effective Perceived Noise Level

Although PNL and A-weighted sound level can be used to monitor the peak noise level of an aircraft passby (or flyover or flyby), neither measure takes into account the variation of the noise and the time it lasts. Experiments have shown that annoyance and noisiness increase both with the magnitude and with the duration of a noise event. Figure 3-12 shows a PNL time history of a passby of a typical fanjet aircraft. As the airplane approaches, the discrete frequency whine caused by fan and compressor noise radiated from the engine inlets is very evident. When the airplane is overhead the noise is dominated by that from the fan exit, and is again mostly whine. When the plane has passed, the low frequency jet rumble is heard. The peaks for each source occur at different times, since each source is very directional. The inlet noise is mainly "beamed" forward in the flight direction, while the jet noise is mainly radiated backwards about 45° to the jet exhaust direction.

The Effective Perceived Noise Level (EPNL) takes into account the duration and the level of the noise by integrating the PNL over the duration of the event.

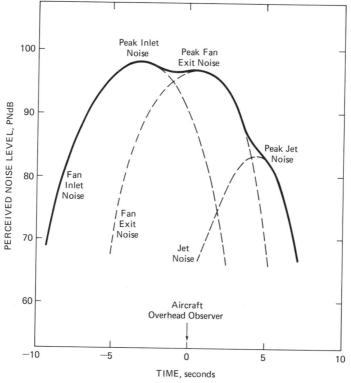

Fig. 3-12 Typical noise history of a fanjet aircraft flyover.

The method is somewhat complicated and is only briefly summarized here. Fuller details are given in FAR Part 36 or ICAO Annex 16 [3, 18]. For each of a number of time intervals k (usually of $\frac{1}{2}$ sec), the Perceived Noise Level PNL(k) is calculated. A tone correction, $C(k)$, is then added, the calculation of which is also too complicated for the space available here [3, 18], giving

$$\text{PNLT}(k) = \text{PNL}(k) + C(k),$$

which has the unit TPNdB.

The maximum instantaneous value of PNLT(k) is found and denoted PNLTM. EPNL is given by the sum of PNLTM and a duration correction factor D:

$$\text{EPNL} = \text{PNLTM} + D.$$

Figure 3-13 is an example of the PNL values of the aircraft passby in Fig. 3-12 after they have been tone-corrected. The EPNL is then the area under the PNLT curve summed on an energy basis and normalized by a time constant T, usually taken to be 10 seconds. PNLT in reality is a continuous function of time and thus EPNL is defined:

$$\text{EPNL} = 10 \log_{10} \left[\frac{1}{T} \int_{t_1}^{t_2} \text{antilog (PNLT/10)} \, dt \right].$$

Thus, from the two previous equations,

$$D = 10 \log_{10} \left[\frac{1}{T} \int_{t_1}^{t_2} \text{antilog (PNLT/10)} \, dt \right] - \text{PNLTM}.$$

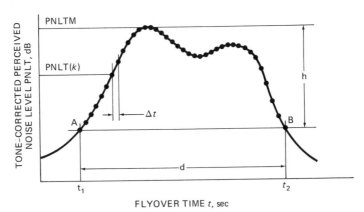

Fig. 3-13 How Tone-Corrected Perceived Noise Level may vary in an aircraft flyover, showing some of the labeling used in the calculation of Effective Perceived Noise Level (Section 3-4.2).

Note that the following restrictions apply in calculating D. If PNLTM > 100 TPNdB, then t_1 is the first point in time after which PNLT $>$ PNLTM - 10 and t_2 is the point in time after which PNLT remains constantly less than PNLTM - 10. If PNLTM < 100 TPNdB, then t_1 is the first point in time after which PNLT > 90 TPNdB and t_2 is the point in time after which PNLT remains constantly less than 90 PNdB. (See the points A and B, and the variable h in Fig. 3-13.) However, if PNLTM < 90 TPNdB, the duration correction is assumed zero.

Since, in practice, only k + 1 discrete values of PNLT are measured, the integrals in the above equations must be replaced by the numerical values of PNLT(k) and a summation. Thus

$$D = 10 \log_{10} \left\{ \frac{1}{T} \sum_{k=0}^{d/\Delta t} \Delta t \text{ antilog } [PNLT(k)/10] \right\} - PNLTM,$$

where Δt is the length of the equal increments of time and d (see Fig. 3-13) is the time interval to the nearest second during which PNLT(k) equals or exceeds either PNLTM - 10 or 90 TPNdB, as discussed above. Half-second intervals of Δt are normally sufficient to obtain a satisfactory PNLT history; however, a shorter time interval may be selected provided approved values of T and Δt are used. ICAO specifies that $T = 10$ sec and $\Delta t = 0.5$ sec, and the above equation becomes

$$D = 10 \log_{10} \left\{ \sum_{k=0}^{2d} \text{antilog } [PNLT(k)/10] \right\} - PNLTM - 13,$$

where the integer $2d$ corresponds to the duration time. The EPNL is then given by substitution for PNLTM and D in the equation EPNL = PNLTM + D.

3-4.3 Noise Exposure Forecast

EPNL is useful to describe the noisiness or annoyance of an individual aircraft, and may be used in aircraft certification and in comparing one aircraft with another. However, to describe the effect of aircraft noise in a community near an airport, a noise measure must make allowance not only for the annoyance of each noise event but for the number of such events and the time of day each occurs. Noise Exposure Forecast (NEF) takes these factors into account [19, 20], as do most other environmental community noise measures in use around the world. Some of these other noise measures, such as the Noise and Number Index (NNI), the Composite Noise Rating (CNR), and the Community Noise Equivalent Level (CNEL), will be reviewed in subsequent sections. Besides these American and British noise measures, there are several rather similar ones found in Europe and South Africa that are no longer much used.

The NEF for aircraft type i on flight path j is defined [20] as:

$$\text{NEF}_{ij} = \text{EPNL}_{ij} + 10 \log_{10} (n_d + 16.67 n_n) - 88,$$

where EPNL_{ij} is the EPNL for aircraft type i on flight path j, n_d is the number of day flights (0701 to 2200 hours), and n_n is the number of night flights (remaining hours) for that particular type of aircraft on that flight path. A penalty of 10 dB is applied to night flights. The nighttime adjustment is chosen so that for the same number of flights *per hour* the nighttime contribution is 10 dB higher than the daytime one [19]. For example, 30 flights/hour during the day would be equivalent to 3 flights/hour at night. The factor of 16.67 arises from the fact that there are 15 daytime hours so defined, and only 9 night-time hours. Thus $10 \times 15/9 = 16.67$.

The constant -88 is chosen for the following reasons [19]: to ensure that the NEF value is different enough from the EPNL value to obviate confusion; and so that a zero or very small value of NEF will correspond to no noise impact on land used even for the most sensitive activities.

It is necessary to compute NEF_{ij} for each type of aircraft for each flight path used. Takeoff and approach are counted as separate operations. Finally, the total NEF at a given location is obtained from an energy summation:

$$\text{NEF} = 10 \log_{10} \sum_{i=1}^{I} \sum_{j=1}^{J} \text{antilog} \, (\text{NEF}_{ij}/10),$$

where I represents the total number of contributions from different types of aircraft and J the total number of flight paths. For example, if there are 10 different aircraft types and 3 different flight paths in use at an airport, then $I \times J$ would be 60, remembering that takeoff and approach contributions must be considered separately, even though many locations are affected by only one or the other.

An example of the NEF calculations for *one* ground position is given in Table 3-2. Though complex, it is obvious that this example is considerably simplified over that for a real airport. To calculate the NEF at many locations around an airport is extremely laborious unless a computer is used, as is often the case [21]. The input data used in these computer calculations should include [21]: airport altitude, runway description in cartesian coordinates, ground tracks of aircraft (i.e., points supposedly directly below the flight path of departing and arriving aircraft), approach procedures (pattern altitudes and descent points), takeoff restrictions (area ceiling altitudes, noise abatement procedures, safety considerations), aircraft operations (numbers, types, day–night distribution, ground track assignments, trip length, gross weight), takeoff profiles (altitude, velocity, thrust from brake release), and last, but not least, noise data. The noise values produced at various slant distances must be known for each aircraft. Values of PNLT and EPNL against slant distance can normally be obtained from

TABLE 3-2. Example of NEF Calculation for a Single Ground Position [19].

Consider a ground location where information on flight tracks, number of operations and EPNL values for each aircraft type and flight track is known, as follows:

Aircraft Type	Flight Track	EPNL	Number of Day Operations	Number of Night Operations
A	1	90	30	4
B	1	95	2	1
C	2	98	5	1

The steps below summarize the calculations for the first aircraft listed above.

1. Find the effective number of operations by multiplying the number of night operations by 16.67 and adding the product to the number of day operations:

$$\begin{array}{lr} \text{Number of night operations (4)} \times 16.67 & 66.68 \\ + \text{ number of day operations} & \underline{30.00} \\ = \text{weighted number of operations} & 96.68 \end{array}$$

2. Determine the total adjustment for number of operations by taking 10 times the logarithm of the effective number of operations:

$$10 \log 96.68 = 19.85$$

3. Add the EPNL value for the aircraft and flight track:

$$19.85 + 90 = 109.85$$

4. Subtract the constant 88 to obtain the NEF contribution for the aircraft:

$$109.85 - 88 = 21.85$$

Calculations for the three aircraft are summarized as:

Aircraft	EPNL	Movements Day	Movements Night	Weighted Number (N)	$10 \log N$	EPNL + $10 \log N - 88$
A	90	30	4	96.68	19.85	21.85
B	95	2	1	18.67	12.71	19.71
C	98	5	1	21.67	13.36	23.36

Finally, the NEF contributions are added on an energy basis to obtain the total NEF:

$$\text{NEF (total)} = 10 \log \left[\text{antilog} \frac{21.85}{10} + \text{antilog} \frac{19.71}{10} + \text{antilog} \frac{23.36}{10} \right]$$

$$= 10 \log [153.1 + 93.5 + 216.8] = 10 \log 463.4 = 26.7$$

the manufacturer. A typical example is given in Fig. 3-14. Note that EPNL does not decrease with distance as rapidly as does PNLT because the noise duration is greater when the distance is great.

An example of NEF contours computed for a major international airport (Sydney, Australia) [22] is given in Fig. 3-15. Also shown are locations from

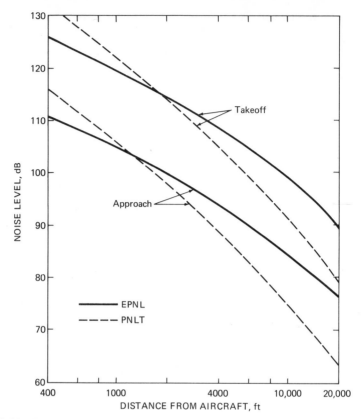

Fig. 3-14   Typical EPNL and PNLT values as they vary with distance from a turbojet aircraft during takeoff and landing [19].

which complaints were received during 1973, not including those from engine runups on the ground. Criteria for acceptable levels of NEF will be discussed later in this chapter.

3-4.4   Noise and Number Index

Noise and Number Index (NNI) is a subjective measure of aircraft noise annoyance developed and used in the United Kingdom. The NNI was the outcome of a survey in September 1961 of noise in the residential districts within ten miles of London (Heathrow) Airport [23]. It is defined by

$$\text{NNI} = \langle \text{PNdB} \rangle_N + 15 \log_{10} N - 80,$$

where $\langle \text{PNdB} \rangle_N$ is the average peak noise level of all aircraft operating during a day, and is given by

$$\langle \text{PNdB} \rangle_N = 10 \log_{10} \left\{ (1/N) \sum_n \text{antilog} \, (\text{PNdB}/10) \right\}.$$

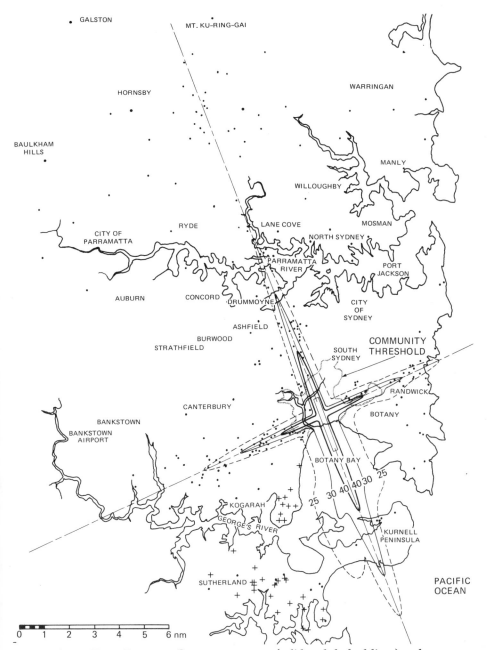

Fig. 3-15   Noise Exposure Forecast contours (solid and dashed lines) and complaint locations (dots and crosses) for Mascot Airport, Sydney, Australia in 1973 [22]. (*Reproduced with the permission of the Australian Department of Transportation, Air Transport Control Group.*)

Here PNdB is the peak PNL produced by individual aircraft during the day, and N is the number of aircraft operations over a 24 hour period. Since no annoyance appeared to occur at levels below 80 PNdB (presumably for one flight), a constant of 80 is used, so that a value of NNI of zero corresponds to no annoyance.

In the U.K. there has been debate as to whether the NNI should be revised [24]. A second survey was carried out around London (Heathrow) Airport in 1967 [25]. The main objectives were: to measure changes between 1961 and 1967 in noise levels and reactions to noise; to find if the 1961 findings were still valid in 1967; and to investigate the effects on annoyance of the following—number of aircraft, duration of noise, time of day or night, inbound or outbound aircraft differences, worst runway usage, and the local environment in which annoyance occurs.

The 1967 survey found that for the same noise exposure the reported annoyance was less than in 1961, though it was not clear whether this was because of noise sensitive people leaving the area or noise insensitive people arriving, or because people adapted or became resigned or apathetic to the noise, or because the background noise from traffic, construction and industry had increased enough to mask more of the aircraft noise. The 1967 survey also found that: people react more to the noise from flights using the runway that affects them most, rather than to the noise from flights using all the runways; annoyance from night flights seemed to be lower than from day flights—the opposite to that assumed in the NEF—though it was suggested that this was an artifact produced by the scarcity of nighttime flights, which have been restricted at London (Heathrow) since the early 1960s.

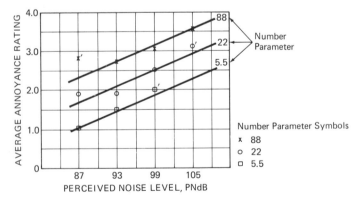

Fig. 3-16   Relationship between average annoyance and Perceived Noise Level in the survey [23] leading to the derivation of Noise and Number Index. The data points originally omitted (see Section 3-5.2) are shown here with a prime.

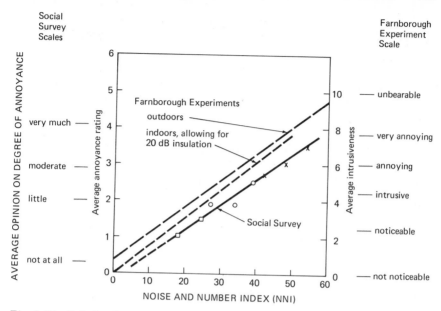

Fig. 3-17 Relationship between annoyance and Noise and Number Index, from experiments reported in Ref. 23. The symbols used correspond exactly with those used in Fig. 3-16.

Although the NNI is still used extensively in the U.K., it makes no allowance for duration of noise events, presence of pure tones, and relative annoyance of day and of night operations. As a consequence, the NNI is simpler than NEF to use, but probably less accurate in its prediction of community annoyance.

The weighting for number of aircraft movements used in NNI is different from that used in NEF and most other aircraft noise measures. In the original NNI study [23], three data points were omitted because the authors felt that not enough people had contributed noise judgments to them. This resulted in NNI being derived from only eight data points. As Schultz [26] has pointed out, if these three points are included (as has been done in Fig. 3-16), then it appears that a curve of PNL + 10 $\log_{10} N$ - 80 fits the data about as well as NNI with its 15 $\log_{10} N$ term. As formulated by its authors, NNI has the relationship to annoyance given in Fig. 3-17.

3-4.5 Composite Noise Rating

The Composite Noise Rating (CNR) has a long history [27–29]; dating back to the early 1950s. The CNR has evolved considerably during this time, as Schultz describes at length [26]. Originally the basic measure it used was the level rank— a set of curves placed about 5 dB apart in the mid-frequency range, rather similar to the NC and NR curves described in Chapter 6. The level rank is obtained by plotting the noise spectrum on the curves and finding the highest zone into which the spectrum protrudes, these being denoted by letters from A to M. The rank found initially, plus the algebraic addition of corrections, gives CNR. The corrections [27] are for: spectrum character, peak factor, repetitive character, level of background noise, time of day, adjustment to exposure, and public relations. The value of CNR obtained is associated with a range of community annoyance categories found from case histories—ranging from no annoyance, through mild annoyance, mild complaints, strong complaints, and threats of legal action, to vigorous legal action.

In the mid-1950s, the CNR was adapted to apply to the noise of military jet aircraft [30], and later of commercial aircraft [31]. Instead of assigning a level rank, the military aircraft noise was converted to an equivalent sound pressure level in the 300–600 Hz range according to a set of curves similar to the level rank curves [26, 30]. The time-varying SPL was time-averaged and then modified by corrections similar to those mentioned above to give the aircraft CNR. The calculation was further modified later when applied to commercial aircraft by using the PNL instead of level rank or the SPL just referred to. The final CNR is of the form

$$CNR = PNL + 10 \log_{10} (N_d + 10N_n) - 12$$

and has a weighting factor applied to penalize the number of nighttime flights, N_n, by 10 dB as compared with the number of daytime flights, N_d. Corrections are no longer made for background noise, previous experience, public relations, or other factors [26], and the new CNR therefore bears little resemblance to its original forms.

CNR is still used, though most people now favor NEF. CNR suffers from the following shortcomings: there is no correction for pure tones, and adjustments for number of flights and different types of aircraft are on a step basis often leading to inaccurate estimates of noise exposure. CNR has not been adopted by the civil aviation community, although it has been widely used by airport and community land use planners, and in military aircraft operations and by the Federal Housing Administration in deciding whether to guarantee loans for new residential construction near airports [26]. CNR is discussed here for completeness, and because it is a base for some of the other descriptors.

3-4.6 Community Noise Equivalent Level

California has adopted a rating scheme for aircraft noise exposure near airports [32, 33]. Known as the Community Noise Equivalent Level, CNEL is somewhat similar to NEF in that a weighting is applied for number of aircraft movements. Unlike NEF, however, CNEL uses the A-weighted sound level as the basic measure, instead of PNL, and besides the night penalty of 10 dB an evening penalty of 5 dB is applied. CNEL is defined [32] as:

$$\text{CNEL} = 10 \log_{10} (1/24) \left\{ 10^{\text{HNLD}/10} + 3 \sum 10^{\text{HNLE}/10} + 10 \sum 10^{\text{HNLN}/10} \right\},$$

where HNLD, HNLE, and HNLN are the day, evening and night hourly noise levels. The hourly noise level is given by

$$\text{HNL} = 10 \log_{10} \left(\sum \int 10^{L/10} \, dt/3600 \right),$$

where L is the instantaneous A-weighted sound level. The integral is calculated for each aircraft movement, and summed. HNL is usually computed electronically.

If the following simplifying assumptions are made: traffic consists of 70% daytime flights (0700 to 1900 hours), 23% evening flights (1900 to 2200), and 7% nighttime flights (2200 to 0700); the flyover A-weighted sound levels are at least 20 dB above the background noise; the average flyover duration is 15 sec; and all aircraft flyover noise levels at a given location have the same maximum value NL, then CNEL becomes [33]

$$\text{CNEL} = \text{NL} - 35 + 10 \log_{10} N$$

where N is the total number of flights per day.

3-4.7 Equivalent Continuous Perceived Noise Level

The International Civil Aviation Organization has recommended that for the purpose of promoting international correlation and communication between countries on community response to aircraft noise, the Equivalent Continuous Perceived Noise Level (ECPNL) should be adopted and referred to as the "international noise exposure reference unit" [18]. ECPNL is given by

$$\text{ECPNL} = \text{TNEL} - 10 \log_{10} (T/t_0),$$

where T is the total period of time under consideration, t_0 is one second, and the Total Noise Exposure Level (TNEL) produced by a succession of aircraft is expressed in terms of the EPNL by

$$\text{TNEL} = 10 \log_{10} \left\{ \sum_{n=1}^{N} \text{antilog} \, [\text{EPNL}(n)/10] \right\} + 10 \log_{10} (T_0/t_0)$$

where T_0 is 10 seconds, and EPNL(n) is the EPNL for the nth event (flyover or ground runup).

ICAO Annex 16 suggests that the period T should be a day or a night, 24 hours, a series of days, or a year. If the night and seasonal conditions are thought to be important, then ECPNL should be weighted as follows to give the weighted ECPNL, known as WECPNL:

$$\text{WECPNL}(2) = 10 \log_{10} \{ \tfrac{5}{8} \text{ antilog } [\text{ECPNLD}(2)/10]$$

$$+ \tfrac{3}{8} \text{ antilog } [\text{ECPNLN}(2)/10]\} + S,$$

where

ECPNLD(2) = ECPNL during daytime, for 2-period rating, 0700–2200
ECPNLN(2) = ECPNL during nighttime, for 2-period rating, 2200–0700

and S is the seasonal adjustment of -5 dB, 0 dB, or $+5$ dB depending on the numbers of hours in the month for which the temperature exceeds certain values, the underlying intention being to predict greater exposure when people are outdoors and without the benefit of the attenuation provided by building structures.

If evening conditions are considered important, a WECPNL(3) is defined on an essentially similar basis in terms of three periods rather than two.

3-5. AIRCRAFT NOISE CRITERIA

The same terminology for criteria and standards defined in Section 2-6 is used in this chapter. Aircraft noise criteria are mostly based on annoyance and supported by various community surveys.

3-5.1 Noise Exposure Forecast

The kind and degree of community response to aircraft noise will be influenced by many community factors in addition to the noise environment. Thus the acceptability of the noise given in Table 3-3 is only an estimate. Examples can be quoted where community reaction has been greater or less than that shown in the table [19]. Recent studies have shown that an *individual's* response to noise can be affected by: fear of aircraft crashing, susceptibility to noise, extent to which the airport and air transportation are considered important, belief in misfeasance of those responsible for the noise problem, dislike of other environmental problems, and belief that the noise affects health. *Community* reaction will be influenced by such factors as: amount of economic and social involvement between community and airport, feelings about the necessity of operations producing noise, and previous handling of airport noise problems [19].

In addition to Table 3-3, Fig. 3-18 can be used to estimate the percentage of people who will be highly annoyed or likely to complain [34]. Note that for an NEF of 20 or less there are likely to be no individual complaints or community reaction, although about 15% of people are still annoyed. As the NEF increases, about 20% more people are highly annoyed for each 10 dB increase.

TABLE 3-3.   Acceptability Related to NEF Values for Residential Areas with Buildings not Specially Constructed for Sound Reduction.

Judgment	NEF Value	Reaction
	45	
Unacceptable		
	40	
Unacceptable		In developed areas, repeated severe complaints and group action may be expected.
	35	
Barely acceptable		In developed areas, individuals may complain, perhaps severely, and group action is possible.
	30	
		Some noise complaints will occur and noise may occasionally interfere with some activities.
	25	
Acceptable		Sporadic complaints may occur.
	20	
Of no concern		No complaints.
	15	

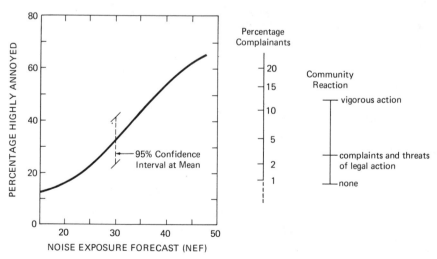

Fig. 3-18 Comparison of individual annoyance and community reaction as a function of Noise Exposure Forecast value [34].

TABLE 3-4.   Land Use Compatibility Table for Various Values of NEF [20].

	Compatibility		
Land use	Zone A (Less than 30 NEF)	Zone B (30–40 NEF)	Zone C (Above 40 NEF)
Residential	Yes	—[b]	No
Hotel, motel, offices, public buildings	Yes	Yes[c]	No
Schools, hospitals, churches, indoor theaters, auditoriums	Yes[c]	No	No
Commercial, industrial	Yes	Yes	—[c]
Outdoor amphitheaters, theaters	Yes[a,c]	No	No
Outdoor recreational (non-spectator)	Yes	Yes	Yes

[a]A detailed noise analysis should be undertaken by qualified personnel for all indoor music auditoriums and all outdoor theaters.

[b]Case history experience indicates that individuals in private residences may complain, perhaps vigorously. Concerted group action is possible. New single-dwelling construction should generally be avoided. For apartment construction, note c applies.

[c]An analysis of building noise reduction requirements should be made and needed noise control features should be included in the design.

Table 3-3 and Fig. 3-18 can be used in planning new airports, or to estimate changes in community reaction or numbers of people annoyed when numbers of aircraft movements increase or decrease, or flight paths are changed at an existing airport. In planning new airports or new residential or commercial developments near existing airports, the land use compatibility table shown in Table 3-4 may also be useful [20]. This land use guide is based on the types of construction which would normally occur when airport noise is not a matter of concern. Increased structural noise attenuation can usually be provided in new construction at relatively low cost, but is more difficult and expensive to provide in existing buildings. Relaxing land use restrictions, if increased noise attenuation is provided, should be used reluctantly, however, especially in the case of residential land uses, where people spend considerable time outdoors.

3-5.2 Noise and Number Index

As previously mentioned, NNI was derived from a 1961 social survey around London (Heathrow) Airport. The noise levels and durations and the number of

TABLE 3-5.   Examples of the Conditions Produced by Various Values of
Noise and Number Index [35].

NNI Value	Example of Conditions
10	Occasional aircraft flying overhead producing noise levels which cause disturbance out of doors but not within houses.
35	Regular overflights by aircraft, noise levels intrusive inside houses.
55	Continual overflying at noise levels which can interfere with conversation within houses—this is the level above which soundproofing grants are available around London (Heathrow) Airport.
65	Incessant overflying by aircraft at noise levels which can interfere with sleep and conversation even within soundproofed houses.
75	This very high value can occur beneath intensively used flight paths when departing aircraft are still low and using full power; no residential accommodation in the U.K. is subjected to such NNI levels.

aircraft to which a sample of the population was exposed were measured and were correlated with the subjective reactions to the noise of each sample. Eighty-five locations were used, and about 100 aircraft were recorded at each. Nearly 2000 people were interviewed. The data were eventually divided into four noise level ranges (84–90 PNdB, 91–96 PNdB, 97–102 PNdB, and 103–108 PNdB) and into three number-of-aircraft-per-day ranges (1–9, 10–39, and 40–110, giving means of about 5.5, 22, and 88 respectively).

The relation between annoyance and PNL is given in Fig. 3-16 [23]. The three data points originally omitted because of small sample populations have been included. Fig. 3-17 shows annoyance score plotted against NNI. Considering this figure and the results in Table 3-4, it would seem that NNI $\doteq$ NEF + 10-15, i.e., for the same annoyance, NNI is about 10-15 units greater. Table 3-5 gives examples of NNI and reaction produced [35].

NNI has been widely used in the U.K. The house insulation grant scheme around London (Heathrow) Airport has been confined to an area within the 55 NNI contour; Surrey County Council has zoned development around London (Gatwick) Airport on the basis of NNI values [35, 36], while the siting of the third London Airport was decided by the Roskill Commission using NNI predictions [24, 37, 38].

3-5.3 Composite Noise Rating

CNR, as computed in Section 3-4.5, has been compared with a number of airport noise case histories [26, 31]. Three zones of community response have been defined, as shown in Table 3-6. Considering the community responses, it is seen that CNR is approximately 75 or 80 dB greater than NEF, or CNR $\doteq$ NEF + 75.

TABLE 3-6. Public Response with Various Values of Composite Noise Rating (for Aircraft) [26, 30].

Takeoffs and Landings	Ground Runups	Zone	Description of Expected Response
Less than 100	Less than 80	1	Essentially no complaints would be expected. The noise may, however, interfere occasionally with certain activities of the residents.
100–115	80–95	2	Individuals may complain, perhaps vigorously. Concerted group action is possible.
Over 115	Over 95	3	Individual reactions would likely include repeated vigorous complaints. Concerted group action might be expected.

3-5.4 Community Noise Equivalent Level

Similar community responses to aircraft noise in terms of CNEL have been determined. Rettinger [33] reports one such result, together with comparisons with NNI and NEF (see Table 3-7). The $>$ signs in his original table, lower row, appear to be in error and have been reversed here. An approximate relationship between CNEL and NEF is CNEL $\doteq$ NEF + 32.

TABLE 3-7. A Summary of Public Response to Aircraft Noise Exposure as Represented by the Major Descriptors of That Exposure [33].

NNI	NEF	CNEL	Meaning of Noise Zone Rating
>50	>40	>72	Noise exposure is not acceptable for residential areas. Many dwellers in this zone will complain vigorously about the "intolerable" noise, and concerted group action is highly probable. Home construction should not be permitted in this area, and airport hotels require substantial sound insulation for acoustic comfort.
40–50	30–40	62–72	Noise exposure is marginal for residential areas. Some complaints about noise are probable. Sound insulation is recommended for all buildings in this zone.
<40	<30	<62	Acceptable for residential use. Few complaints about noise are to be expected.

3-5.5 Other Noise Criteria

Noise criteria in terms of ECPNL or WECPNL do not at present appear to have been developed. Some psychoacoustic researchers have proposed that aircraft noise annoyance should be based on the simple A-weighted sound level of the noisiest aircraft (see, e.g., Rylander *et al.* [39, 40]). Rylander's work is based on the establishment of a noise dose relationship around Swedish airports. He believes that for less than 50 flyovers per day, annoyance is predominantly related to the dB(A) contours of the noisiest such flyover. For more than 50 flights per day he believes his work is incomplete. The advantages of this hypothesis are obvious: it is simpler than NEF, more easily understood by laypersons, and it can easily be used with studies of increasing the noise attenuation of buildings.

The single-event flyover noise concept has several disadvantages: no difference in annoyance is predicted if an airport has 10 or 500 aircraft movements a day; the noisiest aircraft may appear very often—or just once weekly; preferred runway noise abatement procedures and variations in runway use would not produce a change in the noise prediction; day and night operations are not distinguished; and no land use criteria for this type of scheme yet exist.

Rylander's work has not been accepted in the U.S., where the NEF concept is still favored. However, under FAA funding, work has progressed on an alternative scheme using A-weighted sound levels. Termed the "aircraft sound description system" (ASDS), the scheme measures noise exposure in terms of the time in minutes per day that a given threshold sound level is exceeded. At present the threshold has been arbitrarily set at 85 dB(A). The ASDS does not seem to be gaining support at present, and it has the drawbacks that the threshold level has not been properly justified, nor land use criteria established.

The U.S. Environmental Protection Agency is proposing that a variant of the NEF using the A-weighted sound level instead of the PNL should be adopted. Since the prediction computer programs should be readily compatible with this change, and since dB(A) is more easily determined than PNL, such a proposal may have merit.

3-6. EXISTING AND PROPOSED REGULATIONS AND STANDARDS FOR AIRCRAFT NOISE

Aircraft noise regulations vary from country to country. Since space limitations preclude a review of regulations in all countries, this discussion will emphasize U.S. and international legislation. American legislation is, in any case, a major influence on legislation in other countries, mainly because American aircraft manufacturers and airlines dominate the world air transportation industry.

3-6.1   The U.S. Federal Government's Role in Noise Abatement

In the past, the federal government held that noise abatement was a concern for state and local levels of government. However, with certain matters and particularly those where interstate commerce is involved, the federal government resisted state and local efforts to reduce noise [41]. Only in recent years has public pressure forced the federal government to legislate noise abatement in the federally preempted areas of aircraft and interstate motor vehicles.

The power of the federal government to legislate on noise problems is based on Article I, section 8 of the United States Constitution, where the Congress is empowered to regulate interstate commerce. This base enabled seven federal statutes to be enacted [41].

Federal Aviation Act of 1958. [Stat. 749-49 USC Sec. 1348 (c)] This established the federal preemption position to "assign by rule, regulation, or order the use of the navigable airspace under such terms, conditions as he [the Administrator of the FAA] may deem necessary in order to insure the safety of aircraft and the efficient utilization of such airspace." No authority was designated to promulgate noise abatement rules.

Department of Transportation Act of 1966. [PL 89-670] The Department of Transportation was created to develop national transportation policies and programs. The Secretary of Transportation was directed to "promote and undertake research and development relating to transportation, including noise abatement with particular attention to aircraft noise."

Noise Control and Abatement Act of 1968. [PL 90-411] The FAA Act of 1958 was amended to control aircraft noise and sonic boom. "In order to afford present and future relief and protection to the public from unnecessary aircraft noise and sonic boom, the Administrator of the FAA, after consultation with the Secretary of Transportation, shall prescribe and amend standards for the measurement of aircraft noise and sonic boom and shall prescribe and amend such rules and regulations as he may find necessary to provide for the control and abatement of aircraft noise and sonic boom." However, reference was also made to "the highest degree of safety in air commerce," and there was a requirement that "proposed standard, rule, or regulation [be] economically reasonable, technologically practicable, and appropriate for the particular type of aircraft, aircraft engine, appliance or certificate. . . ."

National Environmental Policy Act of 1970. [PL 91-190] The results of this Act have been far-reaching and have affected almost all federal actions and indirectly most new construction. The Act requires "a national policy which will encourage productive and enjoyable harmony between man and his environment;

. . . prevent or eliminate damage to the environment and biosphere, and stimulate the health and welfare of man. . . ." It directs all federal agencies to "utilize a systematic, interdisciplinary approach . . . in planning and in decision-making which may have an impact on man's environment. . . ." The courts have usually interpreted this act to include almost any federal action, and have delayed such actions until environmental impact statements were prepared and reviewed.

Airports and Airways Development Act of 1970. [PL 91-258] The Secretary of Transportation is directed to aid in the development of public airports with financial assistance in airport planning and improvement. Such assistance is dependent on "protection and enhancement of the natural resources and the quality of environment of the Nation" and "the interest of communities in or near which the project may be located." If federal financing is involved in airport development, noise is one factor which must be included.

Noise Control Act of 1972. [PL 92-574] The most comprehensive Act on noise to date, this Act includes a few important provisions on aircraft noise. The FAA Act of 1958 was further amended to include responsibility for protection of "the public health and welfare from aircraft noise and sonic boom." However provisions to consider safety, economic reasonableness, technological practicability and appropriateness to the type of equipment were retained. The act requires consultation by FAA with EPA, and allows EPA to submit to FAA proposed regulations as EPA sees fit to protect public health and welfare.

Airport and Airway Development Act Amendments of 1976. [PL 94-353] This Act authorized for the first time the use of federal airport development funds on projects designed to achieve noise relief. Specifically, Section 11 of the Act now authorizes federal financing of land acquisition to insure compatibility with airport noise levels, and the acquisition of noise suppression equipment [41].

The federal government's rule-making procedures are slow and lengthy and are designed to give the public a chance to comment on proposals and participate in the process before regulations become final [41]. An advanced notice of proposed rule-making (ANPRM) may be published—in the Federal Register and by press release—to announce an agency's intention to propose or amend a regulation. Public comment is invited. Later a notice of proposed rule-making (NPRM) is published to announce details of a proposed regulation. Sufficient details are given to allow interested groups to determine its effect on them, and to allow public study and comment. Based on these comments a final regulation is prepared. Under the National Environmental Policy Act of 1970, an environmental impact statement (EIS) is then prepared and published. After public comments are received and considered, in a minimum of 90 days, a final EIS is written. At last, when an opportunity has been given for a consensus to be reached, a final regulation is published and enforced, with sufficient time allowed for compliance.

A list* of some of the FAA aircraft/airport noise control regulations and notices of proposed rule-making is given below.

FAA AIRCRAFT/AIRPORT NOISE CONTROL ACTIONS

Source Noise Controls Initiated by the FAA

January 11, 1969	NPRM 69-1: Proposed Noise Standards for Aircraft Type Certification.
November 18, 1969	14 CFR 36: Noise Standards; Aircraft Type Certification, effective December 1, 1969.
August 4, 1970	ANPRM 70-33: Proposed Civil Supersonic Aircraft Noise Type Certification Standards.
November 4, 1970	ANPRM 70-44: Proposed Civil Airplane Noise Reduction Retrofit Requirements.
July 25, 1972	NPRM 72-19: Proposed Noise Standards for Newly-Produced Airplanes of Older Type Designs.
January 24, 1973	ANPRM 73-3: Proposed Civil Airplane Fleet Noise (FNL) Requirements.
October 10, 1973	NPRM 73-26: Proposed Noise Standards for Small Propeller-Driven Airplanes.
October 26, 1973	Amendment 2 to 14 CFR 36: Noise Standards for Newly-Produced Airplanes of Older Type Designs, effective December 1, 1973.
December 28, 1973	ANPRM 73-32: Proposed Noise Standards for Short-Haul Aircraft.
March 27, 1974	NPRM 74-14: Proposed Civil Aircraft Fleet Noise Requirements.
January 6, 1975	Amendment 4 to 14 CFR 36: Noise Standards for Small Propeller-Driven Airplanes.
November 5, 1975	NPRM 75-37: Proposed Noise Reduction Stages and Acoustical Change Provisions.
August 19, 1976	Amendment 5 to 14 CFR 36: Acoustical Change Approval Procedures.

*W. A. Leasure, Department of Transportation, Private Communication, June 15, 1977.

October 28, 1976	NPRM 76-21: Proposed Changes to Aircraft Noise Measurement and Evaluation Specifications.
October 28, 1976	NPRM 75-37C: Supplemental Proposed Noise Reduction Stages.
December 23, 1976	Amendment 136 to 14 CFR 91: Phased Compliance Requirement for All Civil Aircraft with Noise Standards of 14 CFR 36.
March 3, 1977	Amendment 7 to 14 CFR 36: Increased Stringency of Noise Standards for Aircraft Type and Air-Worthiness Certification.

Operational Noise Controls Initiated by the FAA

April 10, 1970	NPRM 70-16: Proposed Prohibition of Civil Aircraft Sonic Booms.
February 28, 1972	FAA Order 7110, 22A, Arrival and Departure Handling of High-Performance Aircraft ("Keep-Em-High Program").
February 28, 1972	Advisory Circular 90-59, Arrival and Departure Handling of High-Performance Aircraft.
March 28, 1973	14 CFR 91: Civil Aircraft Sonic Boom, effective April 27, 1973.
March 26, 1974	ANPRM 74-12: Proposed Two-Segment ILS Noise Abatement Approach.
January 18, 1974	Advisory Circular 91-39, Recommended Noise Abatement Takeoff and Departure Procedure for Civil Turbojet-Powered Aircraft.
July 9, 1974	Advisory Circular 91-36A, VFR Flight Near Noise Sensitive Areas.

Airport Operations Noise Controls Initiated by the FAA

July 9, 1975	Alternative Airport Noise Policy.
November 18, 1976	DOT/FAA Aviation Noise Abatement Policy Issued.

Source Noise Control Proposals Submitted by the EPA

December 6, 1974	Noise Standards for Propeller-Driven Small Airplanes, published as NPRM 74-39, January 6, 1975.
January 28, 1975	Civil Fleet Noise Retrofit Requirements, published as NPRM 75-5, February 27, 1975.

January 28, 1975	Fleet Noise Level Requirements, published as NPRM 75-6, February 27, 1975.
February 27, 1975	Civil Supersonic Aircraft, published as NPRM 75-15, March 28, 1975.
January 13, 1976	Airplane Noise Requirements for Operation to and from an Airport within the United States (Current Supersonic Aircraft), published as NPRM 76-1, February 12, 1976.
October 1, 1976	Noise Levels for Turbojet-Powered Airplanes and Large Propeller-Driven Airplanes, published as NPRM 76-22, October 28, 1976.

Operational Noise Control Proposals Submitted by the EPA

December 6, 1974	Minimum Altitudes for Turbojet-Powered Aircraft, published as NPRM 74-40, January 6, 1975.
August 29, 1975	Minimum Flaps Noise Abatement Approach Procedure, published as NPRM 75-35, September 25, 1975.
August 29, 1975	Two-Segment VFR Noise Abatement Approach, published as NPRM 75-35, September 25, 1975.
August 29, 1975	Two-Segment IFR Noise Abatement Approach, published as NPRM 75-35, September 25, 1975.

Airport Operations Noise Control Proposals Submitted by the EPA

October 22, 1976	Airport Noise Regulatory Process, published as NPRM 76-24, November 22, 1976.

Source Noise Control Responses to EPA Proposals

November 29, 1976	Decision Not to Prescribe Fleet Noise Level Requirement (EPA proposal of January 28, 1975).
December 23, 1976	Amendment 6 to 14 CFR 36: Noise Regulations for Propeller-Driven Small Airplanes (EPA proposal of December 6, 1974.)
December 23, 1976	NPRM 76-27: Proposed Noise Regulations for Agricultural and Fire Fighting Small Airplanes (EPA proposal of December 6, 1974).

Operational Noise Control Responses to EPA Proposals

November 29, 1976	Decision Not to Prescribe Minimum Altitudes (EPA proposal of December 6, 1974).

November 29, 1976   NPRM 76-26: Proposed Delayed Landing Flap Noise Abatement Approach Procedure (EPA proposals of August 29, 1975).

November 29, 1976   Amendment 134 to 14 CFR 91: Use of Minimum Certificated Landing Flaps, and Decision Not to Prescribe Two-Segment Approach Requirements (EPA proposals of August 29, 1975).

3-6.2 Federal Aviation Regulations Part 36—Noise Standards

Perhaps the most important aircraft noise standards to date have been those setting noise standards for type certification of subsonic airplanes [3]. These were spawned by the Noise Control and Abatement Act of 1968 (see above). Although these and similar noise regulations in other countries have brought about a reduction of 10–15 EPNdB in noise levels of new passenger aircraft, a general improvement in the aircraft–airport noise situation has yet to occur because of the large numbers of older, noisier jets still in service. However, it may be confidently said that, without such regulations for the last several years, the airport noise situation today would be much worse. It is to be hoped that the FAA regulations issued on December 23, 1976 requiring all existing commercial jet aircraft to comply with FAR Part 36 by January 1, 1985 will actually now bring about a reduction in aircraft/airport noise levels.

Fig. 3-19 shows the noise measuring points used [3] in testing for compliance with FAR Part 36. For takeoff, there is a point 3.5 nautical miles (ca. 6500 m) from the start of takeoff roll. For approach, there is a point 1 nautical mile (1852 m) from the threshold (on the runway centerline). There is also a point on the sideline, on a line parallel to and 0.25 nautical miles (ca. 460 m) from the runway centerline where the noise level is greatest, except for airplanes with more than three jet engines, when this distance is 0.35 nautical miles (ca. 650 m). Fig. 3-20 shows the noise levels allowed under FAR Part 36. The levels allowed are seen to be dependent on aircraft maximum weight. The noise levels allowed may be exceeded at one or two of the three locations if (a) the sum of the excesses is not greater than 3 EPNdB, (b) no excess is more than 2 EPNdB, and (c) the excesses are completely offset by reductions at other required measuring points. Fig. 3-21 shows a comparison made by B. N. Mel'nikov in the U.S.S.R. of the noise from many different commercial airplanes [42].

As already mentioned, the FAA proposed* in November 5, 1975 to amend FAR Part 36 to include more restrictive noise limits (see FAA NPRM 75-37

---

*This amendment has now become effective with publication on March 3, 1977, of Amendment 7 to 14 CFR 36. The amendment has increased the stringency of the noise limits for type certification of all new Large Transport Aircraft for which application was made after November 5, 1975.

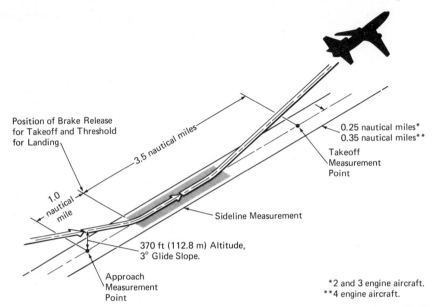

Fig. 3-19 Measurement locations for certification testing of aircraft to FAR Part 36 noise standards.

above). The changes are complicated. Briefly, three stages of noise reduction would be established. Stage 1 would include all airplanes that are noisier than FAR 36 noise limits, including airplanes of older type designs still in service. Stage 2 would deal with all airplanes that comply with present FAR 36 noise

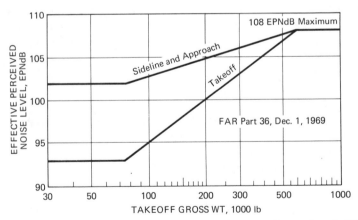

Fig. 3-20 FAR Part 36 noise standards, as introduced (December 1, 1969).

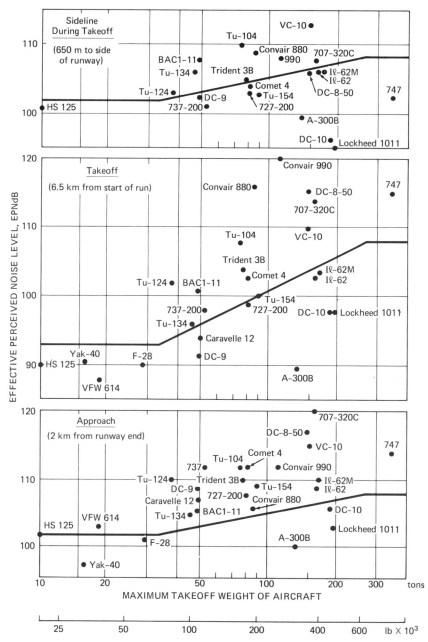

Fig. 3-21 Aircraft noise levels summarized for three measurement positions. FAR Part 36 standards are shown by the stepped curves [42].

limits, but do not comply with the proposed more restrictive noise limits. Stage 3 would cover all airplanes which comply with the proposed more restrictive limits. For new type certificate applications made after the publication date of NPRM 75-37, each airplane would be required to meet the proposed Stage 3 takeoff, sideline, and approach noise limits, these being below existing limits by 10 EPNdB, 9 EPNdB, and 4 EPNdB, respectively. For airplanes with more than three engines, the 0.35 nautical mile sideline measurement point would be reduced to 0.25 nautical miles.

The federal government is proposing to regulate the noise from small airplanes [14, 15]. The levels suggested are similar to those recommended in ICAO Annex 16 [18] and are discussed below.

3-6.3   International Standards and Regulations

The levels recommended for subsonic jet airplanes by ICAO Annex 16 in 1971 [18] are essentially the same as those originally regulated by the FAA in 1969 as FAR Part 36. However, ICAO Annex 16 also recommends the following maximum noise levels for light, propeller-driven airplanes: a 68 dB(A) constant limit up to an airplane weight of 600 kg (1323 lb), ranging linearly with weight from that point to 1500 kg (3307 lb), after which the limit is constant at 80 dB(A) up to 5700 kg (12566 lb). The tests to demonstrate compliance consist of a series of level flights over the measuring station at a height of 300 m (ca. 1000 ft). They should be performed at maximum continuous power and, if possible, at stabilized speed in cruise configuration.

The International Organization for Standardization has issued a recommendation for monitoring aircraft noise around airports [43]. The method recommends either mobile or fixed measuring equipment, although the procedure using the latter is the one primarily described. The use of PNL is recommended.

3-6.4   Future Developments

Likely future developments in regulations are too numerous and varied to discuss in detail.

The noise impact of supersonic transports will not be resolved until considerable in-service experience has been gained. This applies primarily to noise around airports, but there is also the noise impact of sonic booms occurring during supersonic cruise to be considered, an issue extensively reviewed in Ref. 44.

The considerable effort to reduce engine noise in the wide-body fanjets cannot be extended without great penalties in weight, first cost, and operating costs. Further noise reductions will probably have to be sought not at the source (the engine) but in operational techniques such as the two-segment approach, power cutback after takeoff, and minimum altitude regulations. In addition, airport

TABLE 3-8. Proposed CNEL Limits
for California Airports.

Date	CNEL
Until 12/31/75	80
1/1/76–12/31/80	75
1/1/81–12/31/85	70
1/1/86–	65

operations may be planned and regulated to minimize disturbance to the community. As an example of airport noise limits, the California Department of Aeronautics has proposed the noise exposure limits shown in Table 3-8. Better planning and control of construction of buildings near airports will also have an alleviating effect (see, e.g., Ref. 45). However, unless the existing fleet of noisy jets is retired or made quieter, the current aircraft noise problem will continue for many years. It is to be hoped that the FAA regulations issued late in 1976 will bring about a real reduction in aircraft noise in the next few years.

REFERENCES

1. Wyle Laboratories, "Transportation Noise and Noise from Equipment Powered by Internal Combustion Engines." EPA Report No. NTID 300.13, 1971.

2. C. Bartell, L. C. Sutherland, and L. Simpson, "Airport Noise Reduction Forecast, Volume I–Summary Report for 23 Airports." Report No. DOT-TST-75-3, Department of Transportation, Washington, D.C., October, 1974.

3. Federal Aviation Administration, "Federal Aviation Regulations, Part 36, Noise Standards: Aircraft Type and Airworthiness Certification." Washington, D.C., December 1, 1969.

4. J. F. Woodall, "FAA Aircraft Retrofit Feasibility Program." SAE Paper No. 740489, 1974.

5. L. E. Stitt and A. A. Medeiros, "Reduction of JT8D Powered Aircraft Noise by Engine Refanning." SAE Paper No. 740490, 1974.

6. R. L. Frasca, "Noise Reduction Programs for DC-8 and DC-9 Airplanes." Paper presented at the 87th meeting of the Acoustical Society of America, April, 1974.

7. V. L. Blumenthal, R. E. Russel, and J. M. Streckenbach, "A Positive Approach to the Problems of Aircraft Noise." AIAA Paper No. 73-1157, October, 1973.

8. C. C. Ciepluch, "Technology for Low Noise Aircraft—the NASA Quiet Engines." *Noise Control Engineering*, 1 (2), 68–73, 1973.

9. Notice of Proposed Rule-Making, "Civil Aircraft Fleet Noise Requirements." Notice 74-14, *Federal Register*, 11302, March 27, 1974.

10. EPA-recommended Notice of Proposed Rule-Making, "Civil Subsonic Turbojet Engine-Powered Airplanes Noise Retrofit Requirements." December 16, 1974.

11. EPA-Recommended Notice of Proposed Rule-Making, "Fleet Noise Level Requirements." December 16, 1974.

12. EPA Project Report, "Civil Subsonic Turbojet Engine-Powered Airplanes (Retrofit and Fleet Noise Level)." December 16, 1974.

13. D. Gibson, "The Ultimate Noise Barrier—Far Field Radiated Aerodynamic Noise." Inter-Noise 72 Proceedings, Washington, D.C., October 4–6, 1972, 332–337.

14. EPA-Recommended Notice of Proposed Rule-Making, "Noise Standards For Propeller Driven Small Airplanes." November 25, 1974.

15. EPA Project Report, "Noise Certification Rule for Propeller Driven Small Airplanes." November 25, 1974.

16. K. D. Kryter, *J. Acoust. Soc. Amer.*, 31, 1415, 1959.

17. K. D. Kryter and K. S. Pearson, *J. Acoust. Soc. Amer.*, 35, 866, 1963.

18. International Civil Aviation Organization, International Standards and Recommended Practices, "Aircraft Noise." Annex 16 to the Convention of International Civil Aviation, Montreal, Canada, First edition, August, 1971.

19. D. E. Bishop, "Community Noise Exposure Resulting from Aircraft Operations: Application Guide for Predictive Procedure." AMRL-TR-73-105, November, 1974.

20. D. E. Bishop and R. D. Horonjeff, "Procedures for Developing Noise Exposure Forecast Areas for Aircraft Flight Operations," FAA Report DS-67-10, Department of Transportation, Washington, D.C., August, 1967.

21. C. Bartel et al., "Airport Noise Reduction Forecast, Volume II—NEF Computer Program Description and User's Manual." Report No. DOT-TST-75-4, Department of Transportation, Washington, D.C., October, 1974.

22. W. L. J. Bourke, "The Noise Impact of Aircraft Operations at Sydney Airport." *Bulletin Australian Acoust. Soc.*, 3 (2), 20–31, 1975.

23. Committee on the Problem of Noise, "Noise—Final Report." HMSO, London, 1963.

24. Noise Advisory Council, "Aircraft Noise: Should the Noise and Number Index be Revised?" SBN11 7505196, HMSO, London, 1972.

25. "Second Survey of Aircraft Noise Annoyance around London (Heathrow) Airport." HMSO, London, 1971.

26. T. J. Schultz, *Community Noise Ratings*. Applied Science Publishers, London, 1972.

27. W. A. Rosenblith, K. N. Stevens, and the staff of Bolt, Beranek and Newman Inc., "Handbook of Acoustic Noise Control, Vol 2—Noise and Man." WADC TR-52-204. Wright Air Development Center, Wright Patterson Air Force Base, Ohio, 1953, 181–200.

28. K. N. Stevens, W. A. Rosenblith, and R. H. Bolt, "A Community's Reaction to Noise: Can it be Forecast?" *Noise Control*, 1, 63–71, 1955.

29. W. J. Galloway and D. E. Bishop, "Noise Exposure Forecasts: Evolution, Evaluation, Extensions and Land Use Interpretations." BBN Report No. 1862 for the FAA/DOT Office of Noise Abatement, Washington, D.C., December, 1969.

30. K. N. Stevens, A. C. Pietrasanta, and the staff of Bolt, Beranek and Newman Inc., "Procedures for Estimating Noise Exposure and Resulting Community Reactions from Air Base Operations." WADC TN-57-10. Wright Air Development Center, Wright Patterson Air Force Base, Ohio, 1957.

31. Bolt Beranek and Newman Inc., "Land Use Planning Relating to Aircraft Noise." FAA Technical Report, October, 1964; also issued as Report No. AFM86-5, TM-5-365, NAVDOCKS P-38, Department of Defense, 1964.

32. California Administrative Code, Title 4, Subchapter 6, "Noise Standards for California Airports."

33. M. Rettinger, "Acoustic Design and Noise Control." Chemical Publishing Co., Inc., New York, 1973.

34. H. E. von Gierke (chairman), "Draft Report on Impact Characterization of Noise Including Implications of Identifying and Achieving Levels of Cumulative Exposure." Task Group 3, EPA Aircraft/Airport Noise Report Study, June, 1973.

35. Noise Advisory Council, "Aircraft Noise: Flight Routing near Airports." SBN 11 750416 5, HMSO, London, 1971.

36. Surrey County Council, "London (Gatwick) Airport—an Environmental Study," October, 1970.

37. "Report of the Commission on the Third London Airport," HMSO, London, 1972.

38. Noise Advisory Council, "Aircraft Noise: Selection of Runway Sites for Maplin Airport." HMSO, London, 1972.

39. R. Rylander, S. Sörensen, and A. Kajland, "Annoyance Reactions from Aircraft Noise Exposure." *J Sound Vib.*, 24, 419–444, 1972.

40. R. Rylander, S. Sörensen, and K. Berglund, "Re-analysis of Aircraft Noise Annoyance Data against the dB(A) Peak Concept." *J. Sound Vib.*, **36** (3) 399-406, 1974.

41. J. E. Wessler, "Federal Role in Research and Regulation of Aircraft Noise." Paper presented in short course on Airport Noise Developments, University of California at Berkeley, June 20–21, 1975; see also J. E. Wessler, "Approaches for the Reduction of Aircraft Noise." Paper presented in short course on Airport Planning and Design, University of California at Berkeley, June 23, 1977.

42. B. N. Mel'nikov, "Reduction of Aircraft Noise in the Vicinity of Airports." Paper presented at the U.S.S.R./U.S. Aeronautical Technology Symposium, Moscow, July 23–27, 1973; NASA Technical Translation NASA TTF-15237, February, 1974.

43. "Monitoring Aircraft Noise around an Airport." International Organization for Standardization Recommendation ISO/R 1761, June 1970.

44. D. N. May, "Life with the Boom." Part I—*Flight International,* **98**, 519–526a, October 1, 1970; Part II—*Flight International,* **98**, 563–565, 1970.

45. Standards Association of Australia, "Building Siting and Construction against Aircraft Noise Intrusion." Draft Australian Code of Practice DR 74163, October 1, 1974.

4

Recreational vehicle noise to nonusers

EDGAR ROSE*

4-1. INTRODUCTION

In general, the following vehicles are considered motorized
recreational vehicles:

 a. snowmobiles;
 b. pleasure motorboats;
 c. motorcycles;
 d. mobile homes;
 e. dune buggies;
 f. go-karts.

Nonmotorized recreational vehicles, while not soundless, are
rarely a noise problem to the bystander; therefore a discussion
of these vehicles will be omitted from this chapter. Also mobile homes, dune buggies, and go-karts will not be discussed
here; the first falls within the categories discussed in Chapter 2,
and the others have few, if any, regulations with respect

*Outboard Marine Corporation, Waukegan, Illinois 60085

to noise perhaps because they are too rare to warrant such treatment at this time. This chapter will focus on snowmobiles, pleasure motorboats, and motorcycles.

Motorized recreational vehicles, as a group, possess two characteristics (mobility and pleasure potential) which are crucial to any discussion of their noise. To understand why this is so, it is necessary to delve briefly into the recent history of noise.

Early references to noise in the technical literature of this century defined noise as sound of an irregular, nonperiodic, or intermittent nature [1]. During the late twenties, this definition gradually changed to relate noise to sounds possessing large amounts of nonharmonic frequencies. This change in definition was concurrent with the development of the first, rudimentary, sound analyzing equipment. It is significant that during that particular period the problem of noise (at least as discussed in the literature) seemed to exist almost exclusively within the narrow confines established by the limitations of measurement technology. It was not until the early thirties that the present definition of noise as "any undesired sound" appeared in the technical literature [2].

It is also interesting to note that the first attempts at noise abatement centered on early motorized vehicles [3, 4]. Apparently, vehicles have traditionally occupied the role as sources for noise complaints. The transient nature of the noise from moving vehicles is now accepted as an important parameter governing their annoyance. Did the early definition of noise as an intermittent sound derive from the fact that the first sound source causing large-scale annoyance was the motorized vehicle?

Today's definition of noise clearly establishes that while a sound's generation and transmission are physical, its conversion into noise through some form of annoyance mechanism is psychological. In the case of motorized recreational vehicles, one is concerned not only with motorized vehicles' traditional role as a source of all kinds of complaints, but the additional very important consideration that its use may cause annoyance to one segment of the population, the bystander, while generating pleasure to users.

The lack of valid psychological information has so far made it impossible to deal with noise from recreational vehicles in a manner which fully accounts for the separate interests of users and nonusers. This lack of knowledge compounds the difficulties in defining the situations which cause annoyance. Even when the situations are easily recognized as unacceptable, their regulation is difficult. It is, for example, fairly obvious that the very extreme situation of a group of motorcyclists revving their engines at midnight outside an elderly person's country retreat is unwanted behavior—but the translation of our disapproval into regulatory form is a formidable task. Are all the motorcycles producing too much noise, or just some? Are the sources to be limited to certain locations, and who then will define and police them? Is this operation to be controlled by time of day? Is there to be a limit set on the number of machines operating together? Is the noise acceptable for a certain duration only?

Because of these difficulties, noise regulation of recreational vehicles is for the most part limited to a statement of the maximum sound level of a vehicle when tested according to a prescribed, but artificial, operating procedure designed to reveal the noisiest operating mode experienced by a bystander. This approach, as a correlate of real-life recreational vehicle noise nuisance, is only slightly more complete than controlling the nonharmonic content of sounds as per the twenties' concept of noise.

It is hoped that the reader will recognize these limitations. They explain why most if not all SAE documents are phrased in terms of sound levels and avoid mention of noise. As Hillquist states [5], this may seem a semantic ploy by some, but it is consistent with currently accepted acoustical terminology (see, e.g., ANSI S1.1-1960). It further avoids the suggestion of subjective judgment within a procedure for obtaining objective information, which at best has a distant relationship to the annoyance we are ultimately concerned about.

The following sections will deal with specific motorized recreational vehicles, the means to measure their maximum sound levels, and the applicable limits.

4-2. SNOWMOBILES

4-2.1 Background

The rapid gain of popularity of the snowmobile, first introduced in the early sixties, resulted in the astronomical growth of a new manufacturing industry, and the radical restructuring of the previously summer-oriented recreational industry in many snow-belt areas. The changes in lifestyle which resulted from the large-scale introduction of snowmobiles brought forth complaints from nonusers—some unjustified, others having at least a partial basis in fact. Probably the most serious in the latter category were the complaints pertaining to excessive noise.

The industry recognized the need to alleviate the noise problem and formed an SAE committee of Canadian and U.S. engineers from industry and government agencies to take the first step in the abatement of noise through product design: the establishment of a standardized method for the measurement of the maximum sound output of snowmobiles.

4-2.2 Measurement of Snowmobile-Generated Sound Levels

The standardized method which was developed is SAE Recommended Practice J192, published by the Society of Automotive Engineers [6]. A subsequent version of this method, SAE J192a [7], is also frequently used. Both methods are described here.

In the development of SAE J192 great stress was placed on repeatability of results, and on measurement of sound levels under snowmobile operating conditions leading to maximum sound output. Since it was recognized that the widely varying snow conditions normally encountered in the operation of a snowmobile could result in great variations of measured sound levels, it was agreed that the operation of the snowmobile should take place on a grassy surface, with the height of the vegetation limited to 3 inches. The presence of vegetation was required to provide sufficient lateral friction to the skis to permit directional control of the vehicle while running at full throttle, this full power operation (at rated engine speed) resulting in maximum sound output.

The operation on grassy surfaces made it possible to perform development and testing on a 12-month basis. Of course, tests during the winter and early spring months required lengthy and expensive trips from the snow-belt engineering facilities to the southern states of the U.S., principally Florida. SAE J192a, allowing the option of testing on either grassy surfaces or controlled snowy surfaces, obviates these trips, although at a slight loss in the reproducibility of the results.

The test site layout for J192 is shown in Fig. 4-1(a), while that for J192a is shown in Fig. 4-1(b). The microphone is placed 4 ft above ground.

The main difference between test site layouts stems from the fact that in J192 the starting point for the snowmobile is varied for each snowmobile so that it enters the measurement portion of the track at the instant the engine reaches rated speed, whereas in the later J192a the starting point for snowmobile test runs is fixed for the sake of simplicity.

Additionally, J192a places restrictions on the atmospheric conditions acceptable for the conduct of the tests, to reduce the possible variations in sound output due to changes in engine output caused by large variations in ambient pressure or temperature.

Both procedures specify standards of measurement accuracy and call for a number of repeat passbys before establishing an average sound level for each side of the machine. The highest of these average sound levels is the one which is then used to describe the sound output of the machine.

At this time, some government regulatory agencies in Canada and the U.S. use J192; others use J192a.

4-2.3 Snowmobile Sound Level Limits

One of the first regulations pertaining to snowmobile sound levels was formulated by the Department of Parks and Recreation of the State of New York and became effective November 1, 1971 [8]. These regulations stated, in part, that:

The sound level produced by a new snowmobile manufactured after June 1, 1972 shall not exceed 82 dB on an A-weighted network at 50 ft when mea-

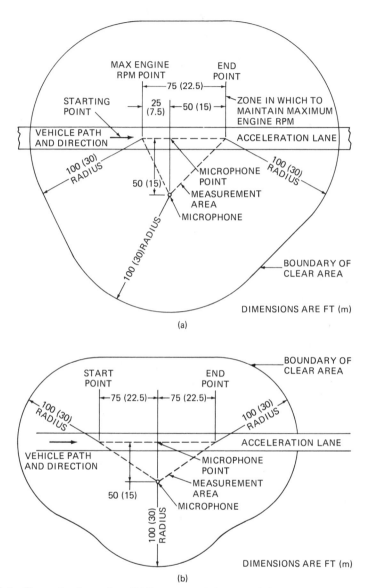

Fig. 4-1 Ground plans for SAE snowmobile sound level test procedures: (a) SAE J192; (b) SAE J192a. The start and end points are shown for left-to-right vehicle passby; they should be reversed for right-to-left passby. (*Reproduced with acknowledgment to the Society of Automotive Engineers*)

sured in accordance with the procedure described herein. The sound level produced by a new snowmobile manufactured after June 1, 1974 shall not exceed 73 dB on an A-weighted network at 50 ft when measured in accordance with the procedure described herein.

The purpose of the first stage of these regulations, the 82 dB(A) limit, was to eliminate the sale in New York, and thus discourage the manufacture, of snowmobile models exceeding sound levels achievable within the state of the art. Virtually all manufacturers were successful in reducing the sound level of their machines to the limit prescribed by this New York State regulation.

The second stage, the 73 dB(A) limit which was to become effective two years after the first stage, required a considerable advance in the state of the art as it existed at that time. In fact, only one manufacturer was successful in meeting this requirement. Furthermore, with the increased availability of 82 dB(A) machines, some evidence started appearing that the chief sources of annoyance were machines producing sound levels in excess of 82 dB(A). The 73 dB(A) machines, on the other hand, were so quiet that normal undulations in the terrain were sufficient to mask the snowmobile sounds in the presence of many normal outdoor ambient sounds.

Thus, two factors combined to lead to New York's cancellation of the 73 dB(A) limit: (1) the inability of most snowmobile manufacturers to meet it; (2) a growing realization that this requirement might have been excessively severe.

By contrast, the regulations developed by the Department of Transport of the Federal Government of Canada [9], although instituted slightly earlier than the New York State regulations, have stood up very well under the test of time. These regulations rely on SAE J192 for detailed specifications of test procedures. They established February 1, 1972 as the deadline for the manufacture of snowmobiles exceeding a sound level of 82 dB(A) and February 1, 1975 for the manufacture of snowmobiles exceeding a sound level of 78 dB(A). The carefully considered wording and realistically effective limits of the Federal Canadian regulations have served to make these regulations a model for most regulations adopted subsequently in various parts of North America.

While Canada, and most states of the U.S.A., now have a 78 dB(A) limit as measured according to SAE J192 or SAE J192a [10], a few states have either higher limits or unenforced lower limits. Thus, for all practical purposes, snowmobiles must be manufactured to meet the 78 dB(A) requirements of Canada.

A major defect in all these efforts to reduce the exposure of nonusers to snowmobile noise is the lack of regulations which would prevent the user from modifying his snowmobile exhaust or engine intake system with resultant increases in sound level output. Such modifications are at times undertaken in the hope of obtaining an increase in horsepower. Although such increases are frequently illusory, they are nevertheless a temptation to some snowmobilers. As long as such user modifications are tacitly tolerated, it is unlikely that significant reductions in snowmobile sound annoyance can be achieved through further reductions in sound levels designed and built into new machines.

4-3. PLEASURE MOTORBOATS

4-3.1 Background

Motorboats used for recreational purposes are powered by outboard motors, or inboard engines producing thrust through stern drives, jet pumps, or direct drive to propellers. While the former group is powered primarily by 2-cycle spark-ignited engines, the latter utilizes mostly 4-cycle spark-ignited engines. The sounds from the two types of prime movers are quite different in character, and are further influenced in intensity and frequency content by the exhaust system, the air intake system, the engine enclosure, and other design parameters.

Outboard motors, because of their exposed location and necessarily low weight, have been more difficult to silence than inboard engines. Through judicious use of absorptive engine enclosures, underwater exhaust, rubber mounts, intake silencers, and other techniques—most of which were described almost 25 years ago by Conover [11]—outboard motor-powered recreational boats have become rather quiet. Inboard-powered boats, using some of the same techniques, have been similarly silenced.

Occasional user-modified outboard motors and open-exhaust inboards nevertheless give rise to justified noise complaints: a sound level measurement procedure was therefore developed and published as SAE Recommended Practice SAE J34 [12].

4-3.2 Measurement of Sound Levels Generated
by Pleasure Motorboats

The determination of sound output from outboard and inboard boats takes place at wide open throttle, rated engine speed, with the boat traveling in a straight line. The microphone is placed 50 ft from this line, according to SAE J34, although a revision of this standard, which upon approval will be numbered SAE J34a, calls for a microphone distance of 25 meters (82 ft). Microphone height is 1.2–1.5 meters (4–5 ft) above the water and no closer than 0.6 meters (2 ft) from the horizontal surface of the dock or platform on which the microphone stands.

The test site (Fig. 4-2) must encompass a body of calm water large enough to allow safe operation of the boats at maximum speed. Since speeds of 60 mph are not uncommon, and occasionally much higher speeds are encountered, great care

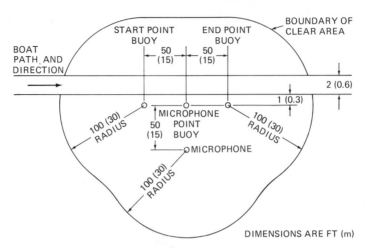

Fig. 4-2 Ground plan for the SAE J34 motorboat sound level test. Start and end points are shown for left-to-right vehicle passby. (*Reproduced with acknowledgment to the Society of Automotive Engineers*)

must be taken in the selection of the test site. Furthermore, the usual restrictions apply to the presence of sound reflective structures and other objects. Tree trunks can be quite effective reflectors; the same is true of river embankments and piers. None of these may be closer than 100 ft to the microphone or the sound measurement portion of the boat path. The latter is marked by three buoys 50 ft apart, one 50 ft from the microphone at the "microphone point," the other two placed either side of this buoy on a straight line perpendicular to the line joining the microphone point to the microphone.

In boat sound level tests special care must be taken to avoid misleading high readings. These can occur when, because of wave action, the underwater exhaust is briefly exposed to the atmosphere, or the engine momentarily overspeeds because of a transient exposure of part of the propeller to the air, or ventilation of the jet pump. Also, since the absence of sound blanking features on bodies of water allows the transmission of natural and other extraneous sounds over large distances, continual vigilance against the intrusion of foreign sounds is required.

4-3.3 Motorboat Sound Level Limits

A few states in the U.S.—most notably California, Oregon, and Arizona—have enacted legislation and regulations relating to the sound emissions from pleasure motorboats. Not all of this legislation has been technically satisfactory. For example, Arizona's regulations state in their entirety [13]:

A. It shall be unlawful for any person to operate a watercraft upon the waters of this state under any condition or in any manner that the watercraft emits a sound level in excess of 86 decibels on the A-weighted scale when measured from a distance of 50 ft *or more* from the watercraft.

B. This section shall not apply to watercraft operated under permits issued in accordance to Section 5-336-C ARS (Arizona Revised Statutes; See Reference Binder 81:3351)

California's regulations [14]—the first to be put into effect—make it illegal to *sell* motorboats having sound outputs in excess of certain values and also forbid the *operation* of motorboats exceeding these sound emission values. The testing procedure specified by these regulations is SAE J34, described above.

The following sound level limits have been established by the California regulations:

86 dB(A) for engines manufactured before January 1, 1976;

84 dB(A) for engines manufactured between January 1, 1976 and December 31, 1977;

82 dB(A) for engines manufactured on or after January 1, 1978.

In effect, the California regulations control the user as well as the manufacturer of motorboats and outboard motors. Thus it becomes possible to eliminate the chief source of noise on waterways: the user-modified exhaust systems, and the special purpose "hot rod" boats with open exhaust stacks or megaphone exhausts. The time-graduated sound levels allow a rational, economically optimized approach to sound level reduction by the manufacturers of motorboats.

Oregon treats all motorboats as off-road recreational vehicles for purposes of sound level regulation [15]. The rule states clearly that "no person shall *operate* any off-road recreational vehicle which exceeds specified noise limits." The sound level limits are gradually reduced over periods of time loosely described by model year designations from before 1976 to after 1978. These rules do not specify exact sound level test procedures; thus, the sound levels listed in these regulations are of little significance.

Several agencies in the U.S. and Canada are actively engaged in the preparation of sound level regulations for motorboats. The Canadian Standards Association is at work on a set of procedures for the measurement of sound levels, which procedures (when finalized) might be adopted by Canadian regulatory agencies.

4-4. MOTORCYCLES

4-4.1 Background

In many cases the motorcycle is used as a utilitarian means of transportation. Nevertheless, its recreational usage is high, and even when it is used for transportation, the choice of a motorcycle in preference to an automobile is often based on the pleasure derived by the motorcyclist from his vehicle. Of course, off-road motorcycle usage is virtually completely recreational.

To measure the sound output of motorcycles, two standard methods have been developed, SAE Recommended Practice J47 [16] and SAE Recommended Practice J331a [17]. The requirements of SAE J47 are, among others, that the motorcycle tests be started with full-throttle accelerations in first gear. If front wheel lift-off should occur, it is permissible to repeat the tests in the next higher gear. Since motorcycles having sufficient torque to raise the front wheels during full-throttle accelerations in first gear are not at all uncommon, the test driver must possess a high degree of skill to avoid injury-causing accidents during this test phase. Because of the obvious hazards involved in SAE J47 testing, it is suggested that SAE J331a be used instead.

The sound levels generated in tests according to SAE J331a are equal to the highest sound levels generated by motorcycles under virtually all conditions other than those entailing high risk factors. Thus this standardized test method should be eminently satisfactory in determining the maximum practical sound generating capability of a motorcycle, and will be the only sound level measuring procedure discussed.

4-4.2 Measurement of Motorcycle-Generated Sound Levels

SAE Recommended Practice J331a is a standardized means of reproducing and measuring the sound level generated at a distance of 50 ft by a motorcycle accelerating at wide open throttle from a moderate cruising speed of 30 mph. This situation will typically occur when a motorcycle is passing other vehicles in residential areas, or on an entry ramp to a high speed highway. Therefore the tests in SAE J331a are generally performed in second gear, with full-throttle accelerations taking place from a steady-state 30 mph for all except very low-powered motorcycles, for which the velocity at "60% rpm" is less than 30 mph. ("60% rpm" is the engine speed at 60% of the engine speed corresponding to the maximum rated net horsepower.)

The test site layout is shown in Fig. 4-3. The surface of the test site, within the isosceles triangle having the microphone location at its apex and a 100 ft base coincident with the vehicle path, must be concrete or asphalt devoid of water, snow, soil, or other materials. The surface outside this triangle must be essentially flat, level, and free of sound-reflecting objects such as billboards, trees, buildings, or vehicles within 100 ft of any part of the previously mentioned isosceles triangle. The microphone height is 4 ft above ground.

The driver, and a saddle bag to hold weight, must be brought to a total weight of 170 ± 5 lb.

Starting from beyond the end point (see Fig. 4-3), the vehicle is accelerated to 30 mph, and driven in second gear up to the first acceleration point. When the

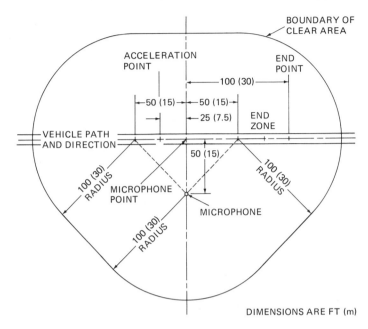

Fig. 4-3   Ground plan for the SAE J331a motorcycle sound level test. Start and end points are shown for left-to-right vehicle passby. (*Reproduced with acknowledgment to the Society of Automotive Engineers*)

acceleration point is reached, the throttle is opened rapidly and fully and the motorcycle is allowed to accelerate until its front reaches the end point, when the throttle is closed. If the engine reaches the speed of maximum rated horsepower before the vehicle arrives at the end point, the throttle is closed.

The motorcycle is run repeatedly in alternate directions, and an average "accelerating" sound level is determined for each side.

There is also a deceleration test. The vehicle proceeds along the vehicle path at the engine speed of maximum rated net horsepower in the gear used for the previous tests. At the end point, after passing the microphone, the throttle is rapidly and fully closed, and the vehicle allowed to decelerate to an engine speed of one-half the rpm at maximum rated net horsepower. An average "decelerating" sound level is determined for each side.

4-4.3 Motorcycle Sound Level Limits

Canada and various sections of the U.S.A. are at this time saddled with a hodgepodge of motorcycle sound regulations. Some agencies utilize regulations which define the sound limits in nonquantitative terms such as "reasonable" or "excessive"; others use quantitative limits, usually 83 dB(A) or 86 dB(A) as measured per SAE J331a. The following is a list of governmental bodies using the quantitative approach for motorcycle sound level limits:

Arlington County, Virginia	Michigan
California	Minnesota
Canada	Montana
Chicago, Illinois	Nevada
Colorado	New Hampshire
Connecticut	New Jersey Turnpike Authority
Detroit, Michigan	New Mexico
Florida	New York
Grand Rapids, Michigan	Oregon
Hawaii	Pennsylvania
Idaho	Rhode Island
Indiana	Washington (State)
Maryland	

Of these, Canada and California may be considered the leaders in effective motorcycle sound emission regulations. Canada specifies test procedures akin to SAE J47, whereas California—like most other governments—favors SAE J331a. Canada has had an 86 dB(A) limit since 1972. Furthermore, 78 dB(A) was proposed to go into effect in Canada on January 1, 1975; so far, this limit has not been adopted.

California's motorcycle noise regulations limit the sound output to 83 dB(A) for motorcycles manufactured on or after January 1, 1975. In general, the SAE J47 procedure can result in sound level readings from 2 dB(A) lower to 9 dB(A) higher than those obtained with the SAE J331a procedure [18].

Neither the Canadian nor the Californian noise regulations go beyond the manufacture and sale of new motorcycles; older motorcycles or user-modified motorcycles are not covered by these regulations. Since virtually all other quantitative noise regulations are based on these two regulations, the same omissions of coverage are usually found in them as well.

The U.S. Federal Government's Environmental Protection Agency established on May 28, 1975 that motorcycles were one of the main causes of vehicular noise. By law, within two years of that date the EPA must establish a set of federal regulations which would regulate motorcycle sound levels on a uniform basis throughout the U.S., but this had not been accomplished by November 1977.

REFERENCES

1. E. H. Barton, *A Text-Book on Sound*. London, 1926.

2. "American Standards Association Acoustical Terminology." *Acoustic Journal*, **2**, 311, 1931.

3. A. B. Wood, *A Textbook of Sound.* English Admiralty, New York, 1930.

4. N. W. McLachlan, *Noise.* London, 1935.

5. W. W. Lang, *Noise-Con 75 Proceedings.* New York, 1975.

6. "Exterior Sound Level for Snowmobiles—SAE J192." In *SAE Handbook,* New York, 1971.

7. "Exterior Sound Level for Snowmobiles—SAE J192a." In *SAE Handbook,* New York, 1974.

8. Official Compilation of Codes, Rules, and Regulations of the State of New York, Chapter II, Subtitle I, 1971.

9. "Statutory Orders and Regulations 1971." *The Canada Gazette,* Part II, 1971.

10. "Summary of Snowmobile Laws and Regulations Compiled by the I.S.I.A." International Snowmobile Industry Association, Falls Church, Virginia, 1974.

11. W. C. Conover, "Noise Control of Outboard Motors." *Noise Control,* March 1955.

12. "Exterior Sound Level Measurement Procedure for Pleasure Motorboats—SAE J34." In *SAE Handbook,* New York, 1974.

13. Arizona Game & Fish Commission Boating Rules, Article 4, Boating and Water Sports, Section R12-482, 1975.

14. California Motorboat Noise Regulations, Harbors & Navigation Code Section 654, 654.05, 654.06 and 668, 1973; amended Chapter 1269, Laws of 1974.

15. Oregon Noise Control Regulations, Environmental Quality Commission, adopted 1974, amended 1974. Portland, Oregon.

16. "Maximum Sound Level Potential for Motorcycles—SAE J47." In *SAE Handbook,* Warrendale, Pennsylvania, 1976.

17. "Sound Levels for Motorcycles—SAE J331a." In *SAE Handbook,* Warrendale, Pennsylvania, 1976.

18. Ross A. Little, "Motorcycle Noise Test Procedure Evaluation." Society of Automotive Engineers, New York, 1974.

5

Noise of transportation to travelers

GILLES CRÉPEAU*

5-1. INTRODUCTION

In this chapter we will analyze the passenger and crew position sound levels of different means of transportation.

The question of vibration is indissolubly linked with transportation noise. However, it has been less studied from a passenger standpoint, and in fact there are no accepted norms for individual means of transport. For this reason vibration is here considered separately from noise in a single section at the end of the chapter, which refers to vibration in transportation generally.

The assessment of hearing damage risk is also briefly covered in an independent section (5-2), which should be read in conjunction with Chapter 11. Hearing damage is an outcome of the physical realities of sound level and exposure time, and is

*Societé d'Habitation du Québec, 255 est, boul Crémazie, Montreal, P.Q. H2M 1L5, Canada. This work was performed while the author was at Centre de Recherche Industrielle du Québec, Dorval, P.Q.

not dependent on the attitudes of travelers toward the benefits and character-istics of the transportation mode (which affects annoyance), nor on such varia-bles as seat spacing (which affects speech interference). The criteria for hearing damage risk need not be expressed in terms of vehicle category, and the subject is therefore treated on its own.

For individual transportation modes, the different noise sources and their acoustic characteristics as sensed inside them are described, and the various sound levels are evaluated according to a number of criteria: speech interference, work and rest interruption, and annoyance. Some limits are tentatively proposed, and the norms and laws which already exist are described.

The assessments of speech interference, of interruption to intellectual activity, and of annoyance must be regarded as tentative. Speech interference is here evaluated in terms of dB(A) without special regard for the particular frequency spectra of the various vehicle categories. Interruption to intellectual activity is evaluated with the use of ranking curves, also translated into dB(A) without particular attention to the spectra involved, and is based moreover on noise inside buildings rather than in vehicles. The evaluation of annoyance has had to make use of a survey of propeller aircraft noise, because information is scant on the annoyance to travelers generated by the other transportation modes. The criteria that have resulted are put forward as suggestions rather than conclu-sions.

5-2. HEARING DAMAGE RISK

The best-known criteria and limits for hearing hazard are those of the EPA [1] , ISO [5] , and OSHA. The sound exposure permitted by each of these bodies is described in Chapter 11; the essential similarity is that the permissible sound level is expressed in dB(A) and is subject to a trading relationship with time, in which a low sound level is permitted for a long period of time and a high sound level for a short period.

To assess hearing damage risk one has, therefore, to know the sound levels in vehicles and the length of time people are exposed to them. A general overview of the sound levels involved is given by the EPA in Ref. 10, from which Fig. 5-1 has been prepared. The range of sound levels occurring on typical journeys is given, and also the energy mean (i.e., the equivalent sound level L_{eq}) on those journeys. These data are a useful supplement to the sound levels given in Sec-tions 5-3 to 5-11. There is, however, little information on the distribution of exposure times for the people using (and operating) the various vehicles. To assess hearing damage risk properly this information is needed; in the remaining parts of this chapter some representative, but not authoritative, exposure times are put forward for each transportation mode. Exposure times are invariably much higher for crew members than for passengers, but the former can more easily be given hearing protection devices where there is such a need.

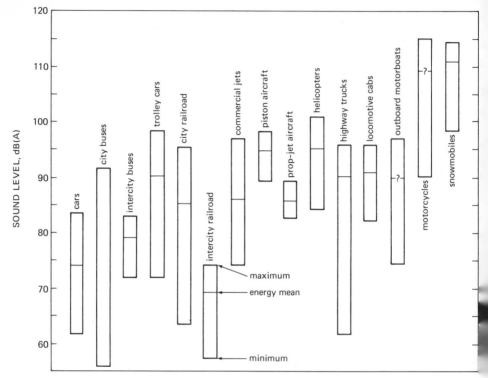

Fig. 5-1 Traveler position sound levels on different types of vehicle: maximum levels, minimum levels, and the energy mean L_{eq} on typical journeys [10].

The remaining difficulty is to choose the correct criterion. Chapter 11 discusses the matter in detail; in the example below we illustrate the principles involved by making use of the EPA criterion [1] of L_{eq} = 70 dB(A), when measured over a year. Figure 5-2 relates L_{eq} measured for various journey times to the L_{eq} over a year.

Example

Problem: A commuter is exposed for 2 hours a day, 5 days a week, 50 weeks a year, to a sound level with an energy mean L_{eq} of 80 dB(A). Using the criterion that the permissible L_{eq} over the year is 70 dB(A), determine whether the commuter's hearing is at risk. Assume that the annual L_{eq} from other sources in his life is 60 dB(A).

Solution: From Fig. 5-2, the journey-time L_{eq} converts to an annual L_{eq} of 67.5 dB(A). To this we must add the 60 dB(A) contributed by other sources. The total L_{eq} is just over 68 dB(A), which is less than the criterion level, so that hearing is not at risk.

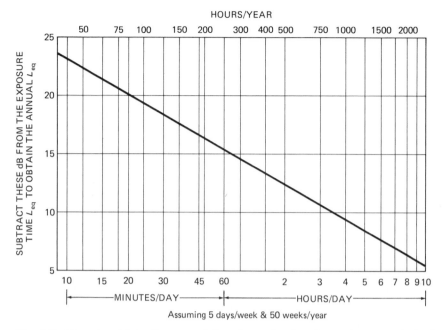

Fig. 5-2   How to obtain the annual L_{eq} from the L_{eq} for a shorter period of exposure.

5-3. AUTOMOBILES

5-3.1 Some Facts

There are at present over 100 million cars in North America (and about 13 million in Britain). The automobile industry in North America, taking into account sales and service as well as manufacture, employs about 5 million people. If one adds to these figures the land devoted to cars (roads, highways, and parking areas), one can readily appreciate the enormous economic importance of the automobile.

Exposure times for car noise are difficult to estimate, and likely cover an enormous range of hours, the heaviest exposure probably occurring among taxi drivers, who, it must be remembered, often drive with their windows open and suffer higher noise levels both from their own vehicle and from others.

5-3.2 Noise Sources

Cars are quieter than buses and trucks. The different noise sources are the exhaust, the carburetor inlet, the cooling fan, the engine, the transmission, the tires, and aerodynamic effects. The noise tends to increase slightly with age because of deterioration, though seldom by more than 2-3 dB(A).

At speeds up to 45 mph, the exhaust is the predominant outside noise source, but it causes no serious problems inside the vehicle, at least while the muffler is adequate. The carburetor inlet noise is also no problem. At these low speeds, the cooling fan, the motor, and the transmission are the significant sources of noise inside the car.

Tire noise increases with speed, with the complexity of the tread pattern, and with the roughness of the ground surface. Snow tires are about 10 dB(A) noisier than ordinary tires. On a smooth surface, with ordinary tires, tire noise becomes significant at speeds above about 40 mph.

Aerodynamic noise, which theoretically increases with the square of the speed, joins tire noise as a major interior noise at high speeds.

For a number of reasons (e.g., differences in engine power and engine cowling, in the design of the muffler, and in the dimensions and insulation of the passenger areas) interior noise levels vary greatly between different types of car. Contrary to popular belief, there is no strong correlation between price and interior quiet. Across the range of cars, sound levels vary between 55 and 90 dB(A). Table 5-1 gives the interior sound levels of a number of mainly 1970 models described in *Popular Science* over the period February, 1970 to January, 1971 [15]. There has been a slight drop in sound levels since those data were amassed, of about a decibel a year [10]; Fig. 5-1 also indicates a contraction in the range of levels.

5-3.3 Criteria

Ease of communication rather than privacy is a design objective in a car. For relatively unstrained conversation between two people seated at the extremes of the same bench seat (say 3 ft apart), the interior sound level should be lower than 65 dB(A), though this probably applies to face-to-face conversation [12]. One can see from Table 5-1 that the noise in vehicles at 60 mph exceeds this criterion, though most cars traveling at 30 mph on smooth surfaces meet it. A similar line of reasoning can be applied to speech and music on the radio, for which the same limit of 65 dB(A) is called for.

Intellectual work rarely has a place in a car, and to a lesser degree the same may be said for rest and sleep. For occasions when cars *are* used as conference rooms and rest areas, one may try to apply (though with doubtful validity) the NC curves of Beranek [3] and the NR curves [4], both of which were derived for daytime noise inside buildings. Bearing in mind the frequency spectrum of noise in cars, they suggest a sound level as low as 50 dB(A). A discussion of this type of curve is contained in Chapter 6.

A further criterion is the ability to hear the warning signal of another vehicle. If the horn of the other vehicle is built to achieve the sound output recommended in SAE J377 [11], which we prefer over the older ISO/R 512 [13], it

TABLE 5-1. Interior Sound Levels in 1970 Model Cars

Automobile	Sound Level, dB(A)		
	30 mph, Smooth Surface	60 mph, Smooth Surface	30 mph, Rough Surface
AMC Gremlin	68	77	75
AMC Hornet	67	74	78
AMC Javelin SST	65	76	76
AMC Rebel	63	75	74
Audi II-LS	62	76	73
Buick Estate Wagon	61	73	68
Buick Riviera	59	68	68
Chevrolet Camaro SS	64	71	74
Chevrolet Malibu	63	69	71
Chevrolet Monte Carlo	61	68	73
Chev. Suburban Carry-All	64	76	78
Chevrolet Vega[a]	68	83	82
Chrysler Town & Country	62	75	75
Citroën D Special	64	75	73
Datsun 510	68	77	77
Datsun 1200	68	79	72
Dodge Challenger 340	65	80	75
Dodge Monaco Wagon	62	75	77
Fiat 1245	67	79	76
Ford Mustang Mach I	67	77	78
Ford Pinto[a]	67	78	84
Ford Thunderbird	58	67	71
Ford Torino	63	72	74
Int. Travelall D-1000	67	71	71
Kaiser Deep Wagoneer	65	73	78
Mercury Colony Park	60	70	69
Mercury Cougar Eliminator	64	75	74
Oldsmobile Toronado	59	72	69
Peugeot 504	62	78	70
Plymouth Barracuda 340	66	75	75
Plymouth Satellite	64	74	77
Pontiac Firebird 400	67	74	74
Pontiac Grand Prix	60	68	72
Renault 16	64	77	74
Saab 96	68	77	78
Simca 1204	73	80	78
Toyota Corolla	66	77	80
Toyota Corona	66	76	76
Volkswagen 1600	72	80	82
Volkswagen Squareback	67	81	78
Volvo 144S	64	78	72

[a]1971 model.

should produce 82–102 dB(A) at a distance of 50 ft, which we can perhaps take as the appropriate warning distance for cars. The audibility of this signal inside the "receiving" car with its windows closed depends on the spectrum of the horn, the insulation qualities of the receiving car and the ongoing noise level inside it. It is difficult to generalize and set an acceptable level for the ongoing interior noise which will allow a horn signal of 82–102 dB(A) to be audible, but it is likely to be in the range 65–85 dB(A). The lower of these levels probably constitutes a strict criterion.

5-3.4 Limits

To our knowledge, there are no formal norms or laws about car interior noise levels. Because of the subject's importance, it would seem desirable to establish some, and also a measurement procedure. There exists at the moment one such norm, SAE J366a, for the sound levels in truck cabs [20]. A norm for cars could be based on this, at least with respect to the measurement procedure (see Section 5-5.4).

An interior sound level of 65 dB(A) should be regarded as a serious short-term design objective by car manufacturers, and an imperative medium-term objective, if ease of communication is an important concept to them.

5-4. BUSES

5-4.1 Some Facts

Because of the overpowering importance of the car in North America, the role of buses is secondary when it comes to surface transportation. There are about 450,000 buses in North America (and about 77,000 in Britain). Three-quarters of these are school buses, and they are responsible for about half the overall mileage.

The North American bus industry employs about 150,000 people and more than 6 billion passengers use their buses each year. It is difficult to establish the average time a passenger spends in a bus annually, but the average for drivers would be somewhat less than eight hours a day.

5-4.2 Noise Sources

Sources of noise in buses are qualitatively the same as for cars, i.e., the exhaust, the intake, the fan, the engine, the transmission, the tires, and aerodynamic effects. Irrespective of speed, the major sources are the engine and exhaust. If its air intake is badly designed, the fan can be an important source, though mainly outside the bus, and then only when the engine idles. At speeds above 50 mph, tire noise (mainly in the frequency range 500–2000 Hz) and aerodynamic noise join engine and transmission noise as the principal sources perceived inside.

TABLE 5-2. Criteria and Suggested Limits for Sound Level in dB(A) on Various Transportation Modes

Vehicle	Criteria						Recommended Limits, dB(A)
	Communication	Privacy (minima)	Radio	Intellectual Effort and Rest	Warning Signal Audibility	Annoyance	
Cars	65	–	65	50	65–85	–	65 (short-term, new production)
Buses	70	65	–	40–50	60–80	77 (long-distance) 87 (urban)	70 (new production) 65 (later)
Trucks	–	–	65–75	–	60–80	87	85 (new production)
Trains	65–70	62	–	40–50	–	77	70 (urban) 65 (intercity) } new rail systems
Aircraft	70 (passenger) 75–80 (crew)	65	75 (crew only)	40–50	–	77–87	commercial jets 70 (cabins) 75 (flight decks) general aviation, helicopters, STOL (new designs) 75 (cabins) 80 (cockpits)
Ships	85–90 (controls) 60 (public areas and cabins)	–	–	40–50 (public areas) 35–45 (cabins)	–	–	45 (cabins, night) 50 (cabins, day, and public areas) } new designs
Hovercraft	65–70	–	75 (crew only)	–	–	87	80 (new designs)
Motorcycles	–	–	–	–	82–102	–	90 (new production)
Snowmobiles	–	–	–	–	87–102	–	90 (new production)
Motorboats	75	–	–	40–60 (cabins)	–	77	85 (outside) 75 (inside) } new production

aDash indicates no information or that the criterion is not applicable.

The length of a bus makes it possible to find standing waves extending to frequencies as low as the auditory lower limit of about 20 Hz. This results in a spectral distribution heavily weighted in the low frequencies. The spectrum drops fairly regularly by about 50 dB per octave at frequencies above the peak.

The resulting sound levels tend to vary between 65 and 90 dB(A), depending on the road surface, the vehicle's speed, the position in the bus, whether the windows are open or not, and whether there is air-conditioning or not.

Rough road surfaces can raise the sound level about 10 dB(A); and 60 mph can be more than 10 dB(A) noisier than 30 mph. With the exception of school buses, most North American buses have rear engines, resulting in sound levels being around 12 dB(A) higher at the back than at the front. Air-conditioning noise affects the front rows of seats more than the back ones. Open windows can increase sound levels 2–3 dB(A) under certain operating conditions and in certain seats.

5-4.3 Criteria

With regard to speech, it seems reasonable for a conversation between two people sitting side by side (i.e., 2 ft apart) to be carried on in a normal voice. For this criterion to be met, the sound level should not exceed 70 dB(A). Since buses are normally public vehicles, one might wish to give equal importance to speech privacy. If one decides that normal speech should not be too intelligible at a distance of more than 3 ft, then the sound level should ideally not be lower than 65 dB(A). However, we believe that emphasis should be given to ease of communication rather than privacy, because privacy can be easily preserved by adjusting voice level.

Buses can be used as an environment for intellectual activity (reading, studying, etc.) and for rest or relaxation, and for these purposes the sound level should not be allowed to intrude. Applying the NC curves of Beranek and the NR curves of Kosten and Van Os, again with doubtful validity, one can propose a level of between 40 and 50 dB(A) as a criterion.

The size of buses probably dictates that an external warning signal should be audible at 100 ft (rather than the 50 ft suggested for cars in Section 5-3.3); this requires an interior noise level, at the driver's position, in the range 60–80 dB(A).

A tentative criterion for annoyance can perhaps be derived from the work of Lippert and Miller on aircraft noise [28]. The level of 77 dB(A) could be a criterion for long distance buses, with 87 dB(A) applicable on urban buses because their journey times are lower.

5-4.4 Limits

To our knowledge, no norms or laws have been formally established on this subject, though we believe they should be. Such norms or laws could be closely related to SAE J336a [20].

Considering all the above, a level of 70 dB(A) or less should, it is suggested, be regarded as a short-term objective for bus constructors, and 65 dB(A) is the logical next step. All these levels refer, of course, to interior noise at any seating position at cruising speed. The generally lower noise levels at the front will then result in lower noise levels for the driver.

5-5. TRUCKS

5-5.1 Some Facts

For the purpose of this chapter, trucks can be classed in three categories: light, medium, and heavy. We are mostly interested in the medium and heavy ones, i.e., those over 6000 lb gross vehicle weight (GVW), mainly because these two categories are the noisiest. Also, the drivers of trucks in these two categories spend relatively more time in their trucks.

There are actually over 20 million trucks on North American highways (and more than 1.6 million in Britain), which translates to more than 210 billion miles a year. Although this gives an annual average of about 10,500 miles per truck, a great proportion of the heavies roll over 50,000 miles per year, averaging four hours each working day. Because this includes time when the trucks are out of service, driver exposure time could be higher—perhaps eight hours a day.

5-5.2 Noise Sources

Overall, the truck has the same noise sources as the car and the bus, as described in the two previous sections of this chapter. However, it should be noted that trucks can be gasoline powered (97.5 percent) or diesel powered (2.5 percent). Diesel trucks are about 8–10 dB(A) noisier than gasoline ones. Whatever their powerplant, sound levels increase with age as a result of mechanical deterioration and relaxed standards of maintenance.

The main noise sources are the engine, the exhaust, the cooling fan, and the air intake. At high speed, tire and aerodynamic noises also assume considerable importance.

Engine noise comes from combustion and compression and the ensuing mechanical movements in the running of the engine. These physical mechanisms cause the engine structure to vibrate, which in turn may vibrate everything attached to it. The difference in sound level between gasoline-powered and diesel-powered trucks is mostly due to their different ignition characteristics. A faster pressure increase occurs in the diesel engine cylinder, and causes greater vibration, than in the gasoline-powered cylinder. From this it follows that the noise increase found with a loaded truck compared with an unloaded one is much less marked with a diesel engine. Therefore, although at low speed diesel-powered trucks are significantly noisier than gasoline-powered ones, at high speed the two types are equally noisy, at least for the same power.

For medium and large trucks, even though their engines are often badly insulated acoustically, the exhaust has nearly as much effect on the noise output as the engine. However, at highway cruising speeds of 60 mph or so, even the exhaust noises are drowned by aerodynamic noise and tire noise. The noise from the cooling fan is also relatively high (it can even exceed the exhaust noise when heard outside), but this noise source is of little importance inside the truck, as is also generally true of noise from the carburetor air intake.

Tires are an important source of noise, especially at high speed. Between 30 and 60 mph, each 10 mph speed increase produces a tire sound level increment of about 4–6 dB(A). Tire noise is also a function of the design of the tread: tires with crossrib design (the majority of truck tires) are noisier than tires with lengthwise design. The difference in noisiness is increased if the tread is worn, and greater still if the tire is a recap.

For medium and heavy trucks, interior sound levels, with windows closed, vary from about 65 dB(A) at engine idle to about 100 dB(A) when the engine output is at maximum. On the road at over 60 mph, the sound level inside the cab is generally between 85 and 95 dB(A).

5-5.3 Criteria

Since truck drivers travel mostly alone, verbal communication is not a major criterion of truck noise acceptability. However, it has a certain importance when the driver listens to the radio. Here, it is difficult to establish a precise criterion, but it is probably in the range 65–75 dB(A), the higher figure being more in the nature of a limit of acceptability than a threshold of nuisance.

With regard to warning signals, the same considerations apply as those for buses (see Section 5-4.3), which call for a maximum interior sound level of 60–80 dB(A).

The majority of trucks carry no relief driver or passenger, so the criteria for rest and intellectual work are perhaps not relevant. However, since the driver is at the wheel a good part of the day (say, four hours or so), the prevailing noise level should not be the cause of psychological or physical fatigue. Our best clue to an appropriate criterion level for this is 87 dB(A) for, according to Lippert and Miller [28], it corresponds to the noise level described as comfortable by air passengers. We have taken this level rather than 77 dB(A), which Ref. 28 also mentions, because the truck driver probably has a higher tolerance to the cab noise nuisance by virtue of having chosen this occupation and by virtue of the benefits he draws from it.

5-5.4 Limits

In North America, the truck is the only kind of vehicle for which the inside noise level is delimited by an established norm, SAE J336a [20]. The norm has not been legislated into force, and simply proposes a design criterion for the maximum noise level in a truck's cabin, along with the equipment and procedures to determine this level.

SAE J336a recommends that the following sound spectrum be employed as a reference in the design and development of new vehicles:

Octave Band Center Frequency, Hz	Sound Pressure Level, dB
63	101.5
125	96
250	90.5
500	85
1000	79.5
2000	74
4000	70
8000	70

Trucks satisfy the standard when the inside noise level is equal to or lower than 87.6 dB(A), and when the sound pressure level of any octave band is not more than 3 dB above the values indicated in the table. The procedure requires that the microphone be placed 6 inches to the right of the driver's right ear and level with it, and that it be directed vertically upwards. The windows and the ventilation openings must be closed and all noisy accessories turned off. The tests should be made on a hard surface (concrete or asphalt), but not a rough one. No large sound-reflecting surfaces should be less than 50 ft from the vehicle, and the wind must be below 15 mph. A gear ratio is chosen which will give a vehicle speed of about 50 mph at the maximum speed of the engine. An accelerating driving technique is required, during which the maximum sound pressure levels in each octave are taken.

SAE J336a seems well conceived and worthy of application to other vehicle types, though naturally with a sound level limit appropriate to each case. Some form of legal support for this type of norm should also be enacted. We recommend that the sound level for truck cabs could be lowered by 2–3 dB(A) in the relatively short term, to 85 dB(A).

5-6. TRAINS

5-6.1 Some Facts

The railroad is a mode of transportation which has not grown significantly in the last twenty years, at least in North America. Long distance passenger trains have suffered particular setbacks, while the numbers of subway and suburban trains remain almost static. Considering the traffic growth in big cities caused by the increase of motor vehicles, the last two categories of train may, however, grow in numbers in the future.

About 99 percent of the long distance trains are diesel–electric, while the short distance ones are electric.

It is difficult to be precise about the time spent by a passenger in urban transit, suburban, or intercity trains. A good average would be less than 2 hours daily, with crew exposure being of the order of 6–8 hours a day.

5-6.2   Noise Sources

With diesel–electric locomotives, the main noise sources are the exhaust, the engine, including the air intake and the cooling fan, the transmission, the inter- action between the wheel and the track, and finally, the generator. Some of these sources are more important than others, notably the engine noise and the accompanying low frequency vibration. Sound levels inside the locomotives are in the range 80–95 dB(A) or higher. The whistle produces an interior level as high as 100 dB(A), but for a very short time.

With electric locomotives as well as with the nonmotorized cars of both elec- tric and diesel–electric trains, the main interior noise sources are the interaction of the wheel and the track, the cooling fans, the electrical traction motors, if any, and the auxiliary equipment, especially the ventilation or air-conditioning system. Wheel–rail noise is the most important of these; it reaches the passengers by two different paths: the mechanical vibration from the wheels transmitted through the suspension, and the sound wave directly through the coach body.

The sound level inside nonmotorized cars of intercity trains is between 60 and 75 dB(A); in electrically driven cars the sound level is between 65 and 75 dB(A). When the windows are open, the sound level could be 10 dB(A) higher or more. Although modern cars are equipped with an air-conditioning system and the windows stay closed, these systems are themselves relatively noisy, to the point that they may dominate the sound spectrum.

The sound levels inside urban transit cars, whether underground or on the surface, are significantly higher than those in the above-mentioned categories, for a number of reasons. The many doors are made to be opened and closed easily and rapidly, and after a while become shaky and noisy; a great proportion of the doors and wall surfaces are composed of windows to add to the visibility, detracting from the overall sound insulation; frequent and fast starts and stops require light cars, which tend to transmit sound and vibration and to radiate the interior noise; and, finally, lack of space dictates small air-conditioning and ventilation ducts, which raise airflow speeds and sound levels. The sound levels in transit cars are therefore between 65 and 95 dB(A) in the open. To all these factors, we must add the amplification in tunnels: the interior tunnel surfaces are usually hard and acoustically reflecting, and increase the sound level almost

10 dB(A) compared to the same operating conditions outside the tunnel. The noise level inside the car when it is in a tunnel could be between 80 and 100 dB(A).

The noise spectrum inside a subway car has a particular shape, with a crest between 250 and 2000 Hz, which translates subjectively into a disagreeable noise. The spectrum shape is due mainly to the friction and vibration of the wheel on the track (or tire on pavement), and occasionally on the side of the track in a curve, resulting in the extremely annoying phenomenon of wheel squeal.

The noise inside cars that roll on tires (e.g., in Montreal, Mexico City, and Paris) tends to be 10-15 dB(A) lower than in those with steel wheels.

5-6.3 Criteria

If we consider that people are at a distance of 2 ft when seated side by side and about 3.5 ft when seated face to face, the acceptable background sound level for speech would be, respectively, 70 dB(A) and 65 dB(A). The intimacy (i.e., privacy) of the conversation is also an important factor. If we assume that normal speech should not be heard without difficulty at more than 5 ft, i.e., beyond earshot of a small group of persons traveling together, the noise level should theoretically be no lower than about 62 dB(A).

On intercity trains, intellectual work can be important, and it is logical to keep the noise level low enough to permit it. Use of Beranek's NC curves suggests a level of between 40 dB(A) and 50 dB(A), levels which are obviously very low and should be regarded as conservative criteria. Many people like to rest and/or sleep when they travel by train, particularly on long distances. The results of research by Dyer and also by Bonvallet [3] suggests the level of 50 dB(A) as a criterion for continuous noise.

Finally, it is necessary to consider the annoyance caused by noise inside the train. This is very important for intercity and transcontinental trains, where the passengers travel for many hours if not many days. According to Lippert and Miller [28], 77 dB(A) could be the annoyance criterion.

5-6.4 Limits

To our knowledge, there are no norms or laws concerning the acceptable sound levels inside locomotives, nor inside cars or coaches. For trains, the establishment of norms and, subsequently, of legislation, should follow a path that is different from the one concerning highway vehicles. In fact, both publicly and privately owned trains and their tracks should be considered as infrastructures which do not change much with time, rather than as vehicles that change every few years. Therefore norms and laws have to refer to future rail systems as a whole, taking into account the social and economic effects of the system as much as the noise potential of each car.

70 dB(A) inside transit or suburban trains, and 65 dB(A) in intercity trains, are recommended as appropriate limits for new rail systems.

5-7. AIRCRAFT

5-7.1 Some Facts

In this section, we will divide aircraft into three categories to take account of sound level and exposure time differences. The first category includes commercial jet aircraft (with two, three, or four engines) and commercial propeller aircraft; the second includes helicopters and short take-off and landing (STOL) aircraft; and the third category includes general aviation aircraft, whether jet or propeller.

In North America there are more than 3000 commercial aircraft in service, of which fewer than 15 percent are propeller driven and about 40 percent are comparatively small, short or medium range jets of two or three engines. These aircraft carry more than 175 million passengers each year. The average flight time is 1.4 hour and the average sound level is 82 dB(A), at least at cruising speed. This first aircraft category is expected to show continued growth in the years to come. In the second category, the number of helicopters in service in North America alone is over 4600. STOL airplanes are not much in service, although there is growing development effort. Helicopters are used to do so many different tasks that it is practically impossible to be precise about individual noise exposure times, either for pilots or passengers, but the daily exposure could be anything up to about two hours a day.

In the third category, general aviation, there are more than 130,000 planes in North America alone, of which 85 percent are single-engine propeller types, and about 1 percent are business jets, which now number over a thousand. The number of propeller planes will likely grow too, but at a slower pace. General aviation planes are of private or corporate ownership, or are taxiplanes or chartered; they are used for pleasure, flying instruction, or business. The average use of these planes is half an hour each day, which is not representative of the time of exposure each time the plane is used, a period that could be anything from half an hour to about four hours.

5-7.2 Noise Sources

The principal noise sources for commercial jets, as perceived inside them, are jet and turbine noise and boundary layer noise. Although the landing gear is a source of noise, as is the use of flaps and spoilers, noise from these sources does not last long enough to be a problem (although the very presence of these noises is said to cause occasional concern among first-time air travelers). At cruising speed, the sound level varies between 77 and 92 dB(A) for the two- or three-engine narrow-body jets; between 75 and 85 dB(A) for the four-engine narrow-body jets; and between 72 and 84 dB(A) for the wide-bodies.

These sound levels are exceeded on takeoff, when for two or three minutes the engines run at maximum power, and on landing, when the thrust is reversed for braking. On these occasions, the sound level in the cabin could be higher than the cruising speed level by as much as 20 dB(A).

The noise spectrum inside these aircraft is such that the noise is not specially unpleasant; it has a rather flat spectrum with no perceptible or unpleasant narrow bands and usually no beats.

The sound level inside jet aircraft rises from front to rear by as much as 10 dB(A). This is more marked in planes where the engines are under the wing. Propeller planes are a totally different case; their sound level is higher in the neighborhood of the propeller, and grows fainter as we move away from this point in either direction. The difference between the sound levels at this point and at the rear could be as much as 30 dB(A). Propeller passenger planes have higher inside noise levels than jet planes. Their sound levels are between 75 and 100 dB(A) and the low frequencies are important, producing beats and vibration.

In the second category, helicopters are more important to consider than STOL aircraft, because there is little information on STOL noise and because STOL planes are not yet much employed commercially.

The noise levels of different types of helicopter are reasonably alike. As heard inside, the main noise sources in order of decreasing importance (and in order of rising high-frequency content) are: the main rotor, the tail rotor and the piston engine (if fitted), the transmission, and, finally, the turbine engine (if a jet).

This results in spectra where the low frequencies are specially intrusive. Additionally, the low-frequency modulation due to the main rotor makes helicopter noise especially unwelcome, particularly inside them. This problem is amplified by the difficulty of combining good sound insulation at low frequencies with a light fuselage.

Sound levels inside helicopters range from about 85 to about 100 dB(A), with a tendency for the smaller ones to be the noisier. A sound level average at cruising speed is around 90 dB(A).

The third aircraft category is general aviation. To summarize the noise sources, we will divide this category into two parts: propeller aircraft and jets. The noise sources in the case of propeller aircraft are the propeller noise, which is between 50 and 700 Hz, with peaks which are at the fundamental and harmonic frequencies of the propeller; and the noise of the engine and the exhaust, which both appear principally at frequencies over 200 Hz. To return to the propeller, the first two or three harmonics as well as the fundamental (usually betwen 60 and 100 Hz) are the most important. Vibration is also sometimes significant. The overall sound level inside the aircraft is between 85 and 105 dB(A). Here again, because of the need for a light structure, insulation from the low frequencies is a major problem, although a less serious one than with helicopters.

The characteristics and the noise sources of business jet planes are practically the same as those of commercial jet aircraft. There is a tendency for the sound level to be lower because the engines are less powerful, but the jet exhaust noise is a problem of the small engines these craft are fitted with: the characteristic frequencies are high, which moves the frequency band where jet noise is at maximum towards an area in the sound spectrum where the ear is fairly sensitive. This spectrum shape has at least the advantage that it can be attenuated with light materials. Thus the sound levels inside business jet aircraft are between 80 and 95 dB(A), compared with between 85 and 105 dB(A) for propeller planes of the same category, even though the sound levels outside business jets are more than 10 dB(A) higher than outside propeller types.

5-7.3 Criteria

For speech communication, using the same reasoning as with buses and trains, the sound level for passengers should be no higher than 70 dB(A), nor, for the sake of privacy, any lower than 65 dB(A). Speech intelligibility has a particular meaning for the crew, who, besides talking to one another, must communicate by radio. The understanding of radio messages has a parallel in the understanding of telephone speech, for which one could use Beranek's curves [3] —yet making an allowance for the fact that aircrew become "phonetically efficient" communicators and that the radio volume can be high, the sound level can probably be as high as about 75 dB(A). For direct communication between crew members, a voice slightly higher than normal could be acceptable, particularly if we consider that the conversation takes place in a work setting. The distance between the two pilots or between either of them and the navigator or flight engineer is about 4 ft on commercial jets; if we presume that the speaker may not be able to turn toward the person he is talking to, the noise level should not exceed about 75 dB(A). In smaller aircraft and in helicopters, where the distance between crew members is less, say about 2.5 ft, the sound level should not exceed 80 dB(A).

For the criteria regarding intellectual activity and rest, we can tentatively use similar figures to those established for bus and train. The generally similar range of trip times for these three kinds of transportation, and the common physical and psychological behavior patterns involved, suggest that such an approach is adequate. Using the NR curves of Kosten and Van Os and the NC curves of Beranek, for both activities the acceptable maximum sound level would be between 40 and 50 dB(A).

With regard to annoyance, Lippert and Miller [28] have researched the matter inside transport aircraft. According to them, if the sound level is lower than 87 dB(A), the surroundings could be described as acoustically comfortable, while ideally the noise level should be lower than 77 dB(A). We must, of course, note that these figures were derived in the days of propeller transports. Since the noise spectra inside propeller planes differ appreciably from those inside today's jet aircraft, and since the vibration levels were higher in the earlier craft, we must accept their research with caution.

5-7.4 Limits

There are no norms or laws on the sound levels inside aircraft. It is obvious that the problem is complex not just technically, but economically and even politically. No government and no public organization could at short notice impose a very restrictive requirement.

Considering all the above, we believe it is realistic to propose a sound level at cruise speed of 75 dB(A) as a long-term objective for the sound level inside planes in general aviation and inside helicopters. This might also be the target for new STOL aircraft. In the cockpits of helicopters and private planes, 80 dB(A) should be considered as an adequate norm, and this level could apply in the cabin when it is not separated from the cockpit. For commercial jets, the long-term objective would be 70 dB(A) in the passenger cabin. As far as the flight deck is concerned, it seems to us that a different and less restrictive norm would be acceptable; privacy not being a problem, the crew can afford to speak with somewhat raised voices and a less restrictive norm is also rendered necessary by the difficulty of sound insulating the front part of the plane, where boundary layer noise is a special problem. We believe that for the flight deck of commercial jets, a norm of 75 dB(A) would be justified.

5-8. SPACECRAFT

The length of time that astronauts are exposed to high noise levels is short, and must be seen in the context of the probability that an astronaut will not, on current experience, fly more than two or three times in his life. Inside the ear canal of Apollo astronauts wearing helmets, the sound level never exceeds 115 dB(A), while 110 dB(A) is not exceeded more than one minute at lift-off nor more than two minutes on re-entry, the two noisy phases of space flight. Propulsion noise and boundary layer noise are the two main sources.

Verbal communication is obviously important, although it is well-rehearsed during lift-off and re-entry. In fact, these two flight regimes are so vital that control sequences are preprogrammed and automatic, the astronauts being witnesses rather than actors. This somewhat downgrades the importance of verbal communication, though it is still just possible at the levels indicated and must, of course, take place in an emergency.

There are no norms or legislation on this subject, as far as we know.

5-9. SHIPS

5-9.1 Some Facts

The ship is a means of passenger transportation which is progressively losing its importance because of its low speed and high operating cost.

The merchant fleet of the United States, considering only ships of more than 1000 tons, is slightly over 2000 units, while for Great Britain, it is about 3700. About 80% are cargo boats, while 20% are cargo or mixed cargo and passenger ships. Although most parts of a ship do not have especially high sound levels, at least in terms of hearing hazard, other parts do, and exposure time becomes a factor of importance in assessing risk, even though it is difficult to establish at all precisely. For passengers, the average period of exposure would be between a few hours and a few weeks, occurring perhaps no more than once a decade. For the crew, exposure time may be 8 hours a day or more; it must be remembered that more than 5 days per week may be worked, and that off-duty hours may also be spent in a noisier environment than is usual ashore.

5-9.2 Noise Sources

The main sources of noise on a ship are the engine, the gearing and the propeller, and the ensuing significant structural vibration. The propelling engine and the propeller run at low speed, and much of the sound energy made by these two pieces of machinery is at low frequencies, which can all too easily be transmitted to far-off places on the ship. The noise of the crankshaft, the gearing, and the turbines is predominantly in the high frequencies, while the ventilation systems generate a continuous broad-band noise which is too high only if badly designed. To these noise sources are added minor and occasional ones like plumbing noise, "people noise," and the creaking of the partitions due indirectly to the impact of the waves on the hull.

5-9.3 Criteria

The evaluation of noise on ships is best undertaken for two different areas, the working quarters (including the engine room) and the living quarters (cabins and public areas including decks). It is obvious that the criterion for the cabins will be the most restrictive.

The sound levels referred to below come from measurements made on more than a hundred cargo and passenger ships [31]. In the engine room, near the engines, the sound level varies between 80 and 102 dB(A), with an average of around 89 dB(A), while at the controls the noise level is between 80 and 88 dB(A), with an average of 84 dB(A).

In an unavoidably noisy working place, speech communication should still be possible in a loud voice at a distance of less than a foot, which is achievable with a background noise of 85 dB(A) or lower. This sound level corresponds to the average level in the vicinity of the controls. (When the background noise is close to 90 dB(A), a loud-voiced conversation is not possible at a distance of more than 6 inches, which is unacceptable except perhaps in situations where one seldom has to communicate.)

For the passengers, evaluating annoyance using the NC and NR curves appropriate for the living quarters of a house, a level of 40–50 dB(A) should apply to the public areas, and 35–45 dB(A) to the cabins.

5-9.4 Limits

As far as we know there are no laws or standards, though the purchaser may set sound level limits in his specification.

All things considered, we believe it is logical to recommend a limit of 50 dB(A) for all living quarters other than the cabins, but including indoor traffic areas. For the cabins, a limit of 45 dB(A) during the night would be desirable though it could rise to 50 dB(A) during the day. These limits could only be economically applied to new ships, and there would obviously be a considerable time lag before they achieved results.

5-10. HOVERCRAFT

5-10.1 Some Facts

Hovercraft exist mainly in Europe and especially in Britain. Although they have been operating for nearly two decades, the number of hovercraft in service is limited and it is difficult to generalize about their noise output with any confidence.

The time spent by the crew in hovercraft is invariably less than four hours per day and is in most cases less than two hours; it is far lower than this for passengers.

5-10.2 Noise Sources

The main sources of noise are the propeller and the engine, along with aerodynamic noise related to the air cushion. The average sound level inside the cabin is between 80 and 95 dB(A), depending on model, position in the cabin, and vehicle speed.

5-10.3 Criteria

Speech communication should be evaluated as for the other means of public transportation where people are seated side by side, resulting again in a criterion of between 65 and 75 dB(A). For the crew, radio communication without a headset is similar to the requirements on aircraft, but since single-crew operation is the norm there is no other requirement for speech communication.

For the short passenger exposure times which are involved (seldom more than one hour), we can perhaps neglect consideration of intellectual activity or rest. A criterion for annoyance is relevant, however, and that due to Lippert and Miller [28] is perhaps particularly applicable to hovercraft, since hovercraft have sound levels, frequency spectra, and low frequency vibration that is broadly comparable with the propeller aircraft that were the subject of Lippert and Miller's research, though exposure times are shorter. According to their study, the sound level is comfortable when it is lower than 87 dB(A), which we can take as acceptable bearing in mind the short journey times.

5-10.4   Limits

To our knowledge, there are no norms or laws on the subject, though we would recommend one be set at about 80 dB(A) for all seats inside the hovercraft, whether for passengers or crew. This should apply to new designs.

5-11.   RECREATIONAL VEHICLES

5-11.1   Some Facts

We will discuss three kinds of vehicle: motorcycles, snowmobiles, and motorboats. There are more than three million motorcycles in North America (and about one million in Great Britain), over two million snowmobiles and about six million powered boats, of which 90% have outboard motors. (We will call all powered boats "motorboats" here.)

It is extremely difficult to know the duty cycles of these vehicles, but we will assume average daily exposure times for the motorcycle of about one hour, and for the snowmobile and the motorboat about two hours.

5-11.2   Noise Sources

Motorcycle sound level is a function of speed. The exhaust is the main source, and its spectrum is preponderantly a high frequency one in the case of a two-stroke engine, and a low frequency one in the case of a four-stroke. The second source of noise is the air intake, and the third is the engine. The sound level at the driver's position can vary between 80 dB(A) and 115 dB(A) according to the model and the operating conditions.

Snowmobile noise has been subject to increasingly strict legislation in a great number of American states and in Canada, which tends to explain why snowmobile noise is a function of age. The main source of noise when measured at 50 ft according to SAE J192 or J192a procedures [39] is the exhaust. This source is under the vehicle, however, so that its relative importance is diminished at the driver's position, which is not directly exposed to it. The other important sources are the carburetor air inlet and the engine, which is usualy an air-cooled two-stroke. The spectral shape of snowmobile noise is annoying and irritating. The sound level at the driver's ear position varies between 90 and 115 dB(A).

For motorboats, it is surprisingly not necessary to make a distinction between those with outboard and those with inboard engines. The main noise sources are the same in both cases: the exhaust, the engine, and the carburetor air intake, and the sound level at the driver's position is between 78 and 115 dB(A), although for most of the latest models and in most driving conditions the sound level is below 95 dB(A).

5-11.3 Criteria

For motorcycles and snowmobiles, the speech communication criterion has little importance, but it would be valuable if their drivers could be able to hear a vehicle horn at at least 50 ft. If the horn complies with the SAE J377 norm, its sound level at 50 ft will be 82–102 dB(A). The audibility of this signal depends on its spectral shape relative to that of the vehicle's, but probably calls for the vehicle sound level to be no higher than the horn's. (The driver's helmet should attenuate both sounds about equally, and may be disregarded as a factor in this analysis).

These considerations do not hold for the motorboat, where speech communication rather than warning signal audibility is the criterion of particular importance. In this case, it seems a reasonable objective that a normal conversation should be possible at a distance of 1 ft and loud-voiced conversation at 3 ft. These two objectives can be met with a sound level of 75 dB(A). For motorboats with cabins and living quarters, a criterion could be taken from Beranek's recommendation of a sound level of 40 dB(A) as appropriate in homes during the day, or Kosten and Van Os's recommendation of 45–60 dB(A) (taking into account the relevant correction factors). With respect to annoyance, Lippert and Miller [28] remain the only source of knowledge we know of, and we are inclined to select from their work a sound level of 77 dB(A).

5-11.4 Limits

We know of no norms or laws on the subject, though they are being worked on.

We propose that the sound level of 90 dB(A) be considered as a serious objective at the driver's ear position on snowmobiles and motorcycles. A sound level of 85 dB(A) at the noisiest place outside the cabin should be considered an objective for motorboats of both types, with 75 dB(A) inside in both cases.

5-12.   VIBRATION

5-12.1   Introduction

Noise and mechanical vibration, though similar, are treated separately in this chapter because human response to vibration is so complex a problem that there are no accepted criteria for individual modes of transport, and only one series of partially accepted norms for vibration in general.

The important parameters behind a vibration stimulus are the position of the person; the point at which the vibration is applied and its direction; and the amplitude, duration, and frequency of motion at the forcing point. However, for this chapter people can be regarded as being seated or standing and subjected mainly to vertical displacements. We can then contemplate a criterion or norm which sets amplitude limits as a function only of frequency and exposure time.

The frequencies of interest can be separated into those below 1 Hz, those from 1 to 100 Hz, and those above 100 Hz.

5-12.2   Criteria

In the lowest of these frequency ranges, the predominant physical effect is motion sickness. According to Bender and Collins [40], the problematical frequencies lie between 0.2 and 0.4 Hz, and a criterion of $0.1g$ applies.

For the frequency range 1–100 Hz human vibration response is not well understood, although many researchers have studied the problem. Koffman [45] concluded that for the frequency range 1–20 Hz, subjective judgements changed from "very acceptable" to only "acceptable" as amplitude rose from 8 to 15 cm/sec^2, which is equivalent to the range 0.008–0.015 g. However, according to Dyer [41], this level of acceleration is around the threshold of perceptibility.

Goldman and Von Gierke [42] concluded that in a vertical position, either seated or standing, vibration is most easily transmitted through the human body and is therefore least tolerable when it is below 10 Hz, or more precisely between 4 and 8 Hz. In the frequency range 1–100 Hz, Goldman [43] presents the graph which is reproduced here as Fig. 5-3. It delineates vibration levels in g which people find "perceptible," "unpleasant," and "intolerable."

Vibration effects can be expressed not only in terms of discomfort, but also in terms of their effect on visual acuity, and therefore on tasks like reading and writing. Researchers generally conclude that vibration which is judged just perceptible is no obstacle to reading and writing, but unpleasant vibration may well be. This is, of course, only an approximate and tentative statement on matters being currently researched. A particular concern is the visual acuity of helicopter pilots, who sometimes do not see high-tension wires.

At frequencies above 100 Hz, the main effect is a tickling sensation in hands or feet, whichever are exposed. This can cause physiological damage when ampli-

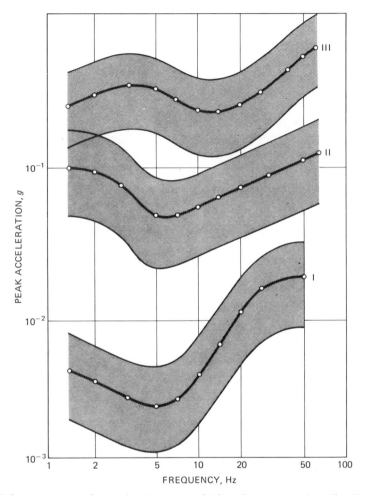

Fig. 5-3 Average peak accelerations at which subjects perceive vibration (I), find it unpleasant (II), or refuse to tolerate it further (III) [43]. Shaded areas are about one standard deviation on either side of the mean.

tudes and exposure times are high enough (e.g., in occupational chain saw users), but is not a problem in transportation. A limit is said to be in the vicinity of $1g$ at 100 Hz, and to rise thereafter at 6 dB per octave.

5-12.3 Norm

The most accepted norm is ISO 2631 "Guide for the evaluation of human exposure to whole-body vibration" [44], which covers only the frequency range from 1 to about 100 Hz, for seated or standing positions only. The document

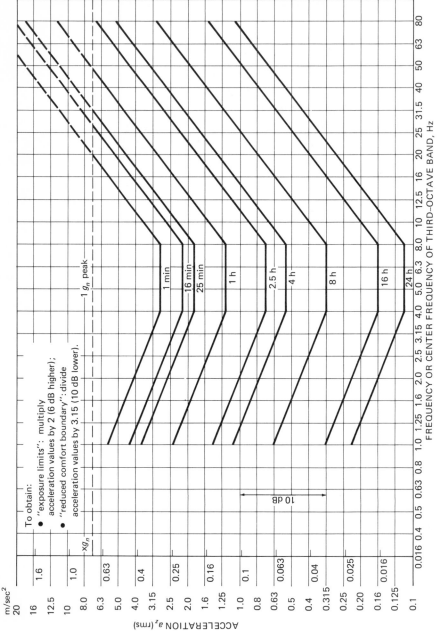

Fig. 5-4 Longitudinal (a_z) acceleration limits as a function of frequency and exposure time; "fatigue-decreased proficiency boundary" [44].

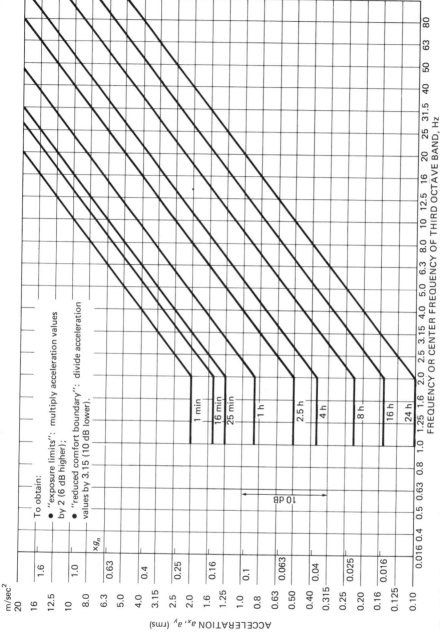

Fig. 5-5 Transverse (a_x, a_y) acceleration limits as a function of frequency and exposure time; "fatigue-decreased proficiency boundary" [44].

refers to three different criteria: the fatigue-decreased proficiency boundary, the reduced comfort boundary, and the exposure limit. The first of these applies to vehicle operators (e.g., pilots, drivers, and conductors.) The second applies to passengers, and is fairly arbitrarily set at one-third the amplitude of the first, i.e., 10 dB lower. The exposure limit is defined as occurring at twice the amplitude of fatigue-decreased proficiency, and is set at about half the limit considered to be the threshold of pain (or limit of voluntary tolerance) for healthy humans while strapped in their seats.

In the ISO curves reproduced as Fig. 5-4 and 5-5, a_z represents vertical vibration, a_x is that in a fore-and-aft direction, and a_y is that from side to side. The accelerations are rms values, except that the horizontal dashed line in Fig. 5-4 marks a peak level of $1g$.

BIBLIOGRAPHY

General

1. "Information on Levels of Environmental Noise Requisite to Protect Public Health and Welfare with an Adequate Margin of Safety." U.S. Environmental Protection Agency Report No. EPA 550/9-74-004, March, 1974.

2. Leo L. Beranek, *Noise and Vibration Control.* McGraw-Hill, New York, 1971.

3. Leo L. Beranek, *Noise Reduction.* McGraw-Hill, New York, 1960.

4. "Assessment of Noise with Respect to Community Response," International Organization for Standardization Recommendation ISO/R 1996, 1971.

5. "Assessment of Occupational Noise Exposure for Hearing Conservation Purposes." International Organization for Standardization Recommendation ISO/R 1999, 1971.

6. Karl D. Kryter, *The Effects of Noise on Man.* Academic Press, New York, 1970.

7. Richard H. Lyon, *Lectures in Transportation Noise.* Grozier Publishing, Cambridge, Mass., 1972.

8. James D. Miller, *Effects of Noise on People.* U.S. Environmental Protection Agency, Washington, D.C., 1971.

9. "Occupational Exposure to Noise." U.S. Department of Health, Education and Welfare, Washington, D.C., 1972.

10. "Passenger Noise Environments of Enclosed Transportation Systems." U.S. Environmental Protection Agency Report No. EPA 550/9-75/025, June, 1975.

11. "Performance of Vehicle Traffic Horns." Society of Automotive Engineers, SAE J377.

12. Wyle Laboratories, "Transportation Noise and Noise from Equipment Powered by Internal Combustion Engines." U.S. Environmental Protection Agency, Washington, D.C., 1971.

13. "Sound Signalling Devices on Motor Vehicles." International Oganization for Standardization Recommendation ISO/R 512, October, 1966.

14. D. T. Aspinall and P. Hunter, "Noise in Commercial Vehicle Cabs; Objective Data." Motor Industry Research Association, Lindley, Warwickshire, U.K., 1969.

Automobiles (5-3)

15. Clifford R. Bragdon, *Noise Pollution*. University of Pennsylvania Press, Philadelphia, 1970.

16. R. D. Ford, G. M. Hugues, and D. J. Saunders, "The Measurement of Noise Inside Cars." *Applied Acoustics,* **3,** 69–84, 1970.

17. R. D. H. Perry and J. A. Raff, "A Review of Vehicle Noise Studies carried out at the Institute of Sound and Vibration Research with a reference to Some Recent Research on Petrol Engine Noise." *J. Sound Vib.,* **28,** 433–470, 1973.

Buses (5-4)

18. G. F. Swetnam and W. S. Murray, "Feasibility Study of Noise Control Modifications for Urban Transit Bus." The Mitre Corporation, Washington, D.C., March, 1973.

Trucks (5-5)

19. James M. Heinen, William A. Leasure, Denzil E. Mathews, and Thomas L. Quindry, "Interior/Exterior Noise Level of Over-the-Road Trucks." U.S. Department of Commerce, Washington, D.C., 1970.

20. "Sound Level for Truck Cab Interior." Society of Automotive Engineers, SAE J336a.

Trains (5-6)

21. Brian H. Aitken and Cyril M. Harris, "Noise in Subway Cars," *Sound and Vibration*, 12–14, February, 1971.

22. T. D. Northwood, "Rail Vehicle Noise." *Engineering Journal*, 30–33, January, 1963.

23. Marshall L. Silver, "Noise and Vibration Control in New Rapid Transit." *Transportation Engineering Journal*, 891–908, November, 1972.

24. G. F. Swetnam, "Rationale for Exterior and Interior Noise Criteria for Dual-Mode and Personal Rapid Transit System." The Mitre Corporation, Washington, D.C., March, 1973.

Aircraft (5-7)

25. William F. Ashe, Lester B. Roberts, and Robert L. Wick, "Light Aircraft Noise Problems." *Aerospace Medicine,* 1133–1137, December, 1963.

26. Leo L. Beranek and Laymon N. Miller, "Noise Levels in the Caravelle During Flight." *Noise Control*, 19–21, September, 1958.

27. W. F. Grimster and C. E. P. Jackson, "Human Aspects of Vibration and Noise in Helicopters." *J. Sound Vib.*, 20(3), 343–351, 1972.

28. S. Lippert and M. M. Miller, "An Acoustical Comfort Index for Aircraft Noise," *J. Acoust. Soc. Amer.*, 23(4), 478, 1951.

29. Jerry V. Tobias, "Cockpit Noise Intensity: Fifteen Single-Engine Light Aircraft." Federal Aviation Administration, Washington, D.C., 1968.

30. Jerry V. Tobias, "Cockpit Noise Intensity: Eleven Twin-Engine Light Aircraft." Federal Aviation Administration, Washington, D.C., 1968.

Spacecraft (5-8)

31. Charles W. Nixon and H. E. Von Gierke, "Noise Effects and Speech Communication in Aerospace Environments." In *Aerospace Medicine*, 2nd ed., Williams and Wilkins, Baltimore.

32. H. E. Von Gierke, "Vibration and Noise Problems Expected in Manned Space Craft." *Noise Control*, May, 1959.

Ships (5-9)

33. Adrian Hagen and Norman O. Hammer, "Shipboard Noise and Vibration from a Habitability Viewpoint." *Marine Technology*, January, 1969.

Hovercraft (5-10)

34. D. Anderton, "Internal Noise Reduction in Hovercraft." *J. Sound Vib.*, 22(3), 343–359, 1972.

35. E. J. Lovesey, "Hovercraft Noise and Vibration." *J. Sound Vib.*, 20(2), 241–245, 1972.

Recreational Vehicles (5-11)

36. Richard A. Campbell, "A Survey of Noise Levels on Board Pleasure Boats." *Sound and Vibration*, 6(2), 28–29, 1972.

37. Jack Curtis and Richard C. Sauer, "An Analysis of Recreational Snowmobile Noise." *Sound and Vibration*, 7(5), 49–50, 1973.

38. National Research Council of Canada, "Snowmobile Noise, Its Sources, Hazards and Control." Report APS-477, Ottawa, 1970.

39. "Exterior Sound Level for Snowmobiles." Society of Automotive Engineers, SAE J192 and J192a.

Vibration (5-12)

40. E. K. Bender and S. M. Collins, "Effects of Vibration on Human Performance: A Literature Review." Bolt Beranek and Newman, Report 1767, February, 1969.

41. I. Dyer, "Passenger Psychological Dynamics—Sources of Information on Urban Transportation." American Society of Civil Engineers, June, 1968.

42. D. E. Goldman and H. E. Von Gierke, "Effects of Shock and Vibration on Man." In C. M. Harris and C. E. Crede, eds., *Shock and Vibration Handbook*, vol. 3, McGraw-Hill, New York, 1961, Chapter 44.

43. D. E. Goldman, "Effects of Vibration on Man." In C. M. Harris, ed., *Handbook of Noise Control*, McGraw-Hill, New York, 1957, Chapter 11.

44. "Guide for the Evaluation of Human Exposure to Whole-body Vibration." International Organization for Standardization Report No. ISO 2631, 1974.

45. J. L. Koffman, "Vibration and Noise." *Automobile Engineer*, 73–77, February, 1957.

6

Interior noise environments

L. W. HEGVOLD* and D. N. MAY†

6-1. INTRODUCTION

This chapter discusses the descriptors used to assess the acoustical suitability of various interior environments, and gives some guidance on the values those descriptors should have. Considered here are interior environments generally; Chapters 5, 7, and 10 give futher consideration to particular situations, namely vehicles, hospitals, and the home.

The descriptors used to assess the suitability of indoor environments are: ranking curves (NC, NCA, PNC, and NR) though these are sometimes abandoned in favor of dB(A); reverberation time, which has particular significance for auditoria and studios; and measures of speech interference (e.g., Articulation Index), which help assess the suitability of open-plan areas.

*Department of Architecture, Western Australia Institute of Technology, South Bentley, Western Australia 6102, Australia.
†Research and Development Division, Ministry of Transportation and Communications, Downsview, Ontario M3M 1J8, Canada.

6-2. RANKING CURVES AND dB(A)

The last twenty years have seen the acceptance of ranking curves as a means of evaluating indoor noise spectra. Although now partly abandoned in favor of simply measuring the sound level in dB(A), the ranking curves are in sufficiently widespread use to warrant some description.

The curves variously proposed all consist of a set of contours on a graph of octave band sound pressure level vs. frequency. When the spectrum of a sound is plotted over them, the maximum value contour reached by the spectrum is taken to describe the acoustical suitability of that environment.

To illustrate the use of these curves, one may take the Noise Criterion (NC) and the Noise Rating (NR) curves; the former (by Beranek [1]) are the most widely used in North America, while the latter (by Kosten and Van Os [2]) have received international endorsement. The NCA and PNC curves differ from the NC curves in that the former allow more low frequency noise (which may have to be accepted as an economic dictate in some circumstances) and the latter allow both less low and less high frequency noise in order to eliminate undesirable "rumble" and "hiss".

The NC curves are shown in Fig. 6-1, and the NR curves in Fig. 6-2. An example demonstrates their use.

Example

Problem: The sound in a room has octave band sound pressure levels in the octave bands centered at 31.5, 63, 125, 250, 500, 1000, 2000, 4000, and 8000 Hz of 50, 55, 60, 60, 55, 50, 45, 40 and 40 dB, respectively. What are its NC and its NR ratings?

Solution: By plotting the spectrum on the NC and NR contours of Figs. 6-1 and 6-2, we find that the maximum incursion into both sets of contours (which occurs at 250 Hz) just reaches the NC = 52 and NR = 52 curves. The answer therefore is that the NC and NR ratings are both 52. (The two sets of curves generally give roughly similar values except for predominantly low frequency sounds.)

Experience has shown that, for most interior environments, the ranking curve values bear a constant relationship to the sound level in dB(A). For example, on many occasions that an NC value is obtained it is found that the sound level in dB(A) $\doteq$ NC + 10. The question therefore arises as to whether the complexity of the ranking curves and the need for an octave analysis of the sound are justified, when a simple sound level measurement may suffice. The answer is that, except where unusual sound spectra are encountered, interior sound environments may be acceptably described in terms of dB(A). In Table 6-1, therefore, the maximum amount of sound accepted as reasonable for various indoor environments is expressed in dB(A). The table is a summary [3] of recommendations from various sources; the determination of these levels is as much a matter of common experience as of authoritative research, and the levels involved are approximate.

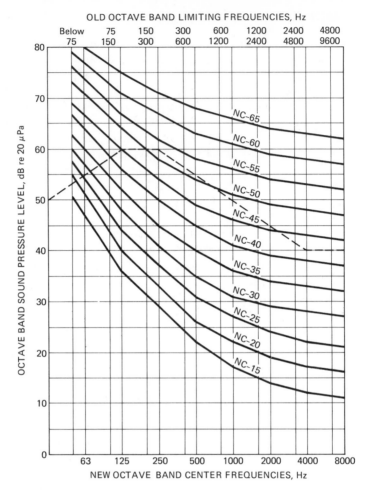

OLD OCTAVE BAND LIMITING FREQUENCIES, Hz

Fig. 6-1   Noise criteria curves [1]. The dashed curve is the spectrum used in the Example in Section 6-2.

This means that when fluctuating levels are encountered, the sound levels in the table may be associated with any statistical descriptor that describes the higher, more annoying levels in the background sound reasonably well. The levels in the table can thus be regarded reasonably accurately as L_{10}, L_{eq}, or L_{50} levels.

Except in instances when interior sound levels must meet a given authority's specification, a perfectly practical approach to setting a limit is to choose a decibel range from Table 6-1 that roughly represents the consensus of opinion of the contributors to the table, and regard that as the desired L_{eq} level. (Where the background noise is steady, its level *is* the L_{eq} level; when it is not, the L_{eq} level must be derived from the sound level history; see Chapter 1.)

In instances where a particular authority has specified the required sound level,

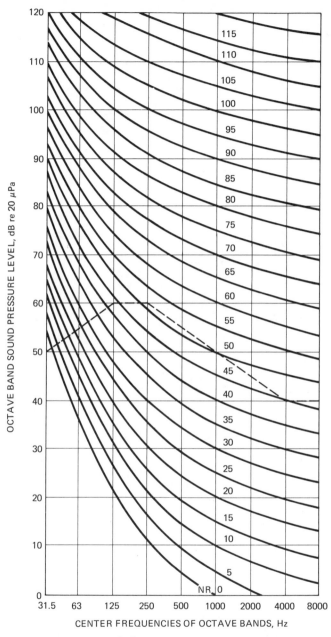

Fig. 6-2 Noise Rating curves [2]. The dashed curve is the spectrum used in the Example in Section 6-2.

TABLE 6-1. Recommended Acceptable Indoor Noise Levels [3].

	Knudsen-Harris 1950	Beranek 1953	Beranek 1957	Lawrence 1962	Kosten-Van Os 1962	Ashrae 1967	Denisov 1970	Kryter 1970	USSR 1971	Beranek 1971	Doelle 1972	Wood 1972	Rettinger 1973	Sweden	Switzerland 1970	Czechoslovakia 1967
	dB(A)	dB(A)	dB(A)	dB(A)	dB(A)	dB(A)	dB(A)	dB(A)	dB(A)	dB(A)	dB(A)	dB(A)	dB(A)	dB(A)	dB(A)	dB(A)
RESIDENT																
Home																
Bedroom	35-45	35	35-45	25	30	25-35		40	35	34-47	35-45	35	34-42	25	35-45	40
Living Room	35-45	35		40	35	30-40		40	35	38-47		40		25	35-45	40
Apartment	35-45		35-40	30		35-45		18		34-47	35-54		38-42		35-50	40
Hotel	35-45		35-40	35-40		35-45		38	35	34-47	30-40		42		35-50	40
COMMERCIAL																
Restaurant	50-55	55	55	40-60	50	40-55		55	55	42-52	45-60	45-50	50		40-50	55
Private Office	40-45	50	30-45	35-45	30-45	25-45		35		38-47	30-45	30-45	46	40		
General Office	45-55		40-55	40-60	60	35-65	40-45	35-40	50	42-52	45-55	45-55	50			
Transport.						35-55	50-60		60							
INDUSTRIAL																
Workshop					70		85						70			
Light		50		40-60						52-61		55-65			45-55	
Heavy		75		60-90						66-80		60-75			50-60	
EDUCATION																
Classroom	35-40	35	35	30-40	30	35-45		35	40	38-47	35-45	35-45	38	35	35-45	
Laboratory				40-50		40-50	40-50			47-46	40-45	45-50	42			
Library	40-45	40	42-45	35-45	35	35-45		40		38-47	40-45	40-45	42			40
HEALTH																
Hospital	35-40	40	42	20-35	35	30-45		40	25	34-47	40-45		38	25-35	25-35	35-40

RECREATION									
Swimpool	60			45–60	60		60	50–60	50
Sports (ampl.)		30		35–45			55–60	45–55	46
Gymnasium				40–50	35				46
AUDITORIUM									
Assembly Hall	35–40	35–40	40–45	30–40	38	30–42	35–45	35–45	38–42
Church	35–40	40	35–40	25–35	40	30–42	35–40	35–40	34
Concert Hall	30–35	25–35	25–35	25–35	28–35	21–30	25–35	30–35	
Court Room	40–45	40–45	40–45	35	40	42	35–40	35–40	30
Record Studio	25–30	25–30	20–30	20	28	21–34	25–30	30	30
TV Studio	25–30	30	25–35	30	28	21–34	30–35	35	34–38
Mot. Pict. Studio	25–30	25–30		25–35	28	21–34	35	25	
Mot. Pict. Theater	35–40	40		35–45	40		40	35–40	38
Lec. Theater	30–35	30–35		30–40	33	30–34	30–35		34

reference must of course be made to that specification. Such authorities could be the building owner, the mortgagor or the lessee. In the U.S. the Department of Housing and Urban Development (HUD) sponsored research leading to the curves in Fig. 1-8 which are intended to provide a basis for gauging the reasonableness of outdoor-measured non-aircraft noise in residential areas. HUD did not officially adopt these curves, however, and the Department's Circular 1390.2 sets out somewhat simpler criteria, as follows:

Unacceptable: Exceeds 80 dB(A) for 60 min. per 24 hours or 75 dB(A) for 8 hours per 24 hours

Normally unacceptable: Exceeds 65 dB(A) for 8 hours per 24 hours

Normally acceptable: Does not exceed 65 dB(A) for 8 hours per 24 hours

Acceptable: Does not exceed 45 dB(A) for more than 30 minutes per 24 hours

The above outdoor-measured criteria were supplemented by the following indoor-measured ones for sleeping quarters:

Acceptable: Does not exceed 55 dB(A) for more than an accumulation of 60 min. per 24 hours, nor 45 dB(A) for more than 30 min. between 11 pm and 7 am, nor 45 dB(A) for more than 8 hours per 24 hours.

It is useful to note that the approximation $dB(A) \doteq NC + 10$ applies to broadband sounds, while $dB(A) \doteq NC + 5$ applies to speech-like sounds. Furthermore, $NC \doteq SIL \doteq PSIL - 3$. SIL and PSIL are measures of a sound's propensity for speech interference, and are described in Chapter 1.

6-3. REVERBERATION TIME

When a sound source is in a room, an observer experiences both direct and reflected sound waves. (When the noise is outside the room, the sound is transmitted through, for example, a wall or a window, which an inside observer may then regard as the "source.") The direct sound wave is prominent close to the source, but gives way to the various reflected waves as the observer moves away. At great enough distances the sound level does not decrease further, because the direct wave contributes little to the sound field, which is now dominated by a series of randomly directed reflections from the room's boundaries. Such positions are described as being in the room's reverberant field. In general, all but a small area of the room near the source is in this field.

The reverberation qualities of a room affect its acoustical suitability. A highly reverberant room, where there are many reflections reaching the observer in addition to the direct wave, is described as "live"; in extreme cases there are perceptible echoes and diminished speech intelligibility. In a room where there is little reverberation, there are not so many reflections because the sound is ab-

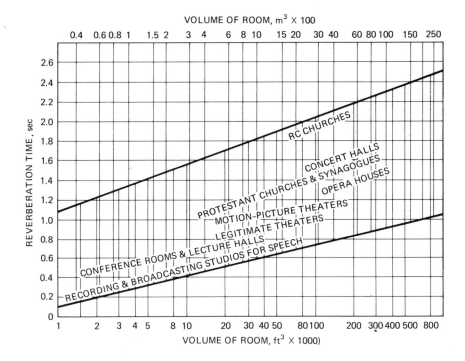

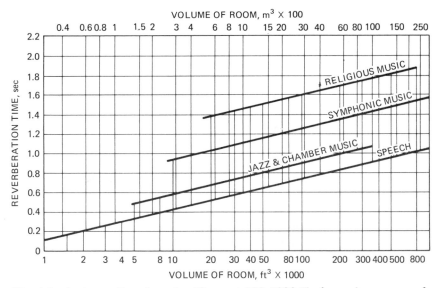

Fig. 6-3 Optimum Reverberation Times at **500–1000 Hz** for various room volumes and usages [4]. The lower of these sets of curves refers to studios.

sorbed at the room's boundaries and to a lesser extent in the air; the room is generally quieter than a live room, and is sometimes described as "dead".

Reverberation time is generally quantified by measuring the time taken for a sound, instantly switched off, to drop in level by 60 dB. This can, of course, be determined for various frequencies. Reverberation time thus defined is usually denoted T_{60} and is measured in seconds.

Acceptable reverberation is found from experience to depend on the use of the room and its volume. The uses for which it is most important to have a proper degree of reverberation are where the sound quality is most valued (e.g., in theaters and in broadcast studios). Figure 6-3 indicates some acceptable reverberation times as a function of room volume [4]. The frequencies 500 and 1000 Hz are typical reference frequencies for reverberation; the only really important characteristics of variation in reverberation with frequency are that the variation should not show particular spikes from room resonances, and that the reverberation time at a frequency below 500 Hz should not exceed the value at 500 Hz by more than about fifty percent.

6-4. OPEN-PLAN OFFICES

Open-plan offices are an indoor environment requiring particular attention to the assessment of acoustical suitability.

The ideal office environment is quiet enough to permit any occupant to talk easily with a visitor or on the telephone, yet is not so amenable to speech communication that he distracts other occupants of the same office. Thus the acoustics of the open office are tied up in two ways with the properties of the speech communication process.

6-4.1 Speech Communication and Privacy

The intelligibility of speech is a function of the signal-to-noise ratio, i.e., the extent to which the speech sound signal exceeds the background noise, taking due account of the frequencies important to speech. The delicate task in an open office is to adjust the signal-to-noise ratio so as to permit adequate local communication together with adequate privacy in relation to points nearby. The concept of signal-to-noise ratio in speech intelligibility is expressed by the Articulation Index (AI). The AI ranges from zero for no communication/perfect privacy, to unity for perfect communication/no privacy. Table 6-2 and Fig. 6-4 summarize the way AI describes ease of communication and extent of privacy. Readers are also referred to Chapter 1 for the calculation of AI. The references in Table 6-2 to Speech Privacy Potential (SPP) are clarified in Section 6-4.5.

6-4.2 Background Sound

It is a common experience that for effective speech communication in a noisy environment the speaker must reduce the distance between himself and the lis-

TABLE 6-2. Relationship Between Articulation Index (AI), Speech Privacy Potential (SPP), and Subjective Impressions.

Communication Condition	AI	SPP	Speech Privacy Condition
Excellent			
	0.65	—	Nil
Good			
	0.50	—	
Fair			Very Poor
	0.30	52	
Poor			Marginal
	0.15	57	
Very Poor			Acceptable
	0.05	62	
Nil			Confidential

tener or increase the level of his voice to maintain a sufficient signal-to-noise ratio and thus the necessary degree of speech intelligibility. Obviously there is an acceptable limit beyond which the increased levels of speech add to the overall noise level. This culminates in the well-known cocktail party effect: with only a few people present, communication using normal voice levels is easy, but as the crowd increases the background noise also increases until speech communication is virtually impossible even with raised voice levels at close range. It is this problem that underlies the success or failure of the open office concept.

It has been shown that most people are unaware of broadband steady sound levels below about 35 dB(A) and will accept levels up to approximately 47 dB(A) [1, 5]. Levels above 47 dB(A) will cause the speaker to raise his voice to compensate for the higher background level; consequently, 47 dB(A) represents the practical upper limit to effective sound masking. Table 6-3 gives typical background sound levels found in various open-plan office environments (without the artificially generated masking sounds which are often introduced). This range of sound levels is consistent with the range shown in Table 6-1 for general offices.

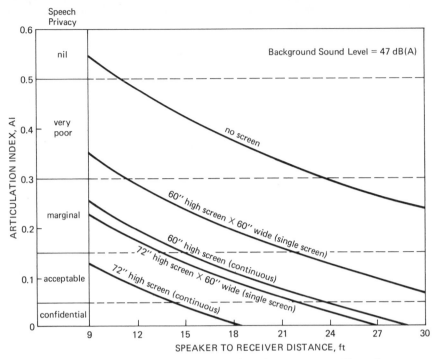

Fig. 6-4   Speech communication with various screen configurations.

6-4.3   Speech Level, Orientation, and Speaker–Receiver Distance

An average person's voice has a range of approximately 20 dB from a quiet to a raised speech effort. Simplifying this range to discrete steps of "quiet," "normal," and "raised," each step of 10 dB results in a change in Articulation Index of 0.34. Merely raising one's voice may therefore destroy the state of acoustical privacy at nearby work locations. Fortunately, the open office environment appears to have the effect of reducing conversational levels.

As the speaker turns from the receiver the received speech level decreases. Considering the receiver as a nondirectional point, each 90° change in speaker orientation results in a decrease in AI of approximately 0.15, so that a maximum decrease of 0.3 is theoretically obtainable when the change in orientation is 180 degrees, i.e., the speaker's back is to the receiver. Thus a 180° change in orientation could change the privacy condition from confidential to very poor. As a speaker normally turns his head through approximately 60° from the direction he faces, these allowances should be reduced by half for practical application.

In these environments the doubling of distance from speaker to receiver can result in a reduction in AI of 0.15 without the use of screens. Thus, if good communication exists at a certain close range (say 3 ft), then without screens a separation of some 20 ft would be required before a condition of acceptable privacy is obtained.

TABLE 6-3. Typical Background Noise for Various Open-Plan Environments[6].

Environment	Typical Background Noise[a]	
Professional/clerical (no machines)	35 dB(A)	(very quiet)
Professional/clerical (some machines)	40 dB(A)	(quiet)
Small drawing office	40 dB(A)	(quiet)
Clerical & numerous machines	45 dB(A)	(acceptable)
Large clerical & numerous machines	50 dB(A)	(noisy)

[a]The levels should be considered to range ±5 dB(A) about the given figure.

The insertion of a single screen or partial height barrier will provide a maximum reduction in AI of 0.3. This will improve an already reasonable situation but will not provide acceptable conditions if the AI is initially high. The use of continuous interlocking screens of 72 in. in height will give an improvement in AI of approximately 0.4 over the no-screen condition.

6-4.4 A Design Guide

Figure 6-4 gives a series of curves that can be used for predicting the privacy conditions in an open plan at any receiver position with respect to any source position. The curves are for a well designed open-plan office environment with a ceiling system having good sound absorptivity [a Noise Reduction Coefficient (NRC) greater than 0.8] and an electronic masking sound system set at 47 dB(A), and utilizing either 72 in. or 60 in. high screens. To be effective, the screens should have an NRC value in excess of 0.8 and a Sound Transmission Class of 15 [6].

The curves in Fig. 6-4 are based on normal speech levels and assume the most critical speaker orientation, i.e., speaker facing receiver. In order to determine the speech privacy condition prevailing at any workplace the following procedure should be used:

a. Determine speaker to receiver distance.
b. Establish screen height and configuration.
c. Determine the AI for the speaker–receiver distance from the appropriate curve in Fig. 6-4.
d. Adjust the predicted AI for actual or measured background sound level as follows: for each 1 dB(A) by which the background level is above 47 dB(A), reduce AI by 0.03; and for each 1 dB(A) by which it is below 47 dB(A), increase AI by 0.03.
e. The curves in Fig. 6-4 being for speaker facing receiver, reduce AI by 0.08 and 0.15 for orientations of 90 and 180 degrees, respectively.

TABLE 6-4.   Assessment of Speech Privacy in an Open-Plan Office. (See Fig. 6-5.)

Step	Reference	Example A to B	B to C	C to D	E to F	A to F
1. Location and interzone distance	Design drawings (Fig. 6-5)	9 ft	12 ft	15 ft	9 ft	20 ft
2. Screen height and configuration	Design drawings	72" continuous	72" continuous	No screen	60" single	72" continuous
3. Determine basic AI	Fig. 6-4	0.13	0.08	0.43	0.35	0
4. Adjust for actual background sound level	Section 6-4.4 [5] [a]	+0.03	+0.03	+0.03	+0.03	+0.03
5. Adjust for speaker-to-receiver orientation	Section 6-4.4 [6]	−0.15	0	−0.08	−0.15	0
6. Prevailing AI	Sum of steps 3, 4, 5	0.01	0.11	0.38	0.23	0.03
7. Privacy condition	Table 6-2 & Fig. 6-4	confidential	acceptable	very poor	marginal	confidential

[a] Assume for the sake of the example that the background level is 46 dB(A).

f. Figure 6-4 being for normal voice levels, add 0.34 to the calculated AI where raised levels are to be provided for.

Example

Figure 6-5 shows a segment of a typical open-plan office, and Table 6-4 gives the calculation of speech privacy conditions for a number of workplaces in the office.

6-4.5 Testing

The only standardized test methods for open office acoustical performance are those described by the Public Buildings Service of the U.S. General Services Administration (GSA) [7]. The tests are performance-oriented and are used to determine whether a condition of acceptable privacy exists for specific ceiling constructions for standardized screen and background sound conditions.

The method uses the concepts of Speech-Privacy Noise Isolation Class (NIC′) and of Speech Privacy Potential (SPP).

NIC′ is basically the attenuation between workplaces at 500 Hz, an important frequency when speech is considered. Its value depends on the distance between workplaces and the presence of screens and other absorptive surfaces. SPP is defined by SPP = NIC′ + NC, where NC is the value of the background noise determined with the help of the NC curves (see Section 6-2). A maximum value of NC = 40 is allowed, this being the upper limit of subjectively acceptable background level, approximately equal to 47 dB(A). The GSA criterion for speech privacy is that SPP must equal or exceed 60. This value of SPP, and the concept of SPP generally, can be better understood by relating it to AI as follows.

If the speech level at one workplace, a, is L_a, and the speech reaches a nearby workplace, b, at a level of L_b, then NIC′ = $L_a - L_b$. The background level L_c and the level of speech at locations a and b are related to the AI (see Section 6-4.3) at each location by

$$\text{AI}_a = (L_a - L_c)/30 \quad \text{or} \quad L_a = L_c + 30\,\text{AI}_a$$

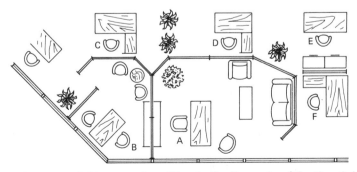

Fig. 6-5 Segment of the open-plan office in the Example of Section 6-4.4 (see also Table 6-4).

and
$$AI_b = (L_b - L_c)/30 \quad \text{or} \quad L_b = L_c + 30 \, AI_b.$$

Then $NIC' = 30(AI_a - AI_b)$, and $SPP = 30(AI_a - AI_b) + NC$, where NC is the NC value corresponding to L_c.

If NC is set at the maximum value of about 40, and AI_b is set low enough for the listener at the nearby workplace not to understand the speaker at the first workplace easily (say $AI_b = 0.10$ from Table 6-2 or Fig. 6-4), then $SPP = 30 \, AI_a + 37$. If AI_a is set at a value making for very good speech intelligibility at the first workplace, say $AI_a = 0.75$, then $SPP = 60$, which is the GSA criterion. Thus SPP provides a useful singe-number value by which to judge speech privacy in an open-plan office.

For $AI_a = 0.75$ and $NC = 40$, the values used above, note that $SPP = 63 - 30 \, AI_b$. This relationship between SPP and AI is represented in Table 6-2. Note also that if SPP must equal or exceed 60, and if the highest acceptable level of background sound is $NC = 40$ [about 47 dB(A)], then from $SPP = NIC' + NC$, the attenuation between workplaces, NIC', should never be less than 20 dB(A); it will have to be greater than this when the background sound is lower than $NC = 40$.

A useful additional description of open-plan office testing by the GSA method is given in Ref. 8.

REFERENCES

1. L. L. Beranek, *Noise and Vibration Control.* McGraw-Hill, New York, 1971, Chapter 18.

2. "Assessment of Noise with Respect to Community Response." International Organization for Standardization Recommendation ISO/R 1996, 1971.

3. "Information on Levels of Environmental Noise Requisite to Protect Public Health and Welfare with an Adequate Margin of Safety." U.S. Environmental Protection Agency Report No. 550/9-74-004, 1974.

4. L. L. Doelle, *Environmental Acoustics.* McGraw-Hill, New York, 1972, Chapters 6 and 10.

5. L. W. Hegvold, "Acoustical Design of Open-Planned Offices," Canadian Building Digest CBD 139, National Research Council, Ottawa, 1971.

6. A. C. C. Warnock, "Acoustical Effects of Screens in Landscaped Offices." Canadian Building Digest CBD 164, National Research Council, Ottawa, 1974.

7. "Standard Methods of Test for the Direct Measurement [PBS(PCD): PBS-C.1] and for the Sufficient Verification [PBS (PCD):PBS-C.2] of Speech Privacy Potential." Public Buildings Service, U.S. General Services Administration, 1975.

8. R. Moulder, "Open Plan Office Elements and Testing." Proceedings of Noisexpo, Atlanta, Georgia, April 30–May 2, 1975, pp. 257–259.

7

Noise in hospitals

J. G. WALKER*

7-1. INTRODUCTION

The problem of noise in hospitals has been appreciated for many years and there is no doubt that it can be particularly trying to both staff and patients. Beginning to 1957, Beranek [1, 2] proposed levels of continuous noise that should not be exceeded in hospital wards. Kosten and Van Os [3] in 1962 also proposed levels of continuous noise that were considered acceptable for sleeping rooms in residential accommodation; these levels were approximately equivalent to the proposals of Beranek for hospitals. Chapter 6 describes the noise contours of these researchers.

The Committee on the Problem of Noise in Britain also recognized the problem [4] and in 1963 pointed out that due attention should be paid to noise reduction in hospital buildings and especially in new hospital building programs. For residential accommodation they restated the main points of the

*Institute of Sound and Vibration Research, The University, Southampton SO9 5NH, England.

proposals of Kosten and Van Os and recommended certain maximum sound levels for rooms in which people sleep. Since 1963, however, it seems that very little coordinated effort has been devoted to noise in hospitals, though it affects not only patients but also staff tasks involving communication, diagnosis, and surgery.

The criteria in general use have considered the hospital noise problem only from the viewpoint of the patient in terms of sleep interference. While this is obviously of great importance, the interference caused to hospital staff by noise cannot be ignored. A survey carried out by Walker and Morgan [5] showed that the staff were more disturbed by, and more concerned about noise than were patients. From their point of view as well as the patients' a misinterpreted instruction or an inaccurate diagnosis can have serious consequences. It is the purpose of this chapter to consider both these sets of people in proposing acceptable noise levels.

7-2. EXISTING CRITERIA

7-2.1 Continuous Noise

Certain criteria already exist which recommend levels of continuous noise acceptable within hospitals. For example, Beranek proposed that a Noise Criteria (NC) rating of 30 would be appropriate. This criteria (see also Chapter 6) specified the maximum sound pressure levels in each of eight octave bands. Beranek's NC ratings were based on the then-accepted octave bands and the Stevens Mark I method of calculating loudness. He has since modified them to give Preferred Noise Criteria (PNC) curves which use the currently preferred octave band frequencies and the latest information on hearing sensitivity, including that used in the Stevens Mark VII method of calculating loudness [6]. Using the PNC curves, Beranek, Blazier, and Figwer [2] recommend that a suitable level for a hospital could range from 25 to 45 PNC.

It is possible to approximate the NC and PNC values to an A-weighted sound level (see Chapter 1), though care should be exercised. NC 30 is approximately equivalent to 40 dB(A), and PNC 25-40 is approximately equivalent to 34-47 dB(A).

Kosten and Van Os proposed a similar system of assessing noise, called the Noise Rating scale [3]. They suggested Noise Ratings of 25 NR for bedrooms and 30 NR for hospitals, which are approximately equivalent to 30 and 35 dB(A). Corrections for the type of area in which a residence is situated can be applied; thus it is desirable that the nighttime noise level in a bedroom in a rural locality be quieter than a bedroom in a busy urban area. The recommended levels range from 23 dB(A) for the rural situation to 40 dB(A) for the busy urban environment.

The foregoing discussion applies to noise that arises from sources both inside and outside the hospital. It appears from surveys that internally generated noise is often more disturbing than noise from outside sources, perhaps because there is a feeling that internal noise could be prevented, while the ability to hear outside noise keeps the patient in contact with the "outside world" so that he feels less isolated.

Measurements carried out in hospital wards suggest that if the hospital is air-conditioned, noise levels can be as high as 60 dB(A) but are more usually around 45–50 dB(A). Questionnaires have shown that patients are often disturbed at night by the noise of air-conditioning systems. In situations where there is no air-conditioning the noise levels can be as low as 30–40 dB(A) and no complaints from the patients occur, which suggests that the criteria of Beranek and Kosten and Van Os would give an acceptable noise environment.

The recommendations of other researchers on acceptable levels of hospital noise are included in Table 6-1 of Chapter 6.

7-2.2 Intermittent Noise

Several units already exist to describe the noise of single events such as an aircraft flyover or a vehicle passby. There are also several methods of expressing the noise environment due to multiple activities. However, in assessing the effects of noise on specially susceptible communities such as those in hospitals, consideration must be given to individual events as well as to the total noise, to ensure that neither the overall effect, nor the individual disturbance is beyond acceptable limits. The acceptability of the hospital environment must be related to task interference as well as to the less clearly definable criteria of general annoyance and patient recovery. Medical staff, for instance, pose fairly strict requirements on the environment for satisfactory stethoscope use. Therefore continuous noise environment measures alone are not entirely valid in the hospital context and an acceptable method must concentrate on the task interference aspects of the single noise event. In the context of patient recovery, if a single noise is loud enough it will wake a sleeping patient. In the case of a normal healthy person, such a disturbance may not be important, but for a sick person it may conceivably impair recovery. Even single noise events must therefore be of a level that will prevent this happening.

The peak level of the intrusion must, however, not be considered alone; account must be taken of the existing ambient level, and it is probably wise to restrict any particular single noise intrusion to a level that is no more than a certain amount above the ambient level. This will be discussed in more detail later.

The available knowledge about the disturbing effects of intermittent noise from the standpoints of sleep and task interference is reviewed in Chapters 13 and 14. A paper on the awakening effects of simulated sonic booms by Ludlow and Morgan [7] showed that more intense noises tended to wake subjects more

TABLE 7-1. Corrections to Be Applied to the Maximum Allowable Noise in Hospitals During Aircraft Flyovers [5, 8].

Number of aircraft flyovers in an 8-hour period	1–9	10–24	25–49	50–99
dB(A) to be added to the level found acceptable for 100 or more flyovers	5	3	2	1

consistently; that awakening was more frequent during the latter part of the night whether or not the subject was exposed to noise; and that the number of awakenings tended to decline as the number of nights of exposure increased; that is, adaptation to the test sleeping environment as well as to noise took place.

Reports on noise in hospitals by Walker and Morgan [5] and by Rice and Walker [8] adapted British Standard BS 4142 [9] to make an allowance for the intrusion of aircraft noise (see Chapter 3) into the hospital environment. Using arguments similar to those of Robinson [10] and assuming a 30-second average flyover duration, they derived corrections based upon the nighttime corrections in the standard. They allowed a limited tradeoff between the allowable peak level during a flyover and the number of events heard in an 8-hour period, as shown in Table 7-1.

The extension of the criterion to other noise sources is difficult, with the possible exception of intermittent traffic noise, where a similar criterion could apply. Busy urban traffic could be treated in the same way, where a high background noise level has occasional even higher levels superimposed by loud vehicles.

The situation of freely flowing traffic might be treated in a different way, perhaps using the L_{10} level. The Wilson Committee [4] tentatively estimated that the following noise levels from any source inside living rooms and bedrooms should not be exceeded for more than 10% of the time:

	Day	Night
Country areas	40 dB(A)	30 dB(A)
Suburban areas	45 dB(A)	35 dB(A)
Busy urban areas	50 dB(A)	35 dB(A)

These levels are taken from the London noise survey of 1961–62 and take account of the sound attenuation through the building structure. The recommended nighttime levels are slightly lower than the background levels measured and recommended in this paper for hospitals, but perhaps represent an ideal situation that should be aimed for. The levels show broad agreement with a great many other similar, more recent recommendations.

One of the greatest sources of annoyance to patients in hospitals seems to arise from internally generated noises. As far as the author is aware, there are no criteria that define maximum noise levels for the type of disturbance resulting from closing doors, drawing curtains and screens, and other daily activities in a hospital. However, many papers and reports from hospital designers, planners, and other authorities point out that such noises are disturbing to patients and must be kept to a minimum.

7-3. EFFECTS OF NOISE ON PATIENTS

Chapter 12 discusses in detail the effects of noise on health. It appears that there is little empirical evidence directly relating noise to physical health or rate of recovery. One study [11] reported a significantly higher rate of admissions to a psychiatric hospital from an area where aircraft noise levels exceeded 100 PNdB or a Noise and Number Index (NNI) of 55, when compared with admissions from areas where the noise intrusion was less. The study found that the type of person most affected is the older woman (of 45 years or more) not living with her husband (either widowed or divorced) who suffers from neurotic or mental illness. The authors of the report pointed out that they have not suggested that aircraft noise itself can cause mental illness; whether it might be assumed that the noise can adversely affect those with some type of mental instability who would otherwise remain stable is not clear.

Jansen [12] has studied sleep interference from noise (see also Chapter 13) and has concluded that all sounds audible at night impair the quality of sleep. Since the depth of sleep is important in determining the refreshment obtained from it, there is therefore some indirect evidence that noise can impair health. In a literature survey on a similar topic, Rice and Lilley [13] found that there was insufficient evidence on which to base precise opinions concerning hazards to health resulting from interference of sleep by noise. However, the Wilson Committee [4] had gone further by stating that "interference with sleep is least to be tolerated because prolonged loss of sleep is known to be injurious to health." It seems, therefore, that at present one way of assessing the effects of noise on health is to consider sleep interference. The particularly susceptible position of patients in hospitals who often find it difficult to sleep under normal circumstances should perhaps receive special consideration.

The level of acceptable noise compatible with undisturbed sleep is extremely difficult to determine, mainly because there are differing opinions as to what constitutes undisturbed sleep. Is a change in sleep stage a disturbance or must behavioral awakening occur? Behavioral awakening is possibly the criterion of sleep disturbance that can most easily be demonstrated to have an effect on health, but the levels of noise which cause it are difficult to assess since it is governed by several variables apart from the intensity of the stimulus. These include the information content of the stimulus, its duration, the sleep stage of the person when the stimulus occurs, and differences in awakening thresholds and in adaptation.

The literature on this topic reviewed in Chapter 13 shows that there are marked inter-individual differences in behavioral awakening thresholds. As yet there is no agreed noise level that is generally known to be compatible with undisturbed sleep, so that any figure given must be regarded as tentative. However investigations suggest that an indoor ambient level of 40 dB(A), rising to 50 dB(A) during intermittent noises which occur on 100 or more occasions during an 8-hour period, should be compatible with undisturbed sleep in the hospital environment. Corrections for other numbers of occurrences may be made according to the system proposed in Table 7-1. These intermittency corrections will probably be acceptable in situations where the background noise level is around 40 dB(A). However, in a rural or very quiet environment, levels of about 30 dB(A) might well be reached during the night; in these circumstances intermittent levels of 50 dB(A) on 100 or more occasions (that is 20 dB(A) above the ambient) may well be unacceptable, and it might be necessary to reduce these allowable peaks. It would probably not be acceptable to raise the ambient artificially, as is sometimes done in offices to mask intrusions, because patients in rural hospitals will normally be accustomed to low-level ambient noises in their home environment.

7-4. EFFECTS OF NOISE ON HOSPITAL STAFF

7-4.1 Speech Interference

Speech interference due to noise has an important effect which is often considered to be distinct from the general annoyance it causes. However, in 1963 McKennell [14] obtained social survey evidence on aircraft noise in the community which indicated that speech interference plays a leading part in the formation of annoyance.

For a continuous ambient noise level, criteria have been formulated that give acceptable listening conditions for speech at normal voice levels. These correspond to an Articulation Index (AI) of 0.6 or a Speech Interference Level (SIL) of about 60 dB. In a study of the acceptability of traffic noise in environments where low voice levels are desirable, as would be preferred in a hospital ward or consulting room, Gordon, Galloway, Nelson, and Kugler [15] suggested that an AI of 0.6 for 50% of the time should be acceptable, while the value of AI should not be allowed to fall below 0.4 for more than 10% of the time. These criteria correspond to traffic noise levels of 46 dB(A) for AI = 0.6 and 52 dB(A) for

AI = 0.4, for a speaker-listener separation of 5 ft. They also suggested that the dB(A) scale could be used as an alternative to AI not only for traffic noise but for aircraft noise. Walker and Morgan [5] used these data in deriving acceptable levels for aircraft within hospitals. They accepted the AI of 0.6 for speaker-listener separations of 4 to 5 ft and low voice levels, and the AI of 0.4 during an aircraft flyover. This is equivalent to an ambient level of about 40 dB(A) rising to about 50 dB(A) during an aircraft flyover for flyover rates of 100 or more in an 8-hour period. Assuming a duration of about 30 seconds each, aircraft fly-overs would occupy approximately 10% of an 8-hour period. A higher noise level is allowed for fewer flyovers.

The SIL corresponding to an AI of 0.06 is approximately 35-37 dB (SIL) with corrections of 5-10 dB (SIL) for different flyover rates.

The speech interference caused by hospital noise deserves special attention because it is evident from a survey by Walker and Morgan [5] in two hospitals near London (Heathrow) Airport that difficulties in consultations with patients due to aircraft noise are a major source of complaint among staff. Psychoanalysis sessions were particularly susceptible to interference.

7-4.2 Interference with Diagnosis

Another major source of staff complaints concerns interference with the use of stethoscopes, where heart murmurs of predominantly low frequency (below 300 Hz) are of great importance in the diagnosis of heart disease but are readily masked by extraneous noises.

Only two studies appear to have examined the effect of background noise on the impairment of auscultatory examination of chest complaints. Groom and Charleston [16] state that a reduction of hospital noise levels to 35 dB(A) affords a very real improvement in auscultatory proficiency, although the attainment of 35 dB(A) in outpatient clinics and wards is probably impractical in the presence of numerous internal as well as external noise sources. Richards, Croome, and Matthews [17] considered a similar problem and found that the majority of doctors in a test involving stethoscope listening to breathing against a background of traffic noise, favored background noise levels less than 46 dB(A). These authors then proposed that levels in consulting rooms should not exceed 40-45 dB(A).

Since the auscultatory examination sets a lower criterion level than the other, it is tentatively suggested that a first criterion should be 35 dB(A) or less. However, because the A-weighting network assigns less importance to frequencies below 500 Hz, these types of disturbance are not sufficiently assessed by a dB(A) reading alone, and a second criterion we suggest is that the overall level should not exceed 50 dB. More details of the frequency spectra of heart murmurs are required before these criteria can be regarded as more than tentative.

7-4.3 Interference with Surgery

The data available for maximum allowable ambient noise levels in operating theaters is extremely limited. The noise levels recommended by Beranek [1, 2] and Kosten and Van Os [3] are clearly aimed at the sleep disturbing effects of noise, and make no allowance for the many activities in a hospital for which criterion levels of about 35 dB(A) are unnecessarily strict.

There is no evidence that noise is a particular concern in operating theaters, but bearing in mind their special function a conservative maximum ambient level seems appropriate, and 40 dB(A) is suggested. Allowances could be made for intermittent noises ranging from 0 dB for noises occurring on 100 or more occasions in an 8-hour period to 5 dB for a noise occurring only once in a similar period (see Table 7-1).

Suitable siting of the operating theater with no outside windows and on lower or basement floors will help in achieving this goal.

7-5. STAFF AND PATIENT ATTITUDES

The survey of Walker and Morgan [5] in two hospitals close to London (Heathrow) Airport showed that staff were more annoyed by aircraft noise than were the patients. Further, patients were annoyed very nearly as much by internal hospital noise as by aircraft noise, even, it seems, in situations where there was no attempt at soundproofing. Probable reasons for these findings were:

a. most staff are exposed to aircraft noise for longer periods than most patients (that is, staff may work several years in one hospital; patients usually stay for only a few weeks at most);
b. staff rather than patients are occupied with exacting tasks;
c. most patients tend to live in the same area as the hospital and are therefore accustomed to, and may even be reassured by, the presence of aircraft noise;
d. patients tend not to complain about conditions when being treated in hospitals.

Staff opinion in the two hospitals, though it may be biased by their cautious attitude to health care, was that aircraft noise could have detrimental effects on health caused by rest or sleep interference, and that this would most affect the acutely nervous patient. Thus patient requirements should be focused on noise levels compatible with undisturbed sleep and rest. The main staff requirements are that noise levels should not interfere with consultations or stethoscope use.

7-6. RECOMMENDATIONS

It is recommended that both ambient and intermittent sound levels should be considered in any criterion proposed for individual areas within a hospital.

A method to correct for intermittent noises which occur on any number of occasions in any given period is tentatively proposed in Table 7-1. It is suggested that this method may be used for other intermittent noise, not only that of aircraft, and for impulse noise as well.

The corrections, which are in the form of additions to the allowable dB(A), will apply to wards, consulting rooms, and operating theaters, but not to diagnostic or examination rooms where even short, infrequent noise intrusions can be seriously disturbing.

The recommendations in the different areas of a hospital are as follows.

a. An ambient noise level of 40 dB(A) in hospital wards is acceptable, rising to 50-55 dB(A) during intermittent noises, depending on the number of occurrences. However, special considerations may be necessary if the ambient level is of the order of 30 dB(A) or less, in order to limit the peak-to-ambient noise ratios.

b. In consulting rooms, an ambient level of 45 dB(A) should be satisfactory, rising to a maximum of 50-55 dB(A) during intermittent noises, again depending on their number.

c. In diagnostic and examination rooms, 35 dB(A) should not be exceeded, with a further restriction of the overall sound pressure level to 50 dB. No correction can be recommended for intermittent noise.

d. In operating theaters, ambient levels of 40 dB(A) should be satisfactory, with maximum levels of 40-45 dB(A) for intermittent noises depending on the number of occurrences.

Example

Problem: The sound level in a hospital ward is measured at 45 dB(A), except during aircraft noise intrusions, when (on 15 occasions during an eight hour period), the sound level rises to about 55 dB(A). Comment on the acceptability of this environment.

Solution: The ambient level of 45 dB(A) is 5 dB(A) above the level recommended in Section 7-6 under subparagraph a. From the same subparagraph (see also Section 7-3), the maximum level of intermittent noise recommended is 50 dB(A) plus 3 dB(A) to correct for the number of flyovers (from Table 7-1). The measured level of 55 dB(A) slightly exceeds the recommended level of 53 dB(A). Both the ambient and the intermittent noise levels are therefore slightly above recommended levels.

REFERENCES

1. L. L. Beranek, "Criteria for Noise and Vibration in Buildings and Vehicles." In L. L. Beranek, ed., *Noise Reduction*, McGraw-Hill, New York, 1960. Chapter 20.

2. L. L. Beranek, W. E. Blazier, and J. J. Figwer, "Preferred Noise Criteria (PNC) Curves and their Application to Rooms." *J. Acoust. Soc. Amer.*, **50**, 1223, 1971.

3. C. W. Kosten and G. J. Van Os, "Community Reaction Criteria for External Noises." Paper F-5, N.P.L. Symposium No. 12, *The Control of Noise*, HMSO, London, 1962.

4. Committee on the Problem of Noise, "Noise—Final Report." HMSO, London, 1963.

5. J. G. Walker and P. A. Morgan, "Effect of Aircraft Noise in Hospitals." Consultation Report 1323, Wolfson Unit, I.S.V.R., Southampton University, 1970.

6. S. S. Stevens, "Perceived Level of Noise by Mk VII and Decibels (E)." *J. Acoust. Soc. Amer.*, **51**, 575–601, 1971.

7. J. E. Ludlow and P. A. Morgan, "Behavioral Awakening and Subjective Reactions to Indoor Sonic Booms." *J. Sound Vib.*, **25**, 479–495, 1972.

8. C. G. Rice and J. G. Walker, "Criteria for the Assessment of Aircraft Noise Nuisance in Hospitals." Paper 72/103, British Acoustical Society Meeting, July, 1970.

9. "Method of Rating Industrial Noise affecting Mixed Residential and Industrial Areas." British Standards Institution, B.S. 4142: 1968.

10. D. W. Robinson, "An Outline Guide to Criteria for the Limitation of Urban Noise." N.P.L. Aero Report Ac39, 1969.

11. I. Abey-Wickrama, M. F. A'Brook, F. E. G. Gattoni, and C. F. Herridge, "Mental Hospital Admissions and Aircraft Noise." *The Lancet*, 1275–1277, December 13, 1969.

12. G. Jansen, "Effects of Noise on the Physiological State." Proc. Conf. on Noise as a Public Health Hazard. Amer. Speech and Hearing Research Assn. Report No. 4, 1968.

13. C. G. Rice and G. M. Lilley, "Effect on Humans (and Animals) of the Sonic Boom." Paper presented at OECD Conf. on Sonic Boom Research, 1970.

14. A. C. McKennell, Central Office of Information Report SS 337, 1963.

15. C. G. Gordon, W. J. Galloway, D. L. Nelson, and B. A. Kugler, "Evaluation of Highway Noise." Bolt Beranek and Newman Inc. Report No. 1861, 1970.

16. D. Groom and S. C. Charleston, "The Effect of Background Noise on Cardiac Auscultation," *Amer. Heart J.*, **52**, 781–790, 1956.

17. E. J. Richards, D. J. Croome, and D. Matthews, "Acceptable Noise Levels in Doctors' Consulting Rooms." Proc. 7th ICA, Budapest, 1971.

8

Exterior industrial and commercial noise

RUPERT TAYLOR*

8-1. INTRODUCTION

Noise is unique among pollutants in several ways. Its effects can be highly subjective, and can produce an immediate reaction; yet at the levels commonly encountered in residential communities it has not been shown to have any significant harmful physical effect. It is therefore extremely difficult to treat noise in the manner some other pollutants are dealt with— by declaring a maximum safe limit. This approach can be attempted, but human response to community noise is so variable and so dependent on other factors—in particular, on people's feelings towards the object or organization that is responsible for the noise—tnat no declared limit set at any practicable level is totally satisfactory.

Noises from fixed sources, as opposed to moving ones like vehicles and aircraft, are commonly assessed by one of three

*Rupert Taylor and Partners Ltd, 113 Westbourne Grove, London W2 4UP, England.

methods. In Britain, British Standard BS 4142 [1] relating to industrial noise may be used with some justification for commercial noise sources as well. The ISO recommendation R 1996 [2] is another method; it is derived from the British standard, but neither Britain nor the United States in fact subscribe to it. Consequently, to include American as well as British and international criteria, we include a normalized L_{dn} method endorsed by the Environmental Protection Agency [3].

8-2. BRITISH, AMERICAN, AND INTERNATIONAL CRITERIA

The approach taken by all three assessment methods to evaluate intrusive noise sources is to compare:

  a. the measured noise from an existing intrusive source (or the predicted noise from a planned one), corrected in ways we shall describe (see Section 8-2.1), with
  b. the background sound level or a criterion level, which may also be corrected (see Section 8-2.2).

The background sound level is the sound level at the point where the noise is to be assessed in the absence of the intruding noise. This point of assessment is generally an outdoor location of a nearby residence.

  The object is to determine how intrusive the industrial or commercial noise is or will be, intrusiveness being measured by an excess of a over b.

8-2.1 The Intruding Noise Description

Measurement of the noise complained of, or which is a potential cause of complaint, can be complex. The essential point is that a noise level is defined which, implicitly or through correction factors, takes account of temporal factors (e.g., the noise's duration or intermittency, or the time of day or year it occurs) and the noise's character (e.g., any tonality or impulsiveness). Tonal noises are those with a distinctive single-frequency content, e.g., whines, screeches, hisses, squeals, and hums. Impulsiveness is generally the presence of bangs, clanks, or thumps.

  In Britain [1] the intruding noise is measured or predicted in dB(A); the measurement is corrected by an addition of 5 dB(A) if it is tonal or 5 dB(A) if it is impulsive, and is subject to a correction for intermittency amounting to $-2.1 (\log_{10} d) (\log_{10} t)$, where t is the total on-time as a fraction of eight hours, and d is the typical on-time duration as a fraction of eight hours by day and of eight minutes by night; this correction is actually in graphical form in the standard.

  The international method [2] also starts with the dB(A); and it has 5 dB(A) added for tonality or 5 dB(A) for impulsiveness. A table of intermittency corrections is given in the standard; alternatively, the equivalent sound level L_{eq} may be used to describe the varying level (see Chapter 1 for its definition).

The American method [3] starts with the use of L_{dn} to describe the intruding noise. L_{dn}, the day–night level, is also defined in Chapter 1; it is a diurnally weighted L_{eq} insofar as it gives extra decibels to sounds occurring at night. The L_{dn} is corrected to arrive at a normalized L_{dn} by 5 dB(A) for tonality and/or impulsiveness (only one 5 dB(A) correction may apply), and also by the other factors set out in Table 1-7 of Chapter 1. These include matters like the nature of the community, which are considered in the British and international methods in their treatment of background noise—see Section 8-2.2.

8-2.2 Background Noise or Criterion Level

The principle involved is to measure the background level, or to derive a notional background level where this is not possible.

In Britain [1] the background level is defined as L_{90}, the sound level exceeded for 90% of the time, measured in the absence of the noise complained of. Where the background level is unobtainable, a notional background level may be established, which consists of a basic level of 50 dB(A) to which is added: 5 dB(A) for established industrial or commercial premises existing in areas with which they are not truly compatible or 10 dB(A) for long-established premises completely in keeping with the locality; the range of corrections given in Table 8-1 for the type of district; 5 dB(A) for weekday noise occurring by day, -5dB(A) for nighttime noise; and 5 dB(A) if the noise occurs only in winter.

The international method [2] defines the background level as L_{95} in the absence of the intruding noise, or an approximation of this from a sound level meter reading. Where a measured level is unobtainable, a criterion level is estab-

TABLE 8-1. Corrections to the Basic Level Recommended in British Standard BS 4142 [1].

Type of District	Correction to Basic Level,[a] dB(A)
1. Rural (residential)	-5
2. Suburban, little road traffic	0
3. Urban (residential)	+5
4. Predominantly residential urban but with some light industry or main roads	+10
5. General industrial area intermediate between (4) and (6)	+15
6. Predominantly industrial area with few dwellings	+20

[a]Select only one.

TABLE 8-2.   Corrections to the Basic Criterion Level Recommended in ISO/R 1996 [2].

Type of District	Correction to Basic Criterion, dB(A)
1. Rural residential, zones of hospitals, recreation	0
2. Suburban residential, little road traffic	+5
3. Urban residential	+10
4. Urban residential with some workshops or with business or with main roads	+15
5. City (business, trade, administration)	+20
6. Predominantly industrial area (heavy industry)	+25

lished essentially similar to the British standard but with different numbers. The basic criterion is 35–45 dB(A)–an invitation to presume 40 dB(A)–to which are added the corrections given in Table 8-2 for the type of district, as well as −5dB(A) for evening noise or −10 to −15 dB(A) for nighttime noise. There is, however, no correction for the age or compatibility of the noise producer in the locality.

In the American method [3], there is no consideration of background noise beyond that implicit in Section 8-2.1. The use of this method will become clear in Section 8-2.3. In Section 8-2.3 we make use of the information generated in Sections 8-2.1 and 8-2.2 to assess the acceptability of the noise.

8-2.3   Assessment of the Intruding Noise

A comparison of the intruding noise with the background level or corrected criterion allows us to estimate the acceptability of the intruding noise.

The British standard [1] suggests that complaints may be expected if the corrected intruding noise level exceeds the background or notional background level by 10 dB(A) or more. Excesses of 5 dB(A) are of marginal significance. If, on the other hand, the corrected intruding level is more than 10 dB(A) below the background or notional background level, there is a positive indication that complaints will not arise. A fuller explanation is, however, given in the standard itself.

The international method [2] predicts an essentially similar response, but gives in addition a table here reproduced as Table 8-3, which is self-explanatory.

The American method [3] involves reference to the graph reproduced in Chapter 1 as Fig. 1-9, in which values of normalized L_{dn} are plotted against a range of community responses of increasing severity. A normalized L_{dn} of 55 dB(A) or lower will generally not result in complaints. Higher values will likely

TABLE 8-3. Community Response Estimated by the ISO/R 1996 [1] Criterion.

Amount by which the Corrected Intruding Sound Level Exceeds the Corrected Criterion, dB(A)	Estimated Community Response	
	Category	Description
0	none	no observed reaction
5	little	sporadic complaints
10	medium	widespread complaints
15	strong	threats of community action
20	very strong	vigorous community action

TABLE 8-4. Community Response Expected in the American Normalized L_{dn} Assessment [3].

Category	Normalized L_{dn}, dB(A)
No reaction or sporadic complaints	50–60
Widespread complaints	60–70
Severe threats of legal action, etc.	70–75
Vigorous action	75–80

result in an increasingly annoyed community. We have summarized this figure in Table 8-4, rounding off the sound level values to the nearest 5 dB(A).

8-2.4 Problems Involved

Background level. It is important that the measured background noise level be typical of the noise exposure of the residents in the area over a period of time. It may be necessary to repeat the measurements several times if there is a possibility that weather conditions affect the results, and it is of course of the greatest importance that the measurements be carried out at the same time of day that the complaints occur. Care is necessary to ensure that temporary noise sources, such as construction work, do not give unrepresentative results.

There are unfortunately many instances when the background noise cannot be so conveniently measured, since it is not always possible to eliminate the noise complained of merely to measure the noise level in its absence. There are two things which can be done in these cases: the background noise can be measured in a similar nearby locality with the same mix of traffic and the same density of housing and located in a similar situation, but which is far enough from the noise complained of for its contribution to the sound level to be insignificant; alternatively, a frequency analysis, say into $\frac{1}{3}$ octave bands, can be carried out and the L_{90} value established for each band. If the noise being studied is of a relatively narrow-band nature, taking into account the normal broadband nature of background noise it may be possible from the spectrum, by eliminating the peaks, to estimate the overall level in dB(A) of the background alone.

If the establishment of the background noise level is being carried out for planning purposes, it is important to research possible changes which could occur between the measurement time and the date of construction of the factory or commercial establishment and cause a permanent change in the background level. An example is the opening of a suitably distant bypass road which could have the effect of reducing traffic flow and lowering traffic sound levels in some locations.

Assessment. There is a possible drawback in using the international method to compare the L_{eq} of the intruding noise with the L_{95} of the background noise. If, when measuring the background noise, not only the L_{95} has been measured but L_{eq} also, the L_{eq} will undoubtedly be higher than the L_{95} by as much as 5-10 dB(A). In other works, one might reach the conclusion that sporadic or even widespread complaints would arise concerning the background noise alone. It is then, on the face of it, somewhat unfair to judge as unacceptable another noise which also has a corrected L_{eq} value 10 dB(A) in excess of L_{95}.

A solution to this problem is to express the background noise level also in terms of L_{eq}, and to use a different table to compare the uncorrected L_{eq} with the background L_{eq}, as suggested by Table 8-5.

TABLE 8-5.   Simplified Assessment Method Using L_{eq} to Offset the Problem Described in Section 8-2.4.

Increase Over Ambient L_{eq}, dB(A)		Estimated Community Response	
Tonal/ Impulsive	Other	Category	Description
0	0	none	no observed reaction
1	3	little	sporadic complaints
3	6	medium	widespread complaints
6	10	strong	threats of community action
6+	10+	very strong	vigorous community action

Because of the nature of the L_{eq} scale, for a non-tonal/impulsive noise complaints start to occur when the total noise energy due to the noise considered is equal to that due to all other environmental noise sources combined. For a satisfactory situation, the noise considered should have no significant effect on the measured ambient L_{eq}.

8-3. INTERNATIONAL NOISE LIMITS

The limitation of intrusive noise by law varies substantially from country to country, and from one state or city to another. At the one extreme are complex sound level limits (such as Chicago's—see below); at the other extreme are general laws of nuisance.

8-3.1   United States

The Noise Control Act of 1972 [4], in Section 5(a) (2), required the Environmental Protection Agency to publish information on the levels of environmental noise which are requisite to protect the public health and welfare with an adequate margin of safety. This information was published in 1974 [3], and it endorsed the normalized L_{dn} method of judging intrusive noise described in Section 8-2.

The Noise Control Act does not itself directly regulate the emission of industrial or commercial noise: legislative control of this kind is exercised at the state and local level. In the case of future revised or new ordinances, the EPA criteria [3] and model ordinance [5] may form a base, but pre-existing ordinances differ substantially.

Chicago. Industrial and commercial noise in Chicago is controlled through a zoning system set out in the 1971 Noise Ordinance [6]. Maximum sound pressure levels in each octave band are prescribed for Restricted Manufacturing Zoning Districts; these levels should not be exceeded at any point on the boundary of a Residence, Business, or Commercial Zoning District due to individual or combined operations of the noise producer's plant. Slightly higher limits pertain to General and Heavy Manufacturing Zoning Districts, measurable at the same locations as before, or at 125 ft from the nearest source property line, whichever is greater. These levels are given in Table 8-6.

Groundborne vibration is controlled by stating that no vibration which is perceptible without the aid of instruments shall be permitted beyond the lot line, except in Heavy Manufacturing Zoning Districts, where the vibration may not create a nuisance or hazard outside the lot.

TABLE 8-6. Chicago's Noise Limits [6].

	Maximum Sound Pressure Levels, dB, Along District Boundaries					
	Restricted Manufacturing Zoning Districts		Manufacturing Zoning Districts			
			Residence		Business (General Manufacturing District)	Commercial (Heavy Manufacturing District)
Octave Band Center Frequency, Hz	Residence	Business–Commercial	(General Manufacturing District)	(Heavy Manufacturing District)		
31.5	72	79	72	75	79	80
63	71	78	71	74	78	79
125	65	72	66	69	73	74
250	57	64	60	64	67	69
500	51	58	54	58	61	63
1000	45	52	49	52	55	57
2000	39	46	44	47	50	52
4000	34	41	40	43	46	48
8000	32	39	37	40	43	45
A-scale levels (for monitoring purposes)	55 dB(A)	62 dB(A)	58 dB(A)	61 dB(A)	64 dB(A)	66 dB(A)

New York State. The State of New York's Model Local Noise Ordinance [7] recommends maximum sound levels that may be created by an industrial, commercial or business operation, as follows:

a. sounds entering Class AA land (especially quiet areas): shall not increase the L_{90} there;

b. sounds entering Class A land (including most residences): during the daytime between 7 am and 7 pm, the sound level should not exceed 65 dB(A) on the slow response, and the L_{10} should not exceed 60 dB(A); at other hours the respective limits are 55 and 50 dB(A);

c. sounds entering Class B land (places where speech communication must be easily possible): the sound level should not exceed 65 dB(A) on the slow response, nor should the L_{10} exceed 60 dB(A).

These limits apply to broadband, continuous sounds; where impulsive or tonal characteristics are present, the quoted limits are to be reduced 5 dB(A). The model ordinance also suggests limits for other receiver zones, and lists certain exceptions.

New York City. A local law enacted by New York City in 1971 [8] provides for the setting up of noise-sensitive zones, so that the administrators can proscribe any activity within the zone which would constitute unnecessary noise, and for the division of the city into ambient noise quality zones. Within each zone maximum allowable sound levels may be laid down.

8-3.2 The European Economic Community

Although important sections of noise control including the limitation of noise from motor vehicles, construction equipment, and machinery are or will be covered by EEC directives, plant noise is a matter for the individual member countries, as given in the following sections.

8-3.3 Great Britain

Common Law Nuisance. Plant noise is controlled in four ways: the long-established law of nuisance can result in a court injunction restraining a person from continuing the emission of noise which is judged to be a nuisance in a legal sense of the word. Specific levels of noise are rarely, if ever, referred to in an injunction, but in some cases the judge may be impressed by expert evidence based on the use of British Standard 4142 [1], which, if it indicates that complaints may be expected, indicates that the person bringing the action is not unusually sensitive to noise. However, the evidence of the plaintiff and of lay witnesses and even a view of the site by the judge can all have importance as great as that of expert evidence.

Statutory Nuisance. Noise is also dealt with as a form of nuisance under the Control of Pollution Act 1974 [9] (replacing the Noise Abatement Act 1960), in which a statutory procedure for local authorities to issue abatement notices having the force of law is laid down. Provided the authority can prove that the noise is a nuisance, i.e., that it causes material discomfort or interference with ordinary living standards, it may specify in the notice the work that must be carried out to abate the noise and may include a specified level to which the noise should be reduced. A fine and a daily fine are the principal remedies, and there is provision for a member of the public to apply directly to a magistrates court if the local authority is unwilling to act.

Noise Abatement Zones. Under Sections 63–67 of the Control of Pollution Act, local authorities are empowered to declare a part or all of their area a noise abatement zone, subject to the confirmation of the Secretary of State for the Environment. The noise emission from classified premises within the zone is then monitored in terms of L_{eq} according to published guidelines about noise abatement zones [10], the results being recorded in a public register. It is then an offense to exceed the registered level without consent, and the authority may issue noise reduction notices requiring that noise from a particular plant be reduced to a specified level by a specified date.

Planning Controls. Under the provisions of the Town and County Planning Act 1971 [11], planning authorities are empowered to refuse or attach conditions to a consent to a proposed new development. Noise can be a leading factor influencing the decision, and is increasingly the subject of special conditions attached to the consent.

The principal guidance available to planning authorities is contained in a Department of the Environment circular, Planning and Noise [12], published in 1973. The basis of the circular is BS 4142, and it states, after conceding that there are circumstances when it is necessary to allow industrial development near houses, that where (by the standards established in BS 4142) the noise from the proposed development is likely to give rise to complaints, it will hardly ever be right to give permission. A warning is given on the danger of a creeping increase in the ambient noise level as a result of a succession of new noise sources being introduced each within a few decibels of the preexisting ambient level. Paragraph 28 states: "Authorities should, as far as possible, operate their development control powers in such a way as to avoid increases in ambient noise levels affecting residential and other noise sensitive development." This will not always be possible; but the main aim must always be to hold down ambient noise levels: and even in urban areas where existing levels are already high it will scarcely ever be justifiable to allow new development which is liable to have the effect of bringing the ambient level affecting existing residential and other noise sensitive development above a Corrected Noise Level of 75 dB(A) by day or 65 dB(A) by night.

At first sight the maximum noise levels quoted above appear exceptionally high, even though in a supplementary table the circular indicates that a "good standard" is 10 dB(A) lower. The reason is that the same circular deals with road traffic noise; and under the Land Compensation Act [13], residents near new or improved roads which create traffic noise outside their houses in excess of 70 dB(A) measured on the L_{10} scale over the hours 6 am to midnight are eligible for sound insulation grants. [Because of an accuracy requirement of ± 2 dB in the method of measurement or prediction, the level which is laid down in the regulations is actually 68 dB(A)].

8-3.4 France

French noise legislation consists mainly of specific noise emission standards. Machines at present covered are confined to construction plant and engines, and also compressors, which must be labeled with their noise level and date of manufacture.

8-3.5 Germany

Planning authorities are empowered to lay down noise emission standards when dealing with applications for permission to modify or build new industrial premises. Standards set out in paragraph 2.321 of the T. A. Lärm [14] include a level of 50 dB(A) by day and 30 dB(A) at night in areas containing residential housing only. Measurements are made 3 m from the source site boundary and 1.2 m above the ground.

8-3.6 Italy, The Netherlands, and The Republic of Ireland

None of these countries appears to have legislation specifically controlling noise. In Italy and the Netherlands noise is included under general nuisance laws. In Ireland local authorities have power to specify noise limits in the granting of planning permission.

8-3.7 Denmark

The Department of the Environment has laid down guidelines for the evaluation of external noise from industry under the Environmental Protection Act [15] of 1973. The guidelines are based on L_{eq} values with a tonal or impulsive character correction, and limits are specified such as 50 dB(A) from a plant situated in an industrial area affecting a nearby multi-storey residential area and 45 dB(A) in a suburban residential area.

8-3.8   Canada

Industrial and commercial noise is under the jurisdiction of the provinces, who may delegate it to the municipalities. In Ontario, the Environmental Protection Act, 1971, as amended [16], controls sound that causes "material discomfort," but provides for municipalities to enact bylaws which meet provincial approval; the Model Municipal Noise Control By-Law [17] proposes sound level limits.

8-3.9   Australia

Legislation relating to plant noise is enacted by the states. Noise policy tends to be based on the Australian standard AS 1055 [18], which is similar in approach to BS 4142. The Code uses an hourly L_{90} as a basis, and corrections are made to produce an adjusted measured noise level similar to the corrected noise level of BS 4142. The city of Richmond has a noise control policy which, although based on AS 1055, is in line with the consensus of international standards [e.g., the acceptable level for daytime in general suburban areas with medium-density transportation is 55 dB(A)].

Example

Problem: The noise from a factory running round the clock, year round, is measured outside a residence and is found to be 65 dB(A) from 7 am to 10 pm on weekdays, and 55 dB(A) at other times. The noise is tonal but not impulsive. Because of the continuous operation, it is impossible to measure the background noise. The factory is a well established one in a noisy urban area close to main roads and with mixed industrial–residential land uses. Use British and American assessment methods to comment on the likelihood of complaints.

Solution:

   a. *Using BS 4142 [1].* The intruding noise, being continuous, has no inter-mittency correction. Since it is tonal, however, add 5 dB(A) (see Section 8-2.1) to the noise level encountered (65 dB(A) for nighttime noise, and 55 dB(A) for daytime noise). The corrected intruding noise levels are therefore 70 dB(A) by day and 60 dB(A) at night. Since we cannot measure the background level, we calculate a notional background level. To a base of 50 dB(A) (see Section 8-2.2), add 5 dB(A) because the area has factories near residences; 15 dB(A) from Table 8-1; 5 dB(A) for the daytime case; and −5 dB(A) for the nighttime case (see Section 8-2.2). There is no correction for seasonal operation. The notional background levels are therefore 75 dB(A) by day and 65 dB(A) at night. These levels are higher than the corrected intruding levels in the same time periods and so complaints are unlikely (see Section 8-2.3).

   b. *Using the U.S. method [3] (see Sections 8-2.2, 8-2.3, and 8-2.4).* We calculate the L_{dn} to be 65 dB(A) from Section 1-5.3. From Table 1-7, subtract 5 dB(A) for the nature of the community, subtract another 5 dB(A) for the long-standing exposure to the noise, and add 5 dB(A) for tonality; the other corrections do not apply. The normalized L_{dn} is, therefore, 60 dB(A). From Fig. 1-9 and Table 8-4, the situation is delicately balanced. The noise is generally noticeable; we can expect sporadic complaints or even widespread ones, though serious legal or political reactions are unlikely.

REFERENCES

1. "Method of Rating Industrial Noise affecting Mixed Residential and Industrial Areas." British Standards Institution, BS 4142 London, 1975.

2. "Assessment of Noise with respect to Community Response." International Organization for Standardization Recommendation R 1996, 1971.

3. "Information on Levels of Environmental Noise Requisite to Protect Public Health and Welfare with an Adequate Margin of Safety." U.S. Environmental Protection Agency Report No. EPA 550/9-74-004, March, 1974.

4. United States Noise Control Act, 1972 (49 U.S.C. Sections 4901 *et seq.*).

5. "Model Community Noise Control Ordinance." U.S. Environmental Protection Agency Report No. EPA 550/9-76-003, September, 1975.

6. Noise Ordinance, City of Chicago, 1971.

7. "Model Local Noise Ordinance." New York State Department of Environmental Conservation, April, 1975.

8. New York City Noise Control Code, Int. No. 66-A, 1971.

9. United Kingdom Control of Pollution Act, 1974.

10. "Control of Pollution Act, 1974. Implementation of Part III—Noise." Circular 2/76, Department of the Environment, HMSO, London, 1976.

11. United Kingdom Town and County Planning Act, 1971.

12. "Planning and Noise." Circular 10/73, Department of the Environment, HMSO, London, 1973.

13. United Kingdom Land Compensation Act, 1973.

14. Technische Anleitung zur Schutz gegen Lärm." Bundesministerium für Arbeit, Bonn, 1968.

15. Denmark Environmental Protection Act, 1973.

16. Ontario Environmental Protection Act, 1971, as amended.

17. "Model Municipal Noise Control By-Law." Ministry of the Environment, Ontario, 1976.

18. "Code of Practice for Noise Assessment in Residential Areas." AS 1055, Australian Standards Association.

9

Construction site noise

ROBIN J. ALFREDSON*
and
D. N. MAY†

9-1. INTRODUCTION

With the growing awareness of noise and its effects on man has come recognition of the need and obligation to control noise from construction sites. The number of people affected by this type of noise is estimated, in the U.S., to be as great as the number affected by freeway noise [1] (see Table 9-1). Two different groups of people are involved. Site personnel are subjected to the highest noise levels. For them there exists the possibility of noise-induced permanent hearing loss or, at least, interference with spoken communication. Acceptable levels with regard to hearing loss are discussed in Chapter 11, and those with regard to speech interference in Chapter 1; they are not taken up here. The second group of people affected by

*Monash University, Department of Mechanical Engineering, Melbourne, Victoria 3168, Australia.

†Research and Development Division, Ministry of Transportation and Communications, Downsview, Ontario M3M 1J8, Canada.

TABLE 9-1. Estimated Number of Americans in Residential Areas Subjected on Any One Day to Noise of Different Kinds at or Above the Specified Day-Night Sound Levels Measured Outdoors [1].

Outdoor L_{dn} level, dB(A)	Number of People Subjected, millions			
	Urban Traffic Noise	Aircraft Noise	Construction Site Noise	Freeway Noise
70+	4–12	4–7	1–3	1–4
65+	15–33	8–15	3–6	2–5
60+	40–70	16–32	7–15	3–6

construction noise are those who live or work in adjacent neighborhoods. Their interests will be emphasized in this chapter.

Noise in the community from construction sites is not dissimilar in many ways from that of industrial sites. However, construction sites are of a temporary nature (although for large projects the activity may extend over a number of years) and can on occasion be located very close to noise-sensitive areas such as hospitals. Further, in order to meet contractual deadlines, the site may continue to operate on weekends or during the hours of darkness. The noise from vehicles arriving, unloading, and departing adds to the general annoyance.

The noise from construction sites is now generally described by the A-weighted equivalent sound level to take account of the enormously time-varying nature of the noise, as shown for example in Fig. 9-1 [2]. This is open to criticism, as the annoyance experienced may depend not only on the mean noise level but also the variability of the noise. Other features such as spectral content, impulsiveness, time of day, type of district, personality of the listener, avoidability of the noise, financial loss or gain resulting from the construction, and the length of time the site has been operating will also influence individual, and hence community, reaction.

Noise exposure at construction sites can be related to several characteristics of the construction activity. The EPA's "Report to the President and Congress on Noise" [3] lists the following five consecutive phases in construction:

1. ground clearing, including demolition and removal of prior structures, trees, rocks;
2. excavation;
3. placing of foundations, including reconditioning of old roadbeds and compacting of trench floors;
4. erection, including framing, placing of walls, floors, and windows, and pipe installation;
5. finishing, including filling, paving, and clean-up.

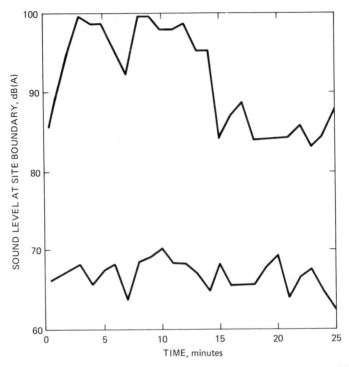

Fig. 9-1 Sound levels at construction site boundaries may vary widely with time. Here the sound levels are averaged over the first four seconds of each minute over a 25 minute period at two different sites [2].

Typical ranges of equivalent sound level on the site for these five phases are given in Table 9-2. The table shows that the initial and final phases are often the noisiest, and the intermediate phases quietest. No class of construction project has notably higher noise *levels* than another when all five phases are considered. However, the different numbers of projects of each type, their size and duration, and the number of people living or working nearby are needed if we are to rank these projects for the noise *exposure* they cause.

Table 9-3 is an estimate of the noise exposure to persons off site to domestic and other construction in urban areas; it differentiates between the exposure of passers-by and people based "permanently" nearby. Of the effects noted in Table 9-3, speech interference is the most widespread nuisance; about 300 million person-hours per week in the U.S. are estimated to involve significant speech interference due to construction noise; another 10 million person-hours would involve severe speech interference. Each year some 30 million people live or work in the vicinity of construction sites, while there are an estimated 24 billion passer-

TABLE 9-2. Noise Levels at Construction Sites in Terms of the Activity in Progress [3].

| | Typical Ranges of L_{eq}, dB(A) | | | | | | | |
| | Domestic Housing | | Office Build-ing, Hotel, Hospital, School, Public Works | | Industrial, Parking Garage, Religious, Amusement & Recreation, Store, Service Station | | Public Works, Roads & Highways, Sewers, Trenches | |
Activity	I	II	I	II	I	II	I	II
Ground clearing	83	83	84	84	84	83	84	84
Excavation	88	75	89	79	89	71	88	78
Foundations	81	81	78	78	77	77	88	88
Erection	81	65	87	75	84	72	79	78
Finishing	88	72	89	75	89	74	84	84

I. All pertinent equipment present at site.
II. Minimum required equipment present at site.

by encounters per year. How this is estimated to increase with the years is shown in Fig. 9-2.

The types of equipment used in construction can be classified as follows [3]:

1. earthmoving equipment (bulldozers, shovels, backhoes, front-loaders) and highway building equipment (scrapers, graders, compactors);
2. materials handling equipment (cranes, derricks, concrete mixers, concrete pumps);
3. stationary equipment (pumps, electric power generators, air compressors).

Noise from these items of equipment differs from model to model, and changes according to the operation involved. The range of levels found for 19 of the more common types of site equipment is shown in Fig. 9-3, based on a limited sample of U.S. equipment but reasonably accurate in both American and British experience. Tables 9-4 and 9-5 convey, in addition, an estimate of the total sound energy from each item (Table 9-4) and how equipment sound energy ranks from one type of site to another (Table 9-5).

The total sound energy is essentially a product of a machine's sound level, the numbers of such machines in service, and the average time they operate. This is the measure which EPA regards as the best indication of the priority they should give to regulating the noise of the various items. Thus, although pile drivers and rock drills produce the highest sound levels when they are operating, it is dump

TABLE 9-3.   Noise Exposure for Various Categories of Person, for Various Types of Construction, in Terms of Its Effects [3].

Noise Source	Exposure, millions of person-hours per week[a]					
	Speech Interference[b]		Sleep Interference[b]		Hearing Damage Risk	
	Moderate 45–60dB(A)	Severe >60dB(A)	Slight 35–50dB(A)	Moderate 50–70dB(A)	Slight 70–80dB(A)	Moderate 80–90dB(A)
Primary (stationary) exposure to domestic construction noise	44	—	—	2	0	—
Primary (stationary) exposure to all other building construction	38	—	—	2	0	—
Primary (stationary) exposure to all other construction in SMSA[c] areas	14	—	—	1	0	—
Secondary (passer-by) exposure of pedestrians to construction in all SMSA[c] areas	—	10	0	—	—	10
Secondary (passer-by) exposure of drivers and passengers to all construction in SMSA[c] areas	—	0.3	0	—	—	0.3

[a] These figures apply to the U.S. population other than construction workers.
[b] Entries in these columns may not be interpreted directly as person-hours of direct speech or sleep interference.
[c] Standard Metropolitan Statistical Area.

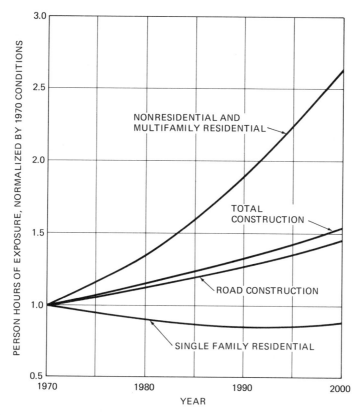

Fig. 9-2 The estimated growth in U.S. construction noise exposure [3].

trucks, air compressors, and concrete mixer trucks that, due to their greater prevalence or longer operating times, produce most sound energy. EPA considers that noise standards for highway trucks will result in a quieting of trucks used in construction, and they have therefore chosen air compressors as their first target of construction equipment legislation.

Two different methods have been proposed for controlling the amount of noise on construction sites. One method lays down acceptable noise levels at the boundary of the site or at the nearest residential, or other sensitive, area. This is the "situation-specific" approach mentioned in Chapter 10. The second, or "source-specific," approach involves specifying acceptable noise levels from each of the various items of plant operating on the site. Both approaches are desirable and have been responsible for producing significant reductions in noise levels. These reductions, the acceptable levels, and the noise from the more important items of plant are discussed below.

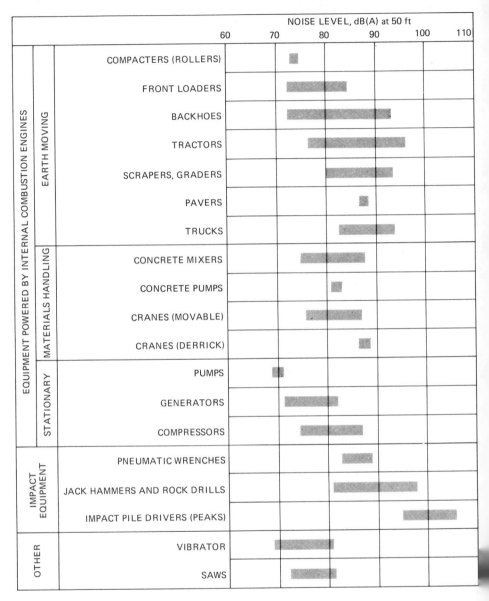

Fig. 9-3 Ranges of noise level for various types of construction equipment [3].

TABLE 9-4. Typical Sound Level and Total Sound Energy from Various Types of Construction Equipment [1].

Construction Equipment	Typical Sound Level, dB(A) at 50 ft.	Estimated Total Sound Energy, kWh/day
1. Dump truck	88	296
2. Portable air compressors	81	147
3. Concrete mixer (truck)	85	111
4. Jackhammer	88	84
5. Scraper	88	79
6. Dozer	87	78
7. Paver	89	75
8. Generator	76	65
9. Pile driver	101	62
10. Rock drill	98	53
11. Pump	76	47
12. Pneumatic tools	85	36
13. Backhoe	85	33

9-2. NOISE FROM INDIVIDUAL TYPES OF EQUIPMENT

The most widely discussed individual items of plant that have received attention to noise are air compressors, rock drills and pavement breakers. It is estimated [4] that in New York City alone there are as many as 5000 air compressors and 20,000 pavement breakers. The noise levels from these and other types of equipment are summarized in Fig. 9-3 and Table 9-4, and we give attention here to their noise sources and their potential for noise reduction.

9-2.1 Air Compressors

The noise of air compressors, which tends to be a function of their air capacity [5], is produced mainly by the engine exhaust, the intake, and the cooling fan [3]. Their noise spectra typically peak at low frequencies and decrease fairly uniformly with increasing frequency. Research in Britain [6] attributes the low frequency noise to the engine exhaust. The cooling fan and structural vibration of the engine and supporting chassis are the likely cause of mid- and higher frequency noise.

The fitting of exhaust mufflers and intake mufflers has effectively reduced the low frequency noise. The designing of more efficient fans and the streamlining of the airflow have helped in reducing fan noise. Chassis and engine struc-

TABLE 9-5. Contribution to Construction Site Noise of Individual Pieces of Equipment on Four Types of Sites in the U.S. [15].

Construction Equipment	Contribution[a] to Construction Site Noise, %			
	Residential	Public Works	Industrial	Nonresidential
Backhoe	5.6	2.2	7.1	3.5
Dozer	10.0	6.8	8.9	4.8
Grader	2.0	1.9	0.3	0.2
Loader	6.3	3.0	4.4	2.5
Paver	2.5	10.8	1.7	0.8
Roller	0.5	1.7	0.2	−
Scraper	3.1	4.8	1.7	1.5
Shovel	2.2	1.0	2.5	1.2
Truck	6.3	21.5	11.3	7.7
Concrete mixer	28.1	10.0	8.9	6.1
Concrete pump	−[b]	−	2.1	2.2
Crane, derrick	−	1.9	1.6	3.1
Crane, mobile	5.6	0.7	1.0	1.9
Air compressor	4.6	6.1	10.0	16.9
Generator	1.8	2.5	1.1	2.5
Pump	1.3	2.7	−	3.5
Paving hammer	0.8	8.5	5.1	2.5
Pile driver	−	−	20.6	24.6
Pneumatic tool	11.3	1.4	6.3	3.1
Rock drill	2.2	13.8	5.1	4.8
Concrete vibrator	4.4	−	0.6	0.4
Saw	−	0.2	0.9	3.1

[a] On an energy basis.
[b] A dash indicates the equipment is not primarily used at the type of site cited or the percent contribution is less than 0.1%.

tural vibration noise has been dealt with by isolating the engine from the chassis and by the fitting of covers over various sections of the engine. The redesign of the engine block is likely to assist further in noise reduction, as is enclosing the engine-compressor combination.

Some 15–20 dB(A) noise reduction is typically possible, at a cost which adds about 20% to the purchase price. The 12 dB(A) average reduction called for by the EPA regulations (see Section 9-3) is estimated to add 12% to the price.

9-2.2 Rock Drills and Pavement Breakers (Jackhammers)

These items are very similar to one another insofar as a reciprocating motion is imparted to a piston by compressed air. The piston is connected to a steel tool, which in turn is used to break or drill such hard materials as rock concrete. The noise levels from these items may change as much as 10 dB(A) depending on the surface being attacked. Soft soil is one extreme and concrete another.

The noise spectra are relatively flat, but tend to peak at both low and high frequencies. The low frequency noise is due to the air exhaust process, while the high frequency noise is due to the ringing of the steel tool. The fitting of exhaust mufflers and the provision of damping on the steel tool have resulted in noise reductions of some 15 dB(A) with a consequent increase in cost of 20–30%.

9-2.3 Pile Driving

Pile driving operations not long ago produced noise levels as high as 100–115 dB(A) at 25 m from the site [7], but recent developments have produced significant noise reductions. These have evolved around quietening the conventional impact hammer method and producing vibratory types of driver. The vibrator approach is very effective in loose and medium density cohesionless soils, particularly where piling extends below the groundwater table. However, it cannot effectively penetrate hard clays, dense cohesionless soils, and rocks.

There has been useful progress in reducing noise from the drop hammer pile driver. Essentially a combined sound absorbent and high transmission loss cladding has been used around the guide which directs the hammer into the pile. Noise levels quoted are of the order of 70 dB(A) at a distance of about 15 m from the site [7]. It is claimed that conversation can be carried out at a few meters from the cladded frame. The cost of this quiet form of piling is some 15% higher than that for the normal drop hammer rig.

9-2.4 Noise from Other Items

The other major sources of noise are generally associated with earthmoving equipment. Such items as bulldozers, scrapers, and trench diggers can produce noise levels of the order of 75–95 dB(A) at 50 ft (15 m). The loudest noise source is usually the exhaust; once this is quieted, the engine, fan, and pumps, and the track are the main sources. There is an awareness of the noise problems of this type of equipment, but they have received less attention than have the problems of air compressor and pavement breaker noise.

9-3. ACCEPTABLE NOISE LEVELS

As mentioned in Section 9-1, two different approaches have been proposed to limit noise from construction sites. The source-specific approach involves specifying acceptable levels from the various items of plant, while the situation-specific one specifies acceptable levels at the site boundary or beyond.

The major advantage of source-specific controls is that the various manufacturers are involved in the noise control process. The manufacturers are in a better position to design quieter equipment than are the contractors. It is obviously more economic for the manufacturers to assume this task than it is for several hundred users.

A difficulty with this approach is that standardized test procedures are required, giving reproducible results which are meaningful in the context of the site operations. Furthermore, even if the levels from individual items are controlled, it is still not possible to guarantee that the noise impact on the community will be at an acceptable level.

The advantage of specifying noise levels at the boundary of the site, or beyond, is that the levels bear a reasonably predictable relationship to the community reaction. Providing acceptable levels and a method of measurement can be agreed upon, the contractor can monitor noise levels and determine his own compliance with regulations. The onus of meeting the regulation lies entirely with the contractor, who obviously must take this factor into account when bidding, as regards increases in both cost and time.

9-3.1 Acceptable Source-Specific Noise Levels

It is important when dealing with noise levels to specify the conditions under which they are to be measured, and the positions of the microphones relative to the noise source.

The measurement procedures which are now becoming accepted in North America are of two types, depending on whether the item is mobile or stationary. Each of these types is discussed below, and an indication is given of currently accepted noise levels.

Mobile Construction Equipment. The Society of Automotive Engineers documents J87 [8] and J88a [9] represent a major effort, which is still continuing, at standardizing test procedures for mobile equipment [10]. These documents are intended to cover such items as track-type tractors, wheel loaders, tractor scrapers, off-highway trucks, mud pumps, excavators, motor graders, etc. Only items whose rated power exceeds 20 bhp are at present considered.

The noise from the equipment operating under two different modes is measured, and the highest of the readings is noted. The modes are "stationary" and "constant-speed moving."

The stationary test involves three different conditions. These are:

1. no load and maximum governed engine speed, with the engine stabilized and component drive systems in neutral;
2. no load, using a cycle "low idle—maximum governed speed (high idle at no load)—low idle," with component drive systems in neutral;

3. no load and maximum governed engine speed, cycling all major components as rapidly as possible.

The constant-speed test requires the equipment generally to be operated in a forward intermediate gear ratio at no load, with the power source operating at

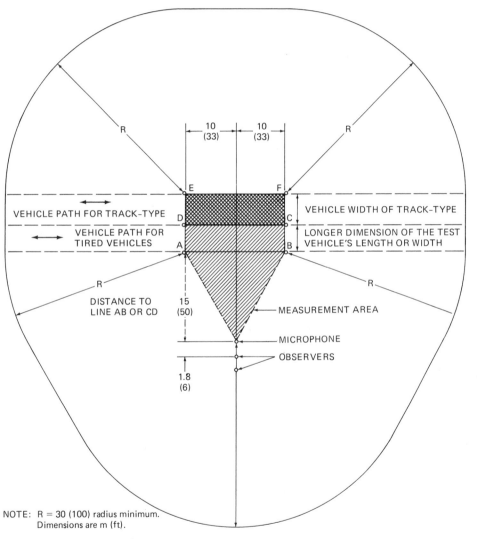

NOTE: R = 30 (100) radius minimum.
       Dimensions are m (ft).

Fig. 9-4 Test site for mobile construction equipment undergoing SAE J88a noise testing [9]. Area ABCD is a hard surface for rubber-tired vehicles; area CDEF is a hard-packed earth area for other (e.g., tracked) equipment. (*Reproduced with acknowledgment to the Society of Automotive Engineers*)

full governor control setting. Machinery with major noise-generating components which could be used while the machine is moving at this speed should have them in operation during the test.

For the stationary test, the machine is located on a hard surface (area ABCD in Fig. 9-4) and measurements are taken at 50 ft (15 m) from the major surfaces of an imaginary box that just drops over the equipment (not including attachment items like buckets and booms), as shown in Fig. 9-5. For the constant-speed test, the machine travels across the hard surface area ABCD in Fig. 9-4 if it has rubber tires. If it has steel wheels, steel drums, or tracks, it passes across the area CDEF, which consists of hard-packed earth. In all cases the microphone is positioned 4 ft (1.2 m) above the ground and is set to read dB(A) on the slow response. The highest sound level noted in the tests is taken to be the sound level of the equipment per this procedure.

SAE J87 [8] recommends a maximum sound level of 88 dB(A) for powered mobile construction equipment of 20–300 rated bhp when tested according to this procedure. The indications are, however, that this level will fall progressively

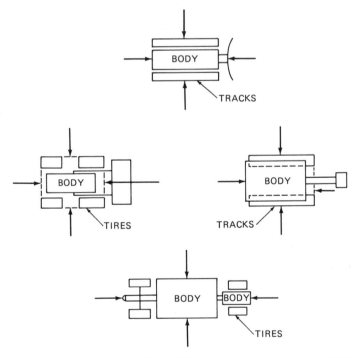

Fig. 9-5 The stationary test in SAE J88a requires measurements 50 ft (15 m) from the major surface outlines as defined above [9]. (*Reproduced with acknowledgement to the Society of Automotive Engineers*)

over a period of a few years to about 80 dB(A) or slightly lower. The Ontario Ministry of the Environment, for example, has proposed levels of 85-88 dB(A) for equipment manufactured in 1976-79, and levels of 83-85 dB(A) from 1980. This applies to excavation equipment, dozers, backhoes, loaders, and similar machinery [11]. In the U.S., the EPA is working on limits to be effective from 1981.

Air Compressors. Until recently, the most comprehensive test procedures for measuring noise from air compressors was that produced by the Compressed Air and Gas Institute [12], known as the CAGI-PNEUROP test method. Essentially ten microphone positions are used, as shown in Fig. 9-6.

Measurements are made 5 ft (1.5 m) above a hard reflecting surface and at distances of 1 m and 7 m from the four major vertical surfaces of the compressor. Two further measurements are made in the direction of maximum noise level, as obtained from a preliminary survey.

A-weighted and octave band sound pressure levels are measured at each position. Three test conditions are specified:

1. compressor operated as for continuous service at its normal rated working pressure;
2. compressor running at idle with the discharge valves from the receiver closed;
3. compressor running at the design full speed under load, delivering its rated output and pressure (the discharge is piped clear of the test area, or silenced).

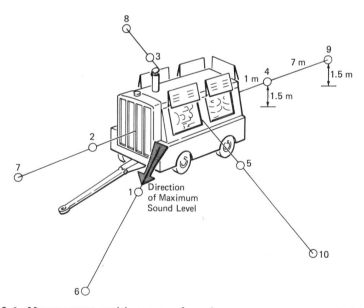

Fig. 9-6   Measurement positions around an air compressor recommended by the Compressed Air and Gas Institute [12].

EPA has specified a modified form of this test in its portable air compressor regulations promulgated in 1976. (The air compressor regulations will later be joined by EPA regulations for loaders and dozers, as well as for rock drills and pavement breakers [13, 14].) This federal legislation will preempt conflicting state and local laws. EPA is careful, however, to point out that extensive participation in construction noise abatement is still possible on the part of states and local authorities: they are able, for example, to set controls on the hours and places or zones in which construction can occur, and on the number of products operated at one time; they can enact legislation that is identical to the federal legislation and thereby share in enforcement; they can also enact nonconflicting legislation governing the noise at receiving points and property lines, or governing the noise output of machinery during operation (see Section 9-3.2). (The federal legislation is aimed at the noise output of individual pieces of equipment at the time of *distribution*.)

Portable air compressors were selected by EPA as the first target for noise abatement on the grounds that they rank with dump trucks and concrete trucks in producing the most sound energy per day. The average noise level of portable air compressors employed at construction sites before the legislation was 88 dB(A) at 7 m. Compliance with the new regulations will reduce this noise by an average of 12 dB(A). On an energy basis, portable air compressors will move from the second most predominant construction site noise, after trucks (see Tables 9-4, 9-5, and 9-6), to the 16th, and contribute only about 1% of construction site sound energy. This will virtually eliminate portable air compressors as a major source of noise [14, 15].

The EPA's air compressor regulations set the trend for future American regulation of construction noise, and we describe in some detail here not only the acoustical but the enforcement aspects.

The measurement procedure adopted is a modified CAGI–PNEUROP one [12]. A-weighted sound levels are measured 7 m from the surface of the compressor,

TABLE 9-6. Contribution of Portable Air Compressor Noise to Construction Site Noise in the U.S. [15].

Site	Contribution to the Construction Site Noise by the Portable Air Compressor,[a] %	Rank at Site[b]
Residential	4.6	7th
Public works	6.1	7th
Industrial	10.0	3rd
Nonresidential	16.9	2nd

[a]On an energy basis.
[b]On an energy basis relative to 20 typical pieces of equipment employed at construction sites.

using the slow sound level meter response, at five microphone locations (to the left, right, front, and back of the machine and above it). The microphone positions for the first four of these locations is 5 ft (1.5 m) above the ground, which should be sealed concrete or asphalt. There should be no other reflecting surfaces within 10 m of a microphone.

The test method differs from the CAGI-PNEUROP method mainly in its inclusion of a microphone location above the compressor. The purpose of this additional microphone is to consider noise received in high-rise buildings adjacent to the site, or at grade when the site is depressed. EPA considered the inclusion of this microphone position to be a guard against compressor design to channel the noise upwards.

The compressor is operated on load at the design full speed, delivering its rated flow and output pressure. The air discharge is provided with a resistive loading such that no significant pressure drop or throttling occurs across the compressor discharge valve. The air discharge itself is ducted clear of the test area, or otherwise silenced.

The sound level reported is an energy average of the sound levels at the five locations, calculated by inserting their values L_a, L_b, L_c, L_d, and L_e in the following equation:

$$\bar{L} = 10 \log \frac{1}{5} \left(\text{antilog} \frac{L_a}{10} + \text{antilog} \frac{L_b}{10} + \cdots + \text{antilog} \frac{L_e}{10} \right)$$

where $\bar{L}$ is the average sound level one wishes to obtain.

This test procedure has similar objectives to those which give the sound power output of the machine. EPA hopes to approve a sound power test method at some time in the future. ANSI S1.23-1976 [16] is such a procedure, and Technical Committee 43 of the International Organization for Standardization is working on one also, as is the National Bureau of Standards. An advantage of a sound power description of noise is that it offers the option of testing in indoor environments, which give good accuracy and allow year-round testing. An advantage of the present procedure, on the other hand, is that it demands little acoustical experience and is similar to the CAGI-PNEUROP procedure already familiar to industry.

Effective January 1, 1978, new compressors with capacities between 75 and 250 ft^3/min shall not produce an average sound level greater than 76 dB(A). The same limit will apply six months later to larger compressors.

The enforcement of these regulations places on the manufacturer the major share of the responsibility for compliance testing. There are three parts to the enforcement strategy: production verification, selective enforcement auditing, and in-use compliance provisions [14].

Production verification is the testing by the manufacturer, or EPA at its discretion, of early production models. The intention is to verify whether a manufacturer has the requisite noise control technology in hand and capably applied to production. Models selected for testing must have been assembled using the manufacturer's normal assembly method and must be units assembled for sale. Models tested must conform with the sound level standard or the manufacturer may be required to cease distribution in commerce (i.e., for sale or hire).

Most of the testing required by the regulations will be performed by the manufacturer at his test site, using his equipment and personnel, although the EPA reserves the right to monitor or perform any tests. Production verification does not involve any formal EPA approval or issuance of certificates, but is nevertheless a mandatory requirement.

The manufacturer must production-verify each model each year, though in some instances this may involve no more than submitting data submitted in previous years. This might occur when there have been no major model changes from one year to the next.

Selective enforcement auditing is testing ordered by the EPA rather than the manufacturer of a statistical sample of production compressors, from a particular compressor category or configuration selected from a particular assembly plant, in order to determine whether production compressors conform to the standards and to provide the basis for further action in the case of nonconformity. EPA is more likely to order such testing when a manufacturer does not undertake this task routinely. If an excessive proportion of a batch of machines is found to be too noisy, EPA may require that every machine of that type be successfully tested before it is distributed.

In-use compliance provisions at present amount to the issuance to purchasers of instructions for proper use and maintenance of the compressor to ensure that they know exactly what is required to minimize the possibility of increases in the noise level. The manufacturer must also issue a list, which EPA has first approved, of the acts which could have a detrimental effect on noise. EPA will then regard any such actions on the part of users as tampering, an offense under federal law. EPA has also announced the possibility that it will amend the regulations to include a requirement that the product be manufactured to meet the required noise level over a period of time corresponding to the product's useful life. (At the moment, data on compressor aging are not sufficient to allow this.)

U.S. cities which have specified acceptable noise levels from portable air compressors include New York [4] and Chicago [17]. The New York level, for air compressors manufactured from January 1, 1976, was a difficult-to-attain 70 dB(A) at a distance of 3 ft (1 m). These and other ordinances are now preempted by the federal legislation.

The proposed levels in Ontario [11] are 75-80 dB(A), depending on compressor size, also measured at 7 m. They are 5-10 dB(A) lower in a "quiet zone."

The projected acceptable levels in several European countries and Japan vary from about 70 to 80 dB(A) at a distance of 7 m [15, 18] (see Fig. 9-7).

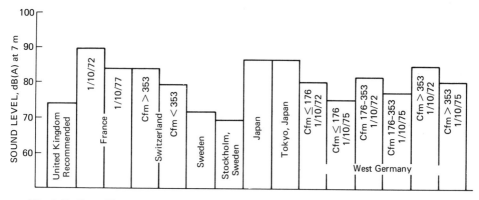

Fig. 9-7   Portable air compressor noise limits in Europe and Japan [15]. Cfm means ft³/min of air flow; some data corrected to 7m; some data corrected for sound level; levels are for any air flow currently available, unless otherwise stated.

Rock Drills and Pavement Breakers. The CAGI–PNEUROP test code [12] includes a measurement procedure for rock drills and pavement breakers. Essentially the device is operated normally and the noise measured at five locations, as shown in Fig. 9-8.

Positions 1 through 4 are 1 m from the machine and in a plane normal to the axis. Position 5 is 1 m from the major outline of the machine and on its major axis. At each position the A-weighted and octave band sound pressure levels are measured.

Legislation dealing with noise from these sources is in preparation at EPA. Meanwhile New York City requires that exhaust mufflers be fitted to pavement breakers manufactured prior to 1974. These mufflers should have a dynamic insertion loss of at least 5 dB(A). Breakers manufactured in 1974 and 1975 must meet an overall noise level of 94 dB(A) at 1 m, while those manufactured since should meet 90 dB(A) [4].

The corresponding values proposed in Ontario are 85–89 dB(A), depending on weight, at a distance of 7 m. A lower limit is applicable in "quiet zones" [11].

Warning Horns. Society of Automotive Engineers SAE J1105 is a document describing the performance and testing of warning horns on mobile construction machinery [19]. Acceptable levels for such horns are not given, but manufacturers and lawmakers would find the document helpful in setting specifications for sound levels, either in procurement or as limits. The requirement that warning horns be audible above the noise of the machine introduces the possibility of specifying a frequency range for horns that lies outside the frequency range of the machine yet in a part of the spectrum that is clearly audible. The document

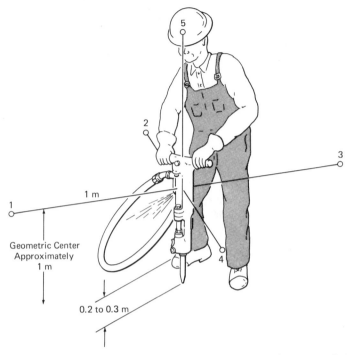

Fig. 9-8 Measurement positions around a pavement breaker or rock drill recommended by the Compressed Air and Gas Institute [12].

attempts to achieve this by requiring that the horn's predominant frequencies should fall within a 710–2800 Hz bandwidth.

Operator Station. Assessment of the generally high noise levels at the operator's station should include consideration of hearing damage (see Chapter 11 and Section 5-2). A test procedure for measuring sound levels at the operator's station on construction and agricultural equipment is SAE J919a [20]. The procedure specifies stationary and moving tests. An assessment of annoyance, speech intereference, or hearing damage risk would require consideration of the duty cycles of the machines and their operators, rather than of a single sound level established by this procedure. The procedure may nevertheless be useful when limits are considered.

9-3.2 Acceptable Situation-Specific Noise Levels

As indicated in Section 9-1, the measurement of noise levels from construction sites is not different in principle from that of industrial sites. A large number of different codes exist for that purpose and in general the techniques used in them can be applied to construction noise. The reader is referred to Chapter 8 and, in particular, to the normalized L_{dn} procedure there described.

One of the earliest references to construction noise sites was made by the Wilson committee [21]. The basis used by this committee was that the noise from construction sites between 7 am and 7 pm should not exceed a level at which conversation in the nearest building would be too difficult with the windows shut. The acceptable level during the day and outside the nearest occupied building was set at 70 dB(A) in rural, suburban, and urban areas away from main traffic and industrial sites. This level was increased to 75 dB(A) in urban areas near main roads and heavy industrial areas.

Acceptable noise levels from construction sites in North America have been set more recently by state and local authorities.

New York City Transit Authority [22] has set a limit of 75 dB(A) for construction between 7 am and 11 pm, and 60 dB(A) at other times.

New York State's Model Local Noise Ordinance [23] limits construction noise as follows:

a. on class A land (essentially residential): at 400 ft from the site boundary, L_{10} must not exceed 70 dB(A) between 7 am and 7 pm, nor 55 dB(A) between 7 pm and 7 am;
b. on class B land (essentially commercial): at 400 ft from the site boundary, L_{10} must not exceed 75 dB(A) during normal business hours, nor 80 dB(A) at other times;
c. on class C land (essentially industrial or agricultural): a hearing conservation criterion applies, essentially amounting to an L_{eq} over 24 hours of 80 dB(A).

The terms L_{10} and L_{eq} are explained in Chapter 1.

The Department of Housing and Urban Development (HUD) in the U.S. has also given some direction regarding acceptable noise levels at construction site boundaries [24], as follows:

Unacceptable: exceeds 80 dB(A) for 60 minutes in 24 hours, or 75 dB(A) for 8 hours in 24 hours;

Normally unacceptable: exceeds 65 dB(A) for 8 hours, or there are loud repetitive sounds;

Normally acceptable: does not exceed 65 dB(A) for 8 hours in 24 hours;

Acceptable: does not exceed 45 dB(A) for 30 minutes in 24 hours.

A study in London, England [25] compared community reaction to construction site noise and traffic noise, and found the former more annoying than traffic noise of comparable levels. A particularly strong reaction was noted in those who considered that, with reasonable effort, much of the construction noise could have been avoided. This is an example of misfeasance (see Chapter 1).

9-4. CONCLUDING COMMENTS

The cost of reducing noise from construction sites should not be taken lightly. Neither should the annoyance caused to those living or working nearby. Control and legislation are obviously highly desirable, but this should not call for unrealistically low noise levels with regard to the currently available technology, for it could not be enforced. Much can be and has been done to identify the main noise sources, and manufacturers have successfully produced worthwhile noise reductions in their products, at a reasonable cost in many cases.

Much can also be done by the contractors to control unnecessary noise without unreasonable hardship. Thus, for example, plant left idling for long periods can be an unnecessary noise source. The use of portable screens and enclosures can produce worthwhile noise reductions. Strategically locating the noisy equipment as far from the boundaries as possible, or to gain the benefit of shielding from other structures, should be considered. Careless loading and unloading should be avoided, and delivery times and routes should be chosen with quiet in mind. Finally the maintenance of equipment, particularly silencers and mufflers, in good working order need hardly be mentioned.

REFERENCES

1. "Identification of Products as Major Sources of Noise." *Federal Register*, **39** (121), 22297–22299, June 21, 1974.

2. F. M. Kessler, "Measurement of Construction Site Noise Levels." Proceedings of International Conference on Noise Control Engineering, Washington, D.C., October, 1972, 52–56.

3. Environmental Protection Agency, "Report to the President and Congress on Noise." Senate document 92-63, February, 1972.

4. R. Gerson and F. C. Hart, "Construction Devices and Construction Site Noises." Proceedings of International Conference on Noise Control Engineering, Washington, D.C., October, 1973, pp. 446–449.

5. B. Dirdal and K. Gjaevenes, Noise from Compressors and Pneumatic Equipment. *Applied Acoustics*, **4**, 23–34, 1971.

6. H. D. Craig, "Noise from Air Compressors and Pneumatic Tools." *Noise Control and Vibration Reduction*, 32–38, January, 1974.

7. "Piling Today." *The Consulting Engineer*, 40–42, November, 1972.

8. "Exterior Sound Level for Powered Mobile Construction Equipment." Society of Automotive Engineers SAE J87, 1972.

9. "Exterior Sound Level Measurement Procedure for Powered Mobile Construction Equipment." Society of Automotive Engineers SAE J88a, 1975.

10. L. D. Bergsten, "Exterior Sound Level Measurement Procedure for Powered Mobile Construction Equipment." Proceedings of International Conference on Noise Control Engineering, Washington, D.C., October, 1973, pp. 237–240.

11. Model Municipal Noise Control By-Law. Ontario Ministry of the Environment, Tables 115-1, -2, -3, May, 1976.

12. "CAGI–PNEUROP Test Code for the Measurement of Sound from Pneumatic Equipment." Compressed Air and Gas Institute, 1969.

13. "EPA Noise Control Program–Progress to Date." Environmental Protection Agency, May, 1976.

14. "Noise Emission Standards for Construction Equipment–Portable Air Compressors." *Federal Register*, **41** (9), January, 14, 1976.

15. "Background Document for Portable Air Compressors." Environmental Protection Agency report No. EPA 550/9-76-004, December, 1975.

16. "Method for the Designation of Sound Power Emitted by Machinery and Equipment." American National Standards Institute ANSI S1.23-1976.

17. Noise Ordinance. City of Chicago Department of Noise Control, 1971.

18. "A Discovery at Bauma." *World Construction*, 41, April, 1973.

19. "Performance, Test and Application Criteria of Electrically Operated Forward Warning Horn for Mobile Construction Machinery." Society of Automotive Engineers SAE J1105, 1975.

20. "Sound Level Measurements at the Operator Station for Agricultural and Construction Equipment." Society of Automotive Engineers SAE J919a, 1971.

21. "Noise–Final Report." HMSO, London, 1963.

22. J. T. O'Neill, "Control of Construction Noise." Proceedings of International Conference on Noise Control Engineering, Washington, D.C., October, 1972, pp. 49–51.

23. Model Local Noise Ordinance. New York State Department of Environmental Conservation, April, 1975.

24. Department of Housing and Urban Development, Circular 1390.2, August, 1971.

25. J. B. Large and J. E. Ludlow, "Community Reaction to Construction Noise." Proceedings of Inter-Noise 75, 1975, pp. 13–20.

10

Noise in and around the home

D. N. MAY*

10-1. INTRODUCTION

While Chapter 6 indicates the sound levels regarded as acceptable in a number of indoor environments which include homes, and other chapters (e.g., 2, 3, 8, and 9) give ways to assess the noise due to sources which may also be experienced in the home, there remain for this chapter to deal with some further sounds which are mainly residential in origin and effect.

Noises in and around the home can be assessed in two ways: in terms of the measured sound exposure occasioned by various appliances and machinery; and in subjective terms covering sound of any origin that disturbs, annoys, is unreasonable, is a nuisance, and so on. Until fairly recently, nearly all regulation of residential noise was accomplished by the latter, subjective type of limitation, but there is now increased effort to obviate the difficulties of subjective controls by setting objective ones.

*Research and Development Division, Ministry of Transportation and Communications, Downsview, Ontario M3M 1J8, Canada.

The many types of residential noise source which cause problems are listed in Table 10-1, a compendium created from numerous miscellaneous reports. By no means all of these sources are regulated, but most are beginning to receive the attention of noise standard committees, legislation drafters, and consumer associations.

TABLE 10-1.   Residential Sounds That Cause Complaints.

Built-in services, major appliances, heavy machinery, vehicles

Furnace	Dishwasher
Plumbing	Washer and spindrier
Doorbell/knocker	Tumble drier
Air conditioner, central	Humidifier, dehumidifier
Air conditioner, room	Refrigerator, freezer
Garage door	Gas fire
Exhaust fan	Alarm, burglar and fire
Telephone	Water fountain
Intercom	Ice cream vendor
Water heater	Vehicle, idling, loading, unloading
Swimming pool filter	Electioneering and other amplified exterior
Heating system pump	noise
Garbage shute, disposer, compactor, truck	Street vacuum
Roadsweeper	Snowplow

Minor appliances and small machinery

Vacuum cleaner	Radio, TV, phonograph, tape player
Floor polisher	Electric organ
Rug shampooer	Saws, rotary, jig, chain etc
Blender	Electric drill
Mixer	Electric sander
Fan	Electric grinder
Fan heater	Typewriter
Can opener	Lawn mower, trimmer
Knife sharpener	Snow blower, thrower
Electric razor	Hedge trimmer
Hair drier	Whistling kettles
Sewing machine	Model airplanes
Clock	Photographic projector
Electric toothbrush	

People and animals

Singing, shouting, "partying"	Door slams, home and car
Dog barks, howls, fights	Hand tools
Musical instruments	Fireworks
Swimming cries, dives	"Children"

10-2.   OBJECTIVE, SITUATION-SPECIFIC LIMITS

Objective limits are generally in the form of a maximum sound level which may be created at, for example, a property line, from sources that are not necessarily specified; or they involve a maximum sound level from a single type of source at a fixed distance.

The first of these approaches, which can be called the situation-specific one, is the most relevant to in-use noise law enforcement, since it applies to the situation actually occurring and complained of. However, it can only be enforced by measurement at the time the noise infringement occurs, not always a simple matter where there is intermittent operation of the source.

The second of these approaches, the source-specific one, is helpful in comparing and purchasing different products for low noise output, and in controlling them at the time of their sale through legislation or labeling requirements. This approach is described in Section 10-3.

10-2.1   Overall U.S. Experience

In its Model Community Noise Control Ordinance [1], the Environmental Protection Agency presents an analysis of the various noise limits for fixed (i.e., non-vehicle) sources, set at residential boundaries in 118 cities. This is reproduced here as Fig. 10-1. The data indicate an average daytime limit of 57 dB(A), and a nighttime limit 5 dB(A) lower. These are, in general, the sound levels that must never be exceeded, rather than the sound levels that must not be exceeded for a certain proportion of the day or night. Some of the individual city ordinances that contributed to this analysis are reviewed in Ref. 2; which also reviews the structure of U.S. noise legislation at other levels.

10-2.2   EPA Model Ordinance

The EPA model ordinance [1] suggests a form of words that may be used in legislation governing noise at residential property lines, as follows. (Note that blanks are left for communities to set their own sound level limits, L_1 and L_2, for the time periods between A am and B pm.)

"No person shall operate or cause to be operated on private property any source of sound in such a manner as to create a sound level which exceeds the limits set forth for the receiving [residential] land use category when measured at or within the property boundary of the receiving use.

$$A \text{ am–} B \text{ pm} \quad L_1 \text{ dB(A)}$$

$$B \text{ pm–} A \text{ am} \quad L_2 \text{ dB(A)}$$

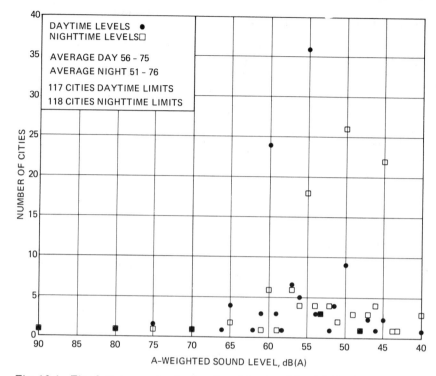

Fig. 10-1 **Fixed source noise levels allowable at residential district boundaries in U.S. Cities [1].**

For any source of sound which emits a pure tone or impulsive sound, the maximum sound level limits set forth [above] shall be reduced by __dB(A)."

Exemptions are construction noise, aircraft and airport noise, explosives firearms and similar devices, motorboats, domestic power tools, refuse collection vehicles, recreational motorized vehicles operating off public rights of way, unamplified human voices, interstate trains, and farming activities. (Many of these sources are covered elsewhere in the model ordinance.)

The EPA model ordinance was prepared in cooperation with the National Institute of Municipal Law Officers (NIMLO). It is not, in itself, an ordinance. However it is available for communities to enact as such, and it also constitutes a body of learned opinion as evidence in civil suits. A comparison of this model ordinance with an earlier NIMLO model ordinance (which is reproduced in Ref. 2) indicates the current tendency to prepare noise legislation with sound level limits rather than subjective limits.

The ordinance suggests that communities enacting the model ordinance could consider the use of sound levels cited in the EPA levels document [3].

10-2.3   EPA Residential Noise Criteria

The abovementioned levels document, which is fully titled "Information on Levels of Environmental Noise Requisite to Protect Public Health and Welfare with an Adequate Margin of Safety" [3], cites the following yearly determined day–night sound levels as criteria for residential land uses:

	indoor	*outdoor*
Residential with outside space, and farm residences	$L_{dn} = 45$ dB(A)	$L_{dn} = 55$ dB(A)
Residential with no outside space	$L_{dn} = 45$ dB(A)	—

These criteria are based on activity interference considerations.

10-2.4   New York State Model Ordinance

The State of New York has a Model Local Noise Ordinance [4] which recommends sound level limits for air-conditioners and air handling devices in residentially zoned areas, as follows:

a. for sounds crossing a property line: 55 dB(A) at any point, 50 dB(A) in the center of a neighboring patio, 50 dB(A) outside a living area window no more than 3 ft from the opening;

b. where multiple dwellings or apartments are involved: 50 dB(A) outside a living area window no more than 3 ft from the opening.

However, these provisions do not apply if operation of the device raises the background sound level less than 5 dB(A).

10-2.5   Ambient U.S. Levels

Any limits should be compatible with the existing ambient levels if they are to be realistic. Limits too far above the ambient will usually be unhelpfully high; those significantly below the ambient will not contribute to increased peace. Table 10-2, taken from Ref. 3, describes the outdoor L_{dn} estimated to exist in U.S. urban areas.

10-2.6   Canadian Limits

In Canada, the Province of Ontario has published a Model Municipal Noise Control By-Law [5] which recommends that the L_{eq} measured outdoors on residential property in any hour should not exceed by more than 6 dB(A) the L_{eq} caused by ground transportation at that point in that hour. The L_{eq} due to ground transportation may be estimated from a knowledge of traffic volume,

TABLE 10-2. Estimated Percentages of U.S. Urban Population (134 Million) Residing in Areas with Various Day–Night Noise Levels Together with Customary Qualitative Description of the Area [3].

Description	Typical Range L_{dn}, dB(A)	Average L_{dn}, dB(A)	Estimated Percentage of Urban Population	Average Census Tract Population Density, Number of People Per Square Mile
Quiet surburban residential	48–52	50	12	630
Normal suburban residential	53–57	55	21	2,000
Urban residential	58–62	60	28	6,300
Noisy urban residential	63–67	65	19	20,000
Very noisy urban residential	68–72	70	7	63,000

distance, and time of day according to a procedure set out in the bylaw. The reasoning is that the traffic noise usually establishes the background level, which—as in the New York ordinance (see the last paragraph of Section 10-2.4)—should not be unduly exceeded.

10-3. OBJECTIVE, SOURCE-SPECIFIC LIMITS

Source-specific sound levels are those which refer to the sounds of sources, usually removed from the acoustically complex situations they operate in in real life. The sound levels may, for example, be measured at specified distances from the source when located in an anechoic or reverberant room. The same procedure could be followed for several different sources in turn, the results allowing their respective sound levels to be easily compared, something which may not be possible in their differing real-life situations.

10-3.1 Measurement Procedures

There are numerous proposed and accepted measurement procedures to describe the sound output of equipment used in the home [9–23]. We paraphrase here the American National Standards Institute S3.17 "Method for Rating the Sound Power Spectra of Small Stationary Noise Sources" [21], which is a new standard designed to give a single number rating (simple enough, therefore, to be easily understood by the consumer) in describing the noise of home equipment like vacuum cleaners, electric mixers, fans, blenders, dishwashers, air-conditioners, hedge clippers, lawn mowers, and so on.

ANSI S3.17 gives a Product Noise Rating (PNR) as follows. In a reverberant room, or in a free field over a reflecting plane (to simulate the floor or ground), sound pressure levels in one-third octave bands are measured, and the sound power level in each band is calculated. (A good explanatory reference to sound power is [24]; the underlying idea is to take measurements that report the source's total acoustic output, which is independent of its directivity pattern, thereby preventing a manufacturer from reporting the sound obtained at just a single favorable position near his product.)

Because the sound power is essentially independent of the distance from the source, its decibel value unfortunately has little meaning to people used to sound levels at given distances from sources. The standard therefore provides a method for obtaining a sound level applicable to reference distances of 1 m (for indoor equipment, and for outdoor equipment tended by an operator) and 4 m (for unattended outdoor equipment). This is done by subtracting 8 dB from the sound power levels in each one-third octave band when 1 m distance applies, and 20 dB when 4 m apply. The levels then obtained are A-weighted, and summated to give PNR. In rough language, PNR is therefore a sound level at a given distance, averaged in space to smooth out the source's directivity pattern.

The standard notes that for rooms with sound characteristics similar to those in typical homes, the PNR obtained for a distance of 1 m will approximate to the sound level actually measured in such a room at a distance from the source of 1 m or greater. Reference 25 is a study of the acoustics of domestic rooms, and Chapter 6 may also be useful in this regard. Reference 26 compares home appliance directivity on reflecting and carpeted floors, which is helpful in translating the source-specific measurements of some appliances to real situations.

10-3.2   Sound Level Limits

No source-specific sound level limits appear to have been set, to date, for the home appliances here described, but this will logically follow acceptance of the measurement standards mentioned in Section 10-3.1. In the U.S., the Association of Home Appliance Manufacturers is working on standards for products such as can openers, food mixers, blenders, dishwashers, clothes washers and driers, refrigerators and freezers, and garbage compactors. The Air Conditioning and Refrigeration Institute publishes sound ratings of air conditioners, which while not constituting a formal limit has the effect of drawing attention to those which are exceptionally noisy (see also Section 10-3.3).

A number of source-specific limits exist for sources other than home appliances.

The State of New York's Model Local Noise Ordinance [4] proposes a maximum sound level from a refuse compacting vehicle of 70 dB(A) at a distance of

10 ft from the compacting unit. It also proposes prohibition of the sale or lease of the following equipment when creating more than the stated sound levels at 50 ft:

Engine-powered equipment of less than 5 rated
bhp for frequent use in residential areas—e.g., lawn } 70 dB(A)
mowers, small garden tools, snowblowers

As above, exceeding 5 hp   } 78 dB(A)

These New York levels are identical to those recommended in SAE J952b [22]. In addition, SAE J1046 [23] cites levels of 70 dB(A) (for rotary mowers), 75 dB(A) (for riding mowers), and 78 dB(A) (for snowblowers and tillers), measured at 50 ft, as representing 1973 "engineering practice" for equipment under 20 bhp.

The Chicago Noise Ordinance [8] limits sound levels as follows, also at 50 ft:

Powered commercial equipment of 20 hp or less,
used infrequently in residential areas—e.g., chain } 88 dB(A)
saws, powered hand tools

Powered equipment for repetitive use in residential
areas—e.g., lawn mowers, small garden tools, } 74 dB(A)
snowblowers

These limits were scheduled to be lowered at various future dates.

10-3.3   Sound Levels Established by Custom

Where no sound level standards are legislated or recommended, they can nevertheless become established by custom. Reference 6 presents sound level measurements of several examples of a number of different types of appliance, measured 3 ft from each at a height of 5 ft, a procedure which, though involving measurements in a number of different homes, is essentially source-specific because 3 ft is close enough to lie within the direct sound field. These measurements are extensive enough to indicate the distribution of sound levels found for each type of appliance quoted. They therefore indicate which sound levels can be regarded as above average, which are below average, and which are the quietest and the noisiest. As well, they allow one to consider buying a category of appliance that is quieter than another (e.g., blenders are generally noisier than mixers). These data are reproduced here as Fig. 10-2. From the same reference we have excerpted the sound levels regarded as typical of various types of outdoor source at 50 ft (see Table 10-3).

In a British context, Jackson and Leventhall [7] present appliance sound levels measured in situ in acoustically simulated domestic rooms, which are given in Table 10-4.

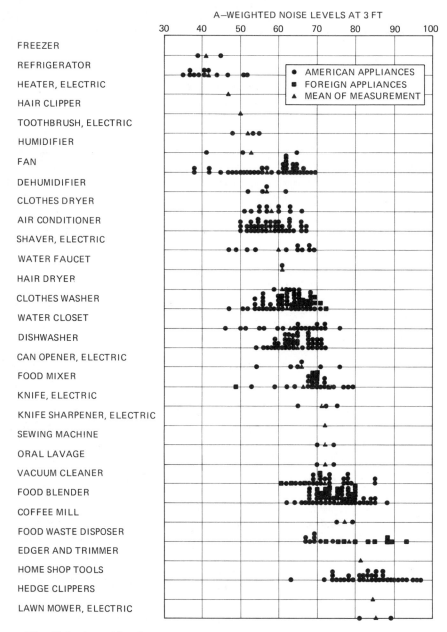

Fig. 10-2 Sound levels from indoor sources in mainly U.S. homes at 3 ft [6].

TABLE 10-3. **Typical Sound Levels from Sources Outdoors at Home [6] .**

Source	Typical A-Weighted Noise Level at 50 ft, dB(A)
Lawn mowers	74
Garden tractors	78
Chain saws	82
Snowblowers	84
Lawn edgers	78
Model aircraft	78
Leaf blowers	76
Generators	71
Tillers	70

Sound levels reported in this way leave unresolved the difficulty of deciding whether it is a bystander one wishes to protect, or the user. The bystander presumably gains less benefit from the source than the user, whose gain may be considerable but whose exposure is greater. In the case of the bystander, the appropriate listening point is likely to be in the reverberant field, which introduces to the question the nature of the room acoustics (see also Section 10-3.1), or it may be in another room altogether, which adds the question of transmission loss. The problem is a little simpler in the case of the user, whose location relative to the source is essentially fixed. Table 10-5 presents data given in Ref. 6 of typical user sound levels, as well as some levels in adjacent rooms.

Section 10-5 discusses the desirability of setting limits for home appliance noise.

10-4. SUBJECTIVE REGULATION OF RESIDENTIAL NOISE

Most residential noise regulation is by ordinances and bylaws which contain few if any objective sound level limits, but instead prohibit noise which disturbs, is unnecessary, is a nuisance, causes annoyance to people of normal sensibilities, is clearly audible after certain hours, and so on. The laws themselves are far too numerous to quote; the most exhaustive summaries of them are available in Refs. 2 and 27–32. In this chapter, we will make use of some of the more influential laws or models [1, 4, 5, 8, 33] to describe the way in which these subjective controls are typically framed. Choosing these particular examples does

TABLE 10-4. Sound Levels from Indoor Sources in British Homes at the Distances Stated [7].

Noise Source	Sound Level, dB(A)			Distance, cm
	Minimum	Average	Maximum	
Food mixers: slow	59	66	71	⎫
medium	62	72	83	
fast	67	77	85	
Food mixer ⎫ slow	57	62	66	⎬ 100
Liquidiser ⎬ medium	70	73	75	
Attachments ⎭ fast	75	78	81	
Purpose-built liquidisers	87	89	90	⎭
Whistling kettles	69	81	93	⎫
Washing machines: washing	54	66	74	
drying	64	72	78	
Hot-air tumble-drier	–	63	–	
Spin driers	69	72	74	
Extractor fans	56	59	60	
Dishwasher	–	71	–	
Waste disposal unit	–	67	–	
Gas cookers	37	44	54	⎬ 150
Gas fires (full on)	28	34	42	
Gas water heaters (wall mounted)	59	63	66	
Vacuum cleaners	67	77	83	
Fan heaters: fan only	41	46	53	
1kW	37	46	53	
2kW	41	47	53	
3kW	47	49	51	⎭
Hair driers: hot	65	71	78	⎫ 10
cold	63	70	79	⎭
Electric tooth brush	–	60	–	⎫ 7.5
Electric razors: rotary	75	80	83	⎬
shuttle	64	68	71	⎭
Flush toilets: high-level	80	82	85	⎫ not stated
low-level	73	76	82	⎭

TABLE 10-5. Sound Levels from Sources in U.S. Homes [6].

Noise Source	Level of Operator Exposure, dB(A)	Level of Exposure of People in Other Rooms, dB(A)
Group I: Quiet major equipment and appliances		
Refrigerator	40	32
Freezer	41	33
Electric heater	44	37
Humidifier	50	43
Floor fan	51	44
Dehumidifier	52	45
Window fan	54	47
Clothes dryer	55	48
Air conditioner	55	48
Group II: Quiet equipment and small appliances		
Hair clipper	60	40
Clothes washer	60	52
Stove hood exhaust fan	61	53
Electric toothbrush	62	42
Water closet	62	54
Dishwasher	64	56
Electric can opener	64	56
Food mixer	65	57
Hair dryer	66	51
Faucet	66	51
Vacuum cleaner	67	60
Electric knife	68	60
Group III: Noisy small appliances		
Electric knife sharpener	70	62
Sewing machine	70	62
Oral lavage	72	62
Food blender	73	65
Electric shaver	75	52
Electric lawn mower	75	55
Food disposal (grinder)	76	68
Group IV: Noisy electric tools		
Electric edger and trimmer	81	61
Hedge clippers	84	64
Home shop tools	85	75

not imply that the noise ordinances of other communities are necessarily less progressive.

10-4.1   General Prohibitions

A general prohibition, intended as a catchall for noise which escapes more specific exclusion, is common in ordinances and bylaws. For example the EPA recommends [1] that "no person shall unreasonably make, continue, or cause to be made or continued, any noise disturbance."

10-4.2   Vocal Sounds

Human and animal vocal sounds are usually also limited by general prohibitions. The EPA [1] outlaws "owning, possessing or harboring any animal or bird which frequently or for continued duration, howls, barks, meows, squawks, or makes other sounds which create a noise disturbance across a residential property line...."

Occasionally, this type of prohibition applies to a certain time of day. For example a venerable British model bylaw [33] requires that "no person shall ... between the hours of 11 pm and 6 am wantonly and continuously shout or sing or otherwise make any loud noise to the annoyance or disturbance of residents."

10-4.3   Prohibition by Zone

Some ordinances provide for an officer concerned with the administration of the ordinance to zone the community according to the areas of noise sensitivity and to altogether prohibit incompatible activities. For example, Ontario [5] suggests that powered models, fireworks, and fire-arms should not be used in a "quiet zone." Other ordinances define minimum distances from, e.g., a hospital, for certain activities.

10-4.4   Prohibition by Time of Day

The prohibition of certain activities by time of day is widely recognized as a valid way of disallowing sounds which are reasonable at one time but not at others. For example the EPA [1] outlaws "operating or permitting the operation of any mechanically powered saw, drill, sander, grinder, lawn or garden tool, snowblower or similar device used outdoors in residential areas between the hours of __ pm and __ am on the following day so as to cause a noise disturbance across a residential real property boundary." (The community is left to set the hours it wishes.)

10-4.5   Prohibition by Duration of Sound

New York State [4] recommends the prohibition of burglar alarms that do not automatically terminate operation within 15 minutes. The prohibition of inordinately long idling of vehicles is another typical control of this type.

10-4.6   Prohibition by Type or Purpose of Activity

Chicago [8] outlaws factory steam whistles when used to signal the beginning or end of work or for any other purpose except an emergency. New York State [4] disfavors the use of emergency warning devices except in emergency or when under test.

10-4.7   Exclusions

Typical exclusions from the terms of ordinances are noise associated with emergencies, and events for which special permits are obtained, such as parades, firework displays, and sports events.

10-5.   SOME PERSPECTIVES

Laws against excessive noise are only useful if there is reasonable success with their enforcement. Experience of this to date [2, 27, 28, 32, 34–48] has not been very satisfactory, which must to some extent reflect upon the adequacy of the laws themselves. In this section are given some perspectives on the scope and structure of suitable legislation, taken mostly from personal experience in Ontario's community noise enforcement.

10-5.1   Objective, Situation-Specific Limits

These should be kept to a minimum to cover what source-specific controls cannot; in a sense the source-specific limits are preventive while situation-specific ones are cures.

Enforcing situation-specific limits is expensive. Because the circumstances are in no way controlled, adequately documenting infractions often requires repeated visits, particularly in the case of intermittent noises (and those continuous noises which the noise producer chooses to switch off when the noise inspector tries to measure them).

Other practical difficulties are: separating the source one is measuring from the ambient sound level and from other problem sources, especially when the source cannot be switched off; deciding how to define a noise excess, either with reference to the actual ambient level (how defined?) or an independently stated limit which balances idealism with realism in taking account of the difficulties of quieting particular sources; accounting for beats, tones, intermittency, and impulsiveness in unambiguous ways, yet ways not requiring extensive (and therefore expensive) measurement; drafting acoustically and morally sensible legislation to deal with difficult property line positions, such as between narrowly separated buildings, on adjoining balconies, between parts of the same building, and between different buildings in commonly owned properties (e.g., condominium townhouses)—all very common noise problem locations.

To face these problems properly when drafting legislation demands extensive field experience in enforcement, something few communities have brought to their legislation. The expense of enforcement should be accepted at the time the legislation is enacted; only communities who wish to act decisively against excessive noise, and enduringly so, should enact comprehensive legislation in this category.

10-5.2   Objective, Source-Specific Limits

For most powered noise sources which can be heard beyond the property line of their users, the best legislation is that which can be applied at the manufacturing level to the various sources isolated from their many and varied real-life installations. This has the advantage of requiring enforcement at relatively few points—from manufacturers, inherently the party unable to afford not to comply.

Thus lawnmower noise, for example, should be enforced by requiring all new lawnmowers sold to conform to a source-specific limit, rather than a situation-specific one requiring a community noise inspector to measure at neighborhood property lines on Sunday mornings. The main disadvantage of source-specific limits of this type are that it takes time for existing products, not covered by the legislation, to go out of service. A gap which must also be covered is in preventing maintenance shortcomings which result in increased noise. All things considered, however, this type of control is probably reasonably cost-effective and result-oriented.

As far as possible, uniformity is required, which demands that higher levels of government assume the responsibility.

A type of noise source which should probably not be the subject of this type of legislation are indoor home appliances. Here free market pressures would probably take due account of public concern for noise if a noise labeling requirement were instituted. This would give the consumer sufficient information to balance his interest in quiet against his interest in price (quiet costs money), speed of operation (quiet may mean slower operation, and therefore a longer noise exposure time), and other factors like appearance and convenience. The requirement therefore is for information rather than prohibition.

10-5.3 Subjective Controls

Legislation of noise excess in qualitative terms is invariably difficult to enforce, even though this type of control is the only one presently possible in cases of "people and animal noise" (where noise measurement is impracticable), e.g., door slams, dog barks, and music practice. The problem is not so much the description of the offence (a difficult task that receives much attention) as the willingness to prosecute and to apply meaningful penalties in circumstances where there is only subjective judgment to rely on.

The obstacle is the seriousness with which the community regards this type of offence in the first place. Thus in one sense the community gets the amount of enforcement it really wants. However, an improvement which is probably called for is to allow private citizens to themselves lay charges under ordinances and bylaws. Of course, this would be subject to the common law requirement that their prosecutions not be malicious.

REFERENCES

1. U.S. Environmental Protection Agency and the National Institute of Municipal Law Officers, Model Community Noise Control Ordinance. EPA Report No. 550/9-76-003, September, 1975.

2. "Laws and Regulatory Schemes for Noise Abatement." Report prepared by George Washington University for the U.S. Environmental Protection Agency. NTID300.4, February, 1971.

3. "Information on Levels of Environmental Noise Requisite to Protect Public Health and Welfare with an Adequate Margin of Safety." U.S. Environmental Protection Agency Report No. 550/9-74-004, March, 1974.

4. Model Local Noise Ordinance. In *Local Noise Ordinance Handbook*, New York State Department of Environmental Conservation, April, 1975.

5. Model Municipal Noise Control By-Law. Ontario Ministry of the Environment, May, 1976.

6. "Report to the President and Congress on Noise." U.S. Environmental Protection Agency, Document No. 92-63, February, 1972.

7. G. M. Jackson and H. G. Leventhall, "Household Appliance Noise." *Applied Acoustics*, 8, 101–118, 1975.

8. Noise Ordinance. City of Chicago Department of Noise Control, 1971.

9. "American National Standard for the Physical Measurement of Sound." American National Standards Institute, ANSI S1.2-1971.

10. "Methods of Testing for Sound Rating Heating Refrigeration and Air-Conditioning Equipment." American Society of Heating, Refrigerating and Air-Conditioning Engineers, ASHRAE 36-72.

11. "Measurement of Sound Power Radiated from Heating, Refrigerating and Air-Conditioning Equipment." American Society of Heating, Refrigerating and Air-Conditioning Engineers, ASHRAE 36-62.

12. "Method of Publishing Sound Ratings for Air Moving Devices." Air Moving and Conditioning Association, AMCA 301-65.

13. H. G. Leventhall and G. M. Jackson, "Assessment of the Noise of Domestic Appliances." Paper 20N2, 7th International Congress on Acoustics, Budapest, 1971.

14. H. G. Leventhall and G. M. Jackson, "The Assessment of the Noise from Domestic Gas Appliances." *Noise Control*, November, 1971.

15. "Sound Rating of Outdoor Unitary Equipment." Air Conditioning and Refrigeration Institute, Standard 270-67.

16. J. R. Schreiner, "Air-Conditioner Sound Rating and Certification." *Noise/News*, 1, 2, 1972.

17. "Room Air Conditioner Sound Rating." Association of Home Appliance Manufacturers, Standard RAC-25R.

18. R. W. Ramsey, "Sound Rating of Outdoor Unitary Air-Conditioning Equipment." Paper presented at 80th meeting of the Acoustical Society of America, Houston, Texas, November 3–6, 1970.

19. H. G. Leventhall, "Determination of Sound Power Levels Through Measurements of Sound Pressure Levels at Prescribed Points Enclosing the Source at a Constant Measuring Distance, and Conforming with the External Shape of the Source." Chelsea College, University of London, 1975.

20. "Methods for the Determination of Sound Power Levels of Small Sources in Reverberation Rooms." American National Standards Institute, ANSI S1.21-1972.

21. "Method for Rating the Sound Power Spectra of Small Stationary Noise Sources." American National Standards Institute, ANSI S3.17-1975.

22. "Sound Levels for Engine Powered Equipment." Society of Automotive Engineers, SAE J952b, 1973.

23. "Exterior Sound Level Measurement Procedure for Small Engine Powered Equipment." Society of Automotive Engineers, SAE J1046, 1974.

24. A. P. G. Peterson and E. E. Gross Jr, *Handbook of Noise Measurement*. GenRad Company, Concord, Mass., 1972.

25. G. M. Jackson and H. G. Leventhall; "The Acoustics of Domestic Rooms." *Applied Acoustics*, 5, 265–277, 1972.

26. K. E. Jeatt and H. G. Leventhall, "Noise Radiation Characteristics of Some Domestic Appliances." Paper 73/03, British Acoustical Society meeting, Chelsea College, University of London, January 4, 1973.

27. C. R. Bragdon "Municipal noise ordinances: 1975." *Sound and Vibration*, 9 (12), 25–30, 1975. Also "Model Noise Ordinances—Model Politics?"—same issue, p. 9.

28. C. R. Bragdon, "Municipal Noise Ordinances." *Sound and Vibration*, 7 (12), 16–22, 1973. Also "City Noise Ordinances—A Status Report"—same issue, pp. 34-35.

29. D. N. May, ed., "A Summary of North American Noise By-Laws." Ontario Ministry of the Environment, November, 1974.

30. "A Survey of Community Noise By-Laws in Canada." Sonic Research Studio, Simon Fraser University, published by Labatt Breweries of Canada, 1972.

31. "Noise Source Regulations in State and Local Noise Ordinances." U.S. Environmental Protection Agency, NTID73.1, March, 1973.

32. "State and Municipal Non-Occupational Noise Abatement Programs." U.S. Environmental Protection Agency, NTID300.8, December, 1971.

33. "Model Byelaws Relating to Noise in Streets and Public Places." Appendix XVI in Committee on the Problem of Noise, "Noise—Final Report," HMSO, London, 1963.

34. D. N. May, ed., "A Summary of Noise Complaint Statistics in Ontario and Elsewhere in Canada." Ontario Ministry of the Environment, November, 1974.

35. "Toward a Quieter City: Report of the Mayor's Task Force on Noise Control." New York City Mayor's Task Force on Noise Control, New York, 1970.

36. H. G. Poertner, "Requirements for Community Noise Control Programs." Purdue Noise Control Conference, 1971 Proceedings, pp. 257-262.

37. H. B. Karplus, "Noise Control Through Legislation." Purdue Noise Control Conference, 1971 Proceedings, pp. 242-245.

38. R. Taylor, "Noise Regulation in Britain." Inter-Noise 72 Proceedings, pp. 28–30.

39. J. T. Howe, "A Balanced Approach—A Lawyer's Viewpoint." International Conference on Transportation and the Environment, 1972 Proceedings, pp. 246–255.

40. J. T. Kaufman, "Control of Noise Through Laws and Regulation." In *Noise as a Public Health Hazard*, American Speech and Hearing Association, 1969, pp. 327–341.

41. C. R. Bragdon, "Guidelines for the Preparation of a Model Noise Ordinance." Inter-Noise 72 Proceedings, pp. 41-43.

42. H. M. Frederikson, "Noise Control on the Local Level." *Archives on Environmental Health*, **20**(5), 651–654, 1970.

43. J. M. Tyler, L. V. Hinton, and J. G. Olin, "State Standards, Regulations, and Responsibilities in Noise Pollution Control" *J. Air Pollution Control Assoc.*, **24**(2), 130-135, 1974.

44. L. F. Yerges, "The Uses and Abuses of Codes and Standards." *Sound and Vibration*, **8** (4), 12–15, 1974.

45. J. S. Moore, "Developing a State-Wide Noise Control Program, the Illinois Experience." Inter-Noise 72 Proceedings, pp. 19-22.

46. V. T. Coates, "Experience of Local and State Governments in Control of Environmental Noise." Inter-Noise 72 Proceedings, pp. 13-17.

47. L. H. Royster and C. E. Scott, "Results of a Survey of Municipal Noise Ordinances in North Carolina." Noise-Con 73 Proceedings, pp. 33-35.

48. C. Caccavari, "Three Community Noise Programs." *Sound and Vibration*, **7**(5), 42–48, 1973.

PART 2

Physical effects assessment

11

Occupational deafness and hearing conservation

A. M. MARTIN*
and
J. G. WALKER*

11.1. INTRODUCTION

Technology has created many environmental pollutants, of
which noise is an immediate and identifiable example. Many
industrial processes since the industrial revolution have generated noise of sufficient sound level to cause deafness. Thus
occupational hearing loss is not a new phenomenon; for example, "boilermakers' deafness" was well known among
foundry workers in the nineteenth century, and gunpowder
explosions are reported to have caused deafness among gun
crews in the Battle of Trafalgar.

*Hearing Conservation Unit, Institute of Sound and Vibration Re-
search, The University, Southampton SO9 5NH, England.

251

Even though it has long been recognized, occupational hearing loss has generally been accepted until recent times as part of the price to be paid for employment and technological progress. Fortunately, a growing awareness of the problem in industry, changing attitudes to employment, and the availability of noise reduction techniques and hearing protectors have resulted in both the quantification of the noise hazard and the reduction of the risk of hearing damage in many instances. Also, the availability of modern electronic apparatus for measuring noise and hearing sensitivity has facilitated the quantitative assessment of damage to hearing caused by noise.

This chapter is concerned with the general subject of hearing conservation and describes the physiology of the ear and the effects of noise-induced hearing loss. The evaluation of the hazard to hearing from noise and its prevention is outlined together with the general requirements of a hearing conservation program.

11-2. THE EAR

11-2.1 General Description

The transmission of sound through the ear and the mechanism of hearing is described briefly in this section. As is illustrated in Fig. 11-1, the ear has three parts, which are generally called the outer, middle, and inner ears. The ear canal (auditory meatus) extends from the pinna to the eardrum (tympanic membrane). The canal itself is approximately oval in cross section and extends for about one inch into the head.

The middle ear is an air-filled cavity which contains a chain of three small,

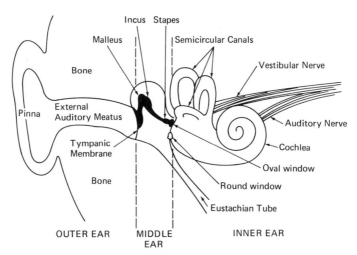

Fig. 11-1 The main components of the ear.

movable bones called the auditory ossicles (the malleus, which is partly within the ear drum, the incus, and the stapes) as well as two minute muscles which may have some importance in helping to protect the ear from intense sounds. The Eustachian tube connects the middle-ear cavity to the pharynx and enables the equalization of air pressure in the middle ear with that in the external ear.

The foot of the stapes is located in the oval window leading to the inner ear. The inner ear is a complex system of fluid-filled cavities positioned deep in the temporal bone of the skull so that it is physically well protected. The system of cavities includes the cochlea and three semicircular canals oriented at right angles to one another. These canals form part of the body's balancing mechanism and need not concern us further.

The cochlea transduces the mechanical vibrations of the ossicles into neural impulses. It is extremely small (about $\frac{1}{4}$ in. in diameter) and similar in shape to a snail's shell, having about $2\frac{3}{4}$ turns coiled round a central, bony pillar, which also carries the nerve fibers from the cochlea to the brain.

The cochlea is split longitudinally into two fluid-filled sections by the cochlea partition, which is itself a tube filled with another fluid. This is illustrated in Fig. 11-2, which is a simplified cross section of the cochlea. The sections of the cochlea are clearly shown. The lower part of the partition is formed by a flexible membrane called the basilar membrane, which supports the hair cells, as illustrated in Fig. 11-3. There are about 30,000 hair cells in the ear. They are the sensors which initiate neural impulses to the brain to produce the sensation of hearing. Immediately above the hair cells is the tectorial membrane; the hairs of the hair cells are probably attached to its underside. The upper part of the cochlear partition is called Reissner's membrane.

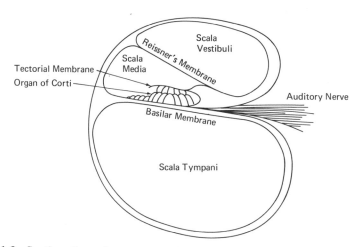

Fig. 11-2 Section through one turn of the cochlea showing its three sections. The basilar membrane, tectorial membrane, and hair cells are not shown in detail.

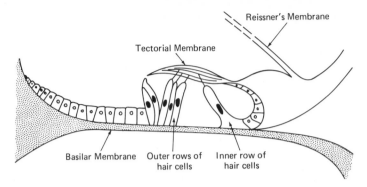

Fig. 11-3 Diagram of the hair cells on the basilar membrane, together with the tectorial membrane.

11-2.2   The Mechanism of Hearing

The chain of events that leads to the sensation of hearing starts with the passage of sound waves down the ear canal. The eardrum vibrates when these pressure fluctuations impinge on it. The movement of the eardrum is transmitted by the ossicles to the fluid in the cochlea via the oval window. There is a small hole called the helicotrema in the cochlea partition at the apex of the cochlea, and another small flexible membrane called the round window in the lower part of the cochlea. These enable the fluid motion to take place. This is best illustrated if the cochlea is considered as a straight tube rather than the spiral it is. The round window and the helicotrema are shown in Fig. 11-4. The basilar membrane vibrates as a result of the fluid motion and the relative motion of the basilar membrane and tectorial membrane bends the hairs of the hair cells. Move-

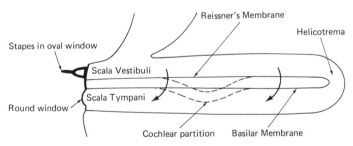

Fig. 11-4 Simplified illustration of the cochlea considered as a straight tube. A positive pressure wave moves the foot of the stapes into the oval window, causing the fluid to move and to bend the cochlear partition. The fluid moves through the helicotrema and displaces the membrane in the oval window. The dashed lines show in an exaggerated manner the deflection caused by a low frequency sound wave.

ment of the hairs causes electrochemical changes to take place in the hair cells; these changes are transmitted as bursts of neural impulses along the auditory nerve to the brain. The ability of the ear to discriminate between different sounds is understood to result partly from analysis of the sound in the cochlea and partly from analysis in the brain.

11-2.3 Hearing Sensitivity

The ear of a normal, healthy young adult is generally regarded as being sensitive to sounds having frequencies in the range from 20 Hz to 20 kHz. The ear is not equally sensitive over this frequency range, the sensitivity being greatest in the 1 kHz to 4 kHz region.

The sensitivity of the ear is normally described in terms of the minimum sound that is audible. This is called the threshold of hearing. Threshold determinations fall into two main categories, free-field measurements and pressure (earphone) measurements. The former are described in terms of the intensity of the sound field in which the observer is placed; the intensity of threshold is called the minimum audible field (MAF). The latter are measures of the sound pressure generated at the entrance to the auditory meatus when an earphone is placed on the ear; the sound pressure at threshold is called the minimum audible pressure (MAP). The majority of hearing sensitivity measurements are of the MAP type, although in circumstances such as the measurement of hearing protector attenuation MAF measurements must be made. Figure 11-5 illustrates the binaural MAF measured at different frequencies for a group of subjects with normal hearing. The MAF is on average about 6 dB more sensitive than MAP. Part of the difference probably arises because MAF measurements are usually made binaurally, whereas MAP measurements are monaural and made with an earphone which occludes the ear, thus increasing the physiological noise generated by internal sources.

Any reduction in hearing sensitivity is generally called "hearing loss." There are many causes of hearing loss, but the main concern here is that which results from exposure to intense noise. It is important to appreciate that noise-induced hearing loss damages the hair cells in the cochlea, while the eardrum is rarely, if ever, damaged as a result of exposure to industrial-type noise. The exact nature of the cellular changes in the hair cells is not clear, but it is certain that the majority of noise-induced hearing loss results from their gradual deterioration, even though it appears from recent research that a substantial proportion of the outer hair cells can be destroyed before any reduction in hearing sensitivity is apparent.

A further cause of deafness in the inner ear is that resulting from the aging process; this is termed presbyacusis. As age increases, hearing sensitivity is reduced, particularly at the high frequencies. Figure 11-6 shows the average effect of aging on hearing sensitivity and includes the range of the most important speech

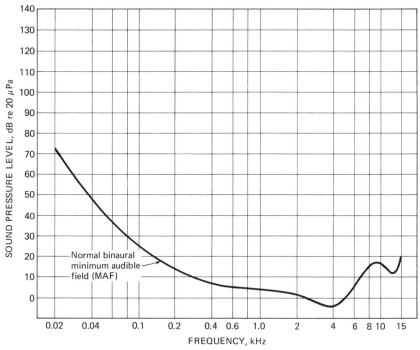

Fig. 11-5 Normal binaural minimum audible field (MAF) at different frequencies.

frequencies. It can be seen that even in old age, the hearing loss in this region is not severe. While presbyacusis may not be a serious disability in itself, when it is added to noise-induced hearing loss, which may have been caused at an early age and was then no great disability, the combined effect can cause a serious handicap in later life (see Fig. 11-9). The causes of presbyacusis are not clearly understood, but it probably results from a general degeneration of the hair cells, stiffening of the basilar membrane, and reduction of the blood supply to the cochlea. Degeneration of the brain cells themselves is probably also responsible for part of the loss of hearing sensitivity.

11-3. MEASUREMENT OF HEARING

In order to examine the effects of noise on the auditory system it is necessary to be able to measure, with some degree of accuracy, the sensitivity of the ear. The technique of measuring the sensitivity of the ear is called audiometry. A method particularly relevant in this context is pure-tone air-conduction audiometry, in

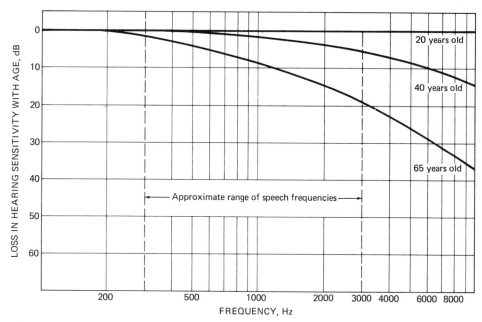

Fig. 11-6 Audiograms showing the typical reduction of hearing sensitivity with increasing age. The approximate range of speech frequencies is also shown.

which a pure-tone signal from an audiometer is presented to the subject by means of an earphone.

Briefly, an audiometer consists of a signal generator with a frequency control, a stepped attenuator, switches for tone presentation and earphone selection, and the earphones. An example of this apparatus is shown in block form in Fig. 11-7, which also shows a typical method by which the subject can indicate his response to the signal. In brief, the audiometer should have stable characteristics and the intensity of the signal should be linear within certain limits throughout its range of attenuation. The test frequencies should include 250, 500, 1000,

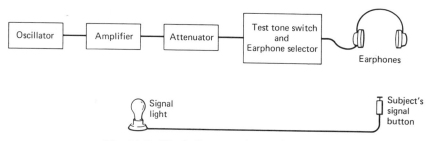

Fig. 11-7 Block diagram of an audiometer.

2000, 3000, 4000, 6000, and 8000 Hz. Current national standards specify the temporal characteristics and intensity tolerances of the signal and stress the need for regular calibration of the audiometer if any value is to be placed on the results.

11-3.1 Audiometry

Since hearing loss often varies with frequency, the threshold of hearing must be determined at a number of frequencies, each ear being tested separately. Since it would be impractical to measure the sound pressure level presented to the ear at threshold on each occasion, an audiometer indicates the difference in decibels between the sound pressure generated at the measured threshold and a standard value. Standard values of the sound pressure generated under several types of earphone at threshold have been specified by a number of national and international standard organizations [1-4]. The audiometer should be calibrated so that it conforms with the recommendations in the particular standard employed, and the measured threshold values can then be quoted with reference to that standard.

The methodology used to determine the threshold has not been standardized universally, but many of the techniques used in practice vary only slightly from one another. It is usual for the subject to indicate in some silent way when he hears the signal presented to him. An important consideration is the difference between ascending and descending thresholds. The ascending threshold is obtained when the signal intensity is increased from inaudibility to audibility. It is sometimes less sensitive than the descending threshold, which is obtained by reducing signal intensity from a clearly audible level to a level below threshold. Normally the threshold of hearing is defined in such a way that both the ascending and descending thresholds are taken into account. In one widely used technique, threshold is defined as the lowest intensity at which at least two out of four signal presentations elicit a response, at least one of which must be on the ascent.

A different audiometric technique, developed originally by Bekesy, is designed to enable the subject himself to adjust the signal intensity to his threshold, and a continuous chart of his hearing sensitivity is obtained at several frequencies. The technique, of which there are a number of variations, is generally referred to as self-recording audiometry.

The measurement of threshold of hearing is not precise for several reasons. The psychological nature of the experiment in which a subject is asked to judge the threshold of hearing leads to loss in accuracy and repeatability of the measurements; and the physiological state of the subject will also affect the measured threshold. The interaction between the earphones and the subject is important, because the position, fit, and pressure of application can affect the measured threshold significantly and should be carefully controlled.

TABLE 11-1. Maximum Permissible Sound Pressure in an Audiometric Enclosure for Measurement of Hearing Level.

Octave band center frequency, Hz		125	250	500	1000	2000	3000	4000	6000	8000
Sound pressure	a:	22	16	18	26	36	39.5	38.5	40	34.5
level, dB	b:	40	40	40	40	47	52	57	62	67

a. Data recommended by Burns [5] when TDH 39 telephones fitted with MX41/AR cushions are used.
b. Data recommended by ANSI [6].

The data differ because Burns's data are designed to allow hearing levels to be measured to −10 dB re BS 2497 [2], while the ANSI data allow masking of thresholds below 0 dB.

Audiometric measurements must be carried out in a room where there is no extraneous noise likely to mask normal thresholds, otherwise the threshold will be artificially elevated and therefore not correct. Examples of noise levels which should not be exceeded in audiometric rooms are given in Table 11-1.

Prior to any audiometric measurement the subject should be examined to see whether there is any wax in the auditory meatus which could affect the hearing sensitivity. Spectacles and similar obstacles will interfere with the proper fit of the earphones and should be removed, and care should be taken to prevent hair covering the ears. Care should also be taken in planning the times at which audiometry is carried out, since the hearing sensitivity of persons working in noisy environments will temporarily worsen during the working day. Hence it is preferable to carry out monitoring audiometry prior to the start of work.

11-4. THE AUDITORY EFFECTS OF NOISE

The earliest manifestation of hearing damage resulting from exposure to intense noise is a change in the threshold sensitivity of hearing of high-frequency sounds. However, it is unlikely that in the early stages of noise-induced hearing loss the person will notice any real changes in his hearing ability.

Before noise-induced hearing loss becomes permanent, exposure to intense noise will result in temporary hearing loss. This is noticeable as a dullness in hearing and it is sometimes accompanied by a ringing sensation (tinnitus) in the ears, although the latter may only be noticeable in quiet surroundings. These effects may disappear a few hours after the noise exposure has ceased, but repeated exposure to noise can result in the hearing loss becoming permanent.

An audiogram taken from a person suffering from noise-induced hearing loss will usually show the greatest loss of hearing sensitivity in the 4 kHz region, which is typically the region most sensitive to damage resulting from many types

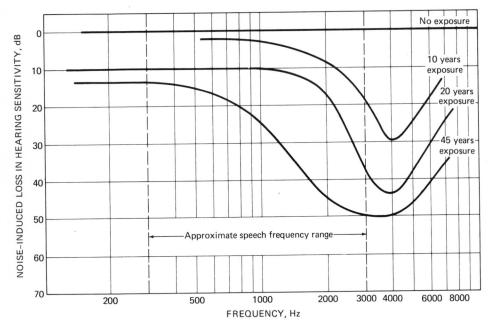

Fig. 11-8 Audiograms showing the typical reduction of hearing sensitivity as a result of long term exposure to industrial noise. The approximate range of speech frequencies is shown.

of industrial noise. The hearing loss starts in this region and with continued exposure, the 4 kHz dip, as it is known, deepens and spreads to lower and higher frequencies. An example of the growth of hearing loss with continued exposure to noise is shown in Fig. 11-8.

The noise-induced hearing loss of the type illustrated in the figure results in difficulties in hearing high-pitched sounds, which can include, for instance, some industrial warning signals. As the 4 kHz dip widens and the lower frequencies are affected, speech reception is increasingly affected. Since most information is borne by the consonant components of speech and many of these contain high-frequency sound energy, speech intelligibility is reduced. Eventually, while the affected person will still be able to hear the speech he will not be able to discriminate sufficiently well to enable him to understand the information it contains. This problem becomes particularly severe when noise-induced deafness is added to that normally encountered in the elderly (presbyacusis); a typical audiogram that results is shown in Fig. 11-9. It can be seen that an elderly person who also suffers from noise-induced hearing loss will have a particularly serious hearing handicap in the speech frequency range.

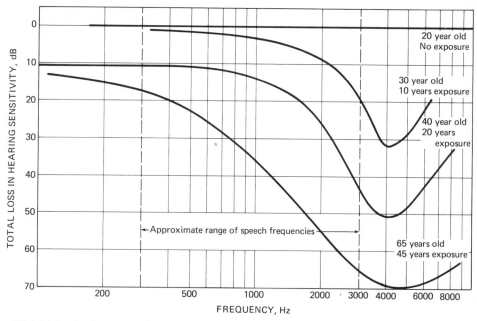

Fig. 11-9 Audiograms showing typical hearing losses resulting from both presbyacusis and noise-induced hearing loss. The approximate range of speech frequencies is shown.

11-5. HAZARDOUS NOISE EXPOSURE

A great deal of effort has been spent over the past three decades in an attempt to formulate methods for quantifying the relationship between noise exposure and hearing loss. Initially much of the work was based on laboratory experiments involving short-term hearing losses. The relationship between temporary hearing loss and persistent hearing loss caused by long-term exposure to noise in industry has formed the basis for a number of criteria for hearing damage risk. Also, in the last few years data have become available that have enabled persistent hearing losses to be correlated directly with long-term noise exposure.

The complex nature of this information has resulted in a divergence of the bases upon which national standards rely. At the present time American and European national standards for limiting noise exposure are founded on different principles. Whereas many European standards assume an energy relationship between sound level and exposure time, i.e., a permissible 3 dB increase in sound level for each halving of exposure duration, this principle is not so well accepted in the U.S., where there is support for a 5 dB increase per halving of exposure duration. The current American and British recommendations are discussed separately in this chapter; note that the EPA recommendations (see Section 11-5.1) are U.S. recommendations that do, in fact, follow an energy relationship, whereas those in the U.S. Occupational Safety and Health Act (see Section 11-6.1) do not.

11-5.1 U.S.A.

The EPA levels document [7] identifies an equivalent sound level (see Chapter 1, Section 1-5.2) of $L_{eq} = 70$ dB(A) as the maximum permissible, when measured on a 24 hours/day basis every day of the year. This level is, however, rounded down from 71.4 dB(A) to provide a margin of safety.

It is possible to apply this criterion to occupational noise only with full knowledge of how a person's exposure is shared between his workplace and other environments. If, however, one assumes that the only significant contribution to his noise dose occurs during a working day of 8 hours, 5 days/week for 50 weeks/year—an assumption which is reasonable for a great many lifestyles—then the exposure at work may rise to an L_{eq} over those hours of 76.4 dB(A) [say, 76 dB(A)] or, adding the margin of safety, to 77.8 dB(A) [say 78 dB(A)].

The criterion is based on the noise level which protects 96% of the population against a noise-induced permanent hearing loss greater than 5 dB at 4000 Hz. It is argued that the people beyond the 96th percentile, who are those with already impaired hearing, are also protected at the criterion level because their hearing is too impaired to be damaged by sounds which, at this level, they cannot hear.

At present there is not complete agreement in the U.S. that the EPA criterion is correct, one point of disagreement being the energy-based trading relationship with time.

11-5.2 United Kingdom

Much of the impetus for the establishment of the equal-energy principle for assessing hearing hazard in Britain is derived from a Medical Research Council and National Physical Laboratory study. The report of this work by Burns and Robinson [8] has since formed the basis for a method of assessing potential hazard to hearing. A number of European national bodies have also based their standard methods of assessment on this type of information.

Equal-Energy Principle. Burns and Robinson [8] showed that for steady-state noise a relationship exists between A-weighted sound energy and persistent hearing loss.

A-weighted sound energy received during a noise exposure may be deduced from the product of the A-weighted sound level and the duration of exposure. A doubling of energy represents an increase in noise level of 3 dB(A). Thus, for example, exposure to 90 dB(A) for a given period is equivalent in terms of sound energy, and therefore hazard, to an exposure to 93 dB(A) for half that period. This is the equal-energy principle. Subsequent research by Atherley and Martin [9] and Rice and Martin [10] has shown that this principle may also be extended to the assessment of risk to hearing from industrial impact noise and from high-level transients such as gunfire. Consequently a single and relatively simple principle may now be said to form the basis for the assessment of most occupational noise. A-weighted sound energy and the equal-energy principle may be considered the unifying factors in what up to the present has been a somewhat confused area of knowledge.

The concept of A-weighted sound energy is embodied in the equivalent sound level, L_{eq}. L_{eq} may be considered a measure of noise dose, a concept familiar in other aspects of occupational medicine but relatively new in the field of noise. Table 11-2 gives examples of values of sound level and exposure duration which, when considered together, represent an L_{eq} of 90 dB(A) for eight hours. This noise dose is considered at present in Britain to be a safe maximum. An L_{eq} of 90 dB(A) over eight hours is equivalent to an L_{eq} of 85 dB(A) over 24 hours, a figure rather higher than the EPA has proposed (see Section 11-5.1).

TABLE 11-2. Permissible Sound Levels for an 8 Hour Workday that Give an L_{eq} for This Period of 90 dB(A).[a]

Sound Level dB(A)	Permissible Daily Exposure
90	8 hours
93	4
96	2
99	1
102	30 minutes
105	15
108	$7\frac{1}{2}$
111	225 seconds
114	112
117	56
120	28
123	14
126	7
129	$3\frac{1}{2}$
132	2
135	1

[a]The table shows the trading relationship between sound level and time that applies to the equal-energy concept. It may be used with other criterion levels based on equal-energy principles by adding or subtracting an appropriate constant to the sound level column.

Standard Methods of Assessment. In Britain at present there are three authoritative documents available that describe methods (based on an energy concept) for the assessment of hazards to hearing from occupational noise exposure [11] :

a. "The Code of Practice for Reducing the Exposure of Employed Persons to Noise." [12]
b. "Hygiene Standard for Wide-Band Noise." [13]
c. "Assessment of Occupational Noise Exposure for Hearing Conservation Purpose." [14]

The Code of Practice is the most suitable document for application to the industrial situation in the United Kingdom at present. It specifies an exposure limit of L_{eq} = 90 dB(A), and describes measurement methods to determine whether the limit is exceeded. The Code makes the assumption that steady-state, fluctuating, intermittent, and impact noise may all be described in terms of L_{eq}. Furthermore, it recommends overriding limits for the unprotected ear, these being a sound pressure level of 135 dB for steady-state noise and 150 dB for impulse noise. The Code is important in Britain because it specifies safe conditions of work and Her Majesty's Factory Inspectorate relies upon its recommendations to define the auditory safety of a working environment under the new Health and Safety at Work Act as discussed in Section 11-6.2.

However, the Code of Practice gives only a single demarcation between acceptable exposures and those which are unacceptable, whereas the second document above [13], by the British Occupational Hygiene Society (BOHS), and the third [14], by the International Organization for Standardization (ISO), provide more detailed information regarding the percentage of persons who would be expected to exceed specified hearing losses after given noise exposures.

The BOHS document refers solely to steady-state noise and is intended to restrict occupational exposure to noise so that handicap does not occur in more than 1 percent of persons exposed during their working lifetime. This objective is considered to be achieved if the noise-induced impairment to hearing at the end of a 30-year working life does not exceed levels of 40 dB calculated as the average loss at the audiometric frequencies 0.5, 1, 2, 3 and 6 kHz in 1 percent of the persons exposed. To meet this criterion, noise levels should not exceed an L_{eq} of 90 dB(A). The standard must not be used for impulse noise or if there are significant pure tones.

ISO Recommendation R 1999 does not quantify absolute levels which must not be exceeded. Rather, it gives a practical relationship between occupational noise exposure, expressed in terms of noise level and duration within a normal working week of 40 hours, and the risk of increase in percentage of persons in specified age groups that may be expected to show hearing impairment as a result of specified exposures. Hearing is considered to be impaired for conversational speech if the arithmetic average of hearing loss for the frequencies 0.5, 1, and 2 kHz is 25 dB or more. The recommendation does not set down limits, but points to levels of 90 dB(A) legislated in some countries. Impulse noises of less than 1 sec duration for a single high-level transient such as gunfire are excluded.

Although the definitions of handicap and impairment are different in the BOHS and ISO documents, they may be assumed to be equivalent in terms of hazard to hearing—i.e., the same noise level assessed by the two methods will probably result in the prediction of the same hearing losses. However, they do not employ the same units of exposure time, nor do they cover precisely the same range of exposures, as shown in Table 11-3.

The BOHS document assumes habitual exposure of 8 hours per day, 5 days per week, 48 weeks per year for a working life of 30 years, and allows for certain exposures of different durations. The ISO recommendation describes noise exposure in terms of a 40 hour week for 50 weeks of the year for a variable time of up to 45 years, again with some provisions for shorter durations. Table 11-3 compares these specifications of exposure time in terms of the maximum permissible sound level in dB(A).

TABLE 11-3. Comparison of Exposure Times Permitted by the BOHS [13] and ISO [14] Documents.

BOHS Permitted Duration, hours/day	Maximum Permissible Sound Level, dB(A)	ISO Permitted Duration, per week
12	88	
11		
10		
9		
8	90	40 hours
7	90.5	35
6	91	30
5	92	25
4	93	20
3	94	15
2	96	10
1	99	5
0.5	102	150 minutes
	105	75
	108	40
	111	20
	114	10

11-6. ILLEGAL NOISE EXPOSURE

The past decade has seen rapid development of the international medico-legal field with respect to occupational hearing loss. The situation at present is still a developing one in many countries. This section reviews the current legal situation in the U.S. and the United Kingdom.

11-6.1 U.S.A.

American federal regulation of noise exposure had its first tangible expression in 1960 regulations accompanying the Walsh–Healey Public Contracts Act [15]. The regulations said that noise must be reasonably controlled to minimize fatigue and the possibility of accidents. Being part of the Act, the regulations were directed toward contractors of the federal government, and not toward employers in general.

In 1969 the Walsh–Healey Act's noise exposure provisions were extended by revisions published in the Federal Register which set out definite limits based essentially on those set out by the American Council of Governmental Industrial Hygienists, but they still applied only to federal contractors. To extend hearing protection to all U.S. employment, the Occupational Safety and Health Act was passed in 1970 [16].

Permissible maximum sound levels under OSHA are given in Table 11-4, which applies to continuous sounds, defined as those for which the variations in sound level involve maxima at intervals of 1 sec or less. When the daily noise exposure is composed of two or more periods of noise exposure of different levels, their combined effect should be considered. If the sum of the following fractions,

$$\frac{C_1}{T_1} + \frac{C_2}{T_2} + \cdots + \frac{C_n}{T_n}$$

exceeds unity, then the mixed exposure should be considered to exceed the limit value. C indicates the total time of exposure at a specified noise level, and T indicates the time of exposure permitted at that level in Table 11-4. Exposure to impulse noise should not exceed 140 dB(C) using the fast response of a sound level meter.

The Act provides that excessive noise should be controlled by any feasible noise control procedures, or by administrative methods such as limiting employees' times of exposure. Only when such methods are unfeasible may the employer turn to the use of hearing protection.

For all employees working regularly or occasionally in areas where the sound level exceeds 90 dB(A), regular audiometric testing must take place, including threshold measurement at 0.5, 1, 2, and 4 kHz. The whole procedure, the audiometric room, the audiometer and the record-keeping should meet prescribed standards (described by Van Atta [17]).

TABLE 11-4. Permissible Exposure Under the Occupational Safety and Health Act [16].

Sound Level, dB(A)	Permissible Daily Exposure, hours
90	8
92	6
95	4
97	3
100	2
102	$1\frac{1}{2}$
105	1
110	$\frac{1}{2}$
115	$\frac{1}{4}$ or less

The Occupational Safety and Health Administration and the Environmental Protection Agency are at present debating the question of acceptable levels for exposure to noise in places of work. The OSHA regulation, based on a limit of 90 dB(A) for an 8-hour continuous exposure with a trading relationship of 5 dB(A) for each halving or doubling of duration, is contested by EPA, who favor a 3 dB(A) trading relationship and the adoption of a limit of 85 dB(A) for 8 hours continuous exposure.

Compensation claims for loss of hearing induced in employment are, in the U.S., subject to the legal precedents set in the various states, but in virtually all states it is accepted that compensation is due. Cudsworth [18] estimates that the numbers of claims over the last twenty years runs into the tens of thousands, and that the average of the maximum compensation for 100% binaural impairment was (in 1971, when he published his paper) close to $8000.

According to Cudsworth, the most commonly accepted way to assess the degree of loss is that recommended by the American Medical Association's Committee on Conservation of Hearing [19]. The pure-tone hearing thresholds at 0.5, 1, and 2 kHz are averaged, 15 dB is subtracted, and the result is multiplied by 1.5 to establish monaural impairment as a percentage of normal hearing capability in that ear. To assess binaural impairment, this percentage is calculated for each ear in turn; then, in the belief that total loss in one ear amounts to only about one-sixth of complete hearing impairment, the percentage for the better ear is multiplied by 5 before it is added to the percentage for the worse ear, and the total is divided by 6.

Example

Problem: The sound levels experienced by an employee in a U.S. plant are analyzed by computing the time he spends in various parts of the factory. He spends (1) one hour a day at 100 dB(A); (2) three hours a day at 90 dB(A); and (3) four hours a day at 60 dB(A). (a) Is his noise exposure permissible according to the Occupational Safety and Health Act? (b) Does his noise exposure exceed the maximum considered safe by the EPA?

Solution: (a) Refer to Section 11-6.1 and Table 11-4. According to OSHA, the total time permitted at 100 dB(A) is 2 hours, the total time permitted at 90 dB(A) is 8 hours, and there is no limit on 60 dB(A) exposure. Then (C_1/T_1) + $(C_2/T_2) = \frac{1}{2} + \frac{3}{8} = \frac{7}{8}$, which is less than unity, so the exposure does not infringe OSHA. (b) The EPA criterion (see Section 11-5.1) calls for maximum L_{eq} over 24 hours of 70 dB(A). To calculate the employee's L_{eq}, refer to Section 1-5.2 (Chapter 1) for the equation defining L_{eq}:

$$L_{eq} = 10 \log_{10} \frac{1}{24} [t_1(10^{L_1/10}) + t_2(10^{L_2/10}) + t_3(10^{L_3/10})]$$

$$= 10 \log_{10} \frac{1}{24} [10^{10} + 3 \times 10^9 + 4 \times 10^6]$$

$$= 82$$

82 dB(A) exceeds the EPA criterion of 70 dB(A) even without considering the employee's off-the-job exposure over a 24 hour period.

11-6.2   United Kingdom

As described by Coles and Martin [20] the present law in Britain can be considered under the headings of legislative law (i.e., statutory law) and common law. The statutory category may be conveniently divided into preventive and compensatory aspects. The situation with regard to preventive statutory law governing noise hazard and its reduction is in a state of rapid development owing to the enactment of the Health and Safety at Work Act [21] in 1974. Prior to this enactment, the Factories Act [22] was the main statutory instrument governing safety at work. This Act has not been repealed. The part of the Factories Act most relevant to noise is Section 29(1) which, with reference to "every place at which any person has at any time to work," states that "every such place shall, so far as is reasonably practicable, be made and kept safe for any person working there." This Section has, however, never been tested in court on the subject of noise.

The Health and Safety at Work Act is designed to replace the Factories Act and specifically includes noise hazard within its scope. It is to be enforced through a number of regulations and codes of practice. One of the first codes of practice to be approved and thus carry the force of law is the Code of Practice for Reducing the Exposure of Employed Persons to Noise [12] (discussed in Section 11-5). Other codes of practice will deal with industrial machinery noise and industrial audiometry. These codes of practice may be either approved or advisory. The Act will result in all employers (not just in industry) taking a much more active role in reducing noise exposure.

Under the terms of the National Insurance (Industrial Injuries) Act [23] financial compensation is payable to injured persons for disability or loss of facility arising from accidents in the course of employment. Some cases of noise-induced tinnitus and deafness have earned compensation under this Act where they have followed some definable single accident or incident of noise exposure. However, slowly developing occupational deafness resulting from repeated noise exposures has only recently become compensatable, as it was only added to the list of over 40 industrial diseases specifically prescribed as to be included under the Act in 1975. For such prescription, the Government has to be satisfied that the condition and its relationship to a particular hazard can be diagnosed with a reasonable degree of certainty and facility. This prescription was based upon a report by the Department of Health and Social Security [23] which recommended that compensation for severe occupational deafness be paid to certain industrial groups. The first awards for compensation for noise-induced deafness were made in February 1976. While the compensatory scheme is limited to certain industrial groups it is intended to be expanded in scope as national facilities for diagnosis of occupational deafness expand. An award under this scheme does not preclude the claimant from suing his employer for damages under common law.

Common law actions whereby a noise-exposed worker sues his employer for damages in respect of occupational hearing loss have only occurred since 1968. The facility has always been available and the knowledge necessary to support a claim has certainly been widespread since 1963, when the British government pamphlet "Noise and the Worker" was first published [24]. Yet the first case was not brought until 1968 and was unsuccessful. Since that time, only a handful of successful cases have been brought, although many have been settled out of court. Although the legal criterion is "on the balance of probabilities," there are real difficulties in proving the three basic factors needed for a successful common law action. Briefly these are:

a. the medical diagnosis of the deafness has to be shown to be that of noise-induced hearing loss;
b. the hearing loss has to be shown to be causally related to the industrial noise exposure;
c. the employer has to be shown to have been negligent in not recognizing the noise hazard and not taking appropriate measures to prevent the damage to hearing.

Furthermore, such cases have to be judged, in respect of negligence, by the knowledge and action to be expected of the reasonable employer at the material time.

11-7. MEASUREMENT OF NOISE EXPOSURE

Whichever national standard specification is employed, the fundamental requirement is the measurement of both sound level and time (i.e., of noise dose) for each individual exposed.

This section reviews briefly the basic instrumentation necessary for the measurement of different types of noise and the techniques of measurement to be applied in practice [25]. Measurement techniques are categorized in terms of the necessary instrumentation as opposed to the types of noise measured. This approach is designed to provide a brief guide for the measurement of noise for those laboratories in industry which already possess certain types of measurement equipment.

11-7.1 The Sound Level Meter

The measurement of noise dose for steady-state noise may be made simply with a precision sound level meter (SLM) conforming to the appropriate national standard, and a timepiece. The meter should be set to the A-weighting network and slow dynamic characteristic. Measurements of the sound level and exposure time should be made in close proximity to the ears of the personnel exposed to the noise. When the noise is steady in level and unbroken in duration, a few measurements of the level should provide a value of noise dose. Short-term fluctuations in sound level of no more than 10 dB can be dealt with by "eye-averaging" the meter reading. However, if the sound level varies by more than 10 dB or varies more slowly with time, as is the case with the noise produced by many cyclic industrial processes, the individual values of A-weighted sound level and on-time should be recorded for each particular sound level. These components may then be summed to give a composite value for noise dose using one of the available standard techniques.

In many practical situations, noise levels fluctuate considerably as various industrial processes, possibly operating in the same area, generate noise with different temporal patterns. It may be necessary in such cases to carry out work-study procedures with an SLM to establish a reliable estimate of noise dose. The situation may be further complicated by the requirement to obtain a measure of the noise dose received by an individual. If an individual remains in the same location during the working day, then his noise dose may be assessed relatively simply. However, if he moves around from quiet to relatively noisy surroundings in a random fashion, both during the day and from day to day (e.g., as maintenance workers do), the measurement of noise dose with an SLM for that individual becomes an extremely complicated procedure. The need to monitor noise dose for a number of individuals probably makes the use of an SLM impractical in such circumstances.

The measurement of dose for impulse noise is not easily achieved with an SLM. This is because measures are required of both sound level and time and it is not usually possible to measure both the peak sound level and the duration of an impulse with an SLM and a simple clock or watch.

11-7.2   The Noise Dosemeter or Dosimeter

This is a relatively new development in the noise field which promises to simplify the measurement of noise dose considerably. The dosemeter measures both A-weighted sound level and duration simultaneously, and hence can provide a direct measure of dose. In principle it should be able to deal with any type of noise, both steady and impulsive, having any type of temporal pattern, and therefore should provide a relatively simple measure of dose in any industrial situation. Care should be taken that the dosemeter employed complies with the particular standard being used to evaluate the noise; for instance, either the energy concept or the OSHA specification of 5 dB per doubling of exposure time.

The dosemeter relies upon long-term integration techniques for the measurement of dose. That is, it sums over a period of time all the A-weighted sound energy incident upon it irrespective of its temporal characteristics. It should not possess limiting dynamic meter characteristics such as "fast" or "slow," and therefore should not in this respect modify the incoming signal. Consequently, it should be equally capable of measuring dose for steady-state noises or for impulses having fast rise times.

The dosemeter is available in two forms: a relatively large, stationary device and a small, personal monitoring instrument worn by the exposed person. The larger instrument usually has a slightly better technical specification than the personal device and may be used to measure the noise dose generated by a particular machine or in a particular area. However, it suffers from a similar drawback to the SLM in that it does not easily provide a measure of the noise dose received by an individual if that individual moves around a factory from day to day in a random fashion. It is in this type of situation that a personal noise dosemeter has an advantage. Being worn, it monitors directly all the sound energy reaching the ears of an individual, wherever he may be, and thus greatly facilitates the measurement of total personal noise dose.

The measurement of dose for steady-state noises, even with complex variations in sound level and temporal pattern, is a relatively simple matter with a dosemeter. However, care should be exercised to ensure that the duration of the measurement period is sufficient to provide a representative measure of a complete day's exposure. Certain types of industrial noise contain both steady-state and impulse noise, where the peak sound levels of the impulses are considerably greater than the background noise levels, although both may be hazardous. In this situation care should be taken that the peaks of the impulses do not overload the dosemeter, thereby introducing errors. A dynamic range of at least 60 dB is often necessary for a dosemeter to give reliable results in such circumstances.

11-7.3 Tape Recorders

The use of a precision tape recorder in the field can greatly facilitate the determination of dose for industrial noise. Tape recordings allow noises having complex sound level and temporal characteristics to be analyzed in the laboratory using a dosemeter or other analytical equipment, such as a graphic level recorder or oscilloscope.

The measurement of steady-state noise by this technique is straightforward, provided that the usual safeguards and standards are maintained. In the case of impulse noise, even greater care is needed. Walker and Behar [26] have shown that the measurement of impulse noise by this method introduces an error of about 1 dB in the peak values of the impulses. Provided that care is taken to ensure that neither microphones nor tape recorders clip the peaks of the impulses and that possible signal-to-noise ratio limitations are noted, a reliable measure of dose should be obtained.

11-7.4 Oscilloscope Techniques

An oscilloscope may be used for the measurement of dose for any type of impulse noise. It can be used to advantage for the measurement of those impulse noises not adequately catered for by a dosemeter or where a dosemeter is not available. However, it is not usually convenient for the measurement of steady-state noise.

The basic extra requirement is for a storage oscilloscope and camera, so that recordings of the pressure–time waveform of the impulse noise can be obtained. The impulse signal may come directly from an SLM or microphone and amplifier, or may be tape recorded and analyzed in the laboratory. Measurements are made of a number of parameters of the waveform of the noise and the noise dose is deduced from these parameters.

For the cases where hearing hazard is assessed by the equal-energy concept, Martin and Atherley [27] have described a simple method for the measurement of equivalent continuous noise level of impulse noise from oscilloscopic recordings of the impulse waveform. The noise is divided into three categories, according to repetition rate, and the technique employed varies according to category.

Oscilloscopic techniques, like the SLM, become rather complex when one is attempting to assess the individual noise dose of a person who moves around a factory. A number of measurements of the waveform may be required, together with work-study procedures, if the noise is generated by several different machines or by a cyclically varying manufacturing process.

11-7.5 Discussion

The calibration of all measurement equipment is essential for reliable results. All instrumentation should be calibrated prior to making the measurements and this should be checked at regular intervals during its use.

If the total workforce remains in one location in a noise with constant sound level and temporal characteristics, the task of evaluating noise dose for each worker is relatively simple. However, as variations in the sound level and temporal pattern become more complex and random, and as individual workers tend to move from one noise environment to another, the determination of dose becomes increasingly difficult. The more variable the noise, the greater the number of measurements and time required to obtain a reliable result. Measurements made with a personal noise dosemeter may require less effort, although in the extreme case of random movements around a factory, an individual may be required to wear a dosemeter for a number of shifts before a representative measure of his noise exposure can be arrived at.

11-8. HEARING PROTECTION

The most obvious and efficient method of reducing exposure to noise is to prevent the generation of the noise in the first place. When it is not possible to reduce noise levels to within safe limits by treatment of the source, using noise control techniques or by replacement with an inherently quieter machine, the problem can sometimes be solved by covering the source with an acoustic hood or by the use of noise barriers. Another means of noise reduction is to remove either the offending machine or the persons exposed to another location; similar results can be achieved by limiting the time that persons are exposed to the noise. There are many situations, however, where such noise control techniques are either impractical or not sufficient. In these cases, personal hearing protective devices must be worn.

The devices provided must be capable of removing the hazard to hearing from the noise environment in which they are being worn. Consequently, knowledge is required of the physical characteristics of the offending noise as well as the acoustic attenuation characteristics of the hearing protectors provided. Furthermore, there is a certain degree of responsibility placed upon the employer to educate the employee about the need to wear hearing protection and to persuade him to do so. Hence other nonacoustical properties, such as comfort and wearer acceptability, have also to be taken into account. Employers and other persons responsible for providing hearing protection should therefore have a good working knowledge of the many aspects and requirements of a hearing protection program.

11-8.1 Types of Hearing Protector

There are many brands and types of hearing protector available on the market and in selecting the most suitable type for any given situation there are several factors to be considered in addition to the protection to hearing they provide. Some of these are comfort, cost, durability, chemical stability, availability, wearer acceptance, and hygiene.

Prefabricated Earplugs. The best prefabricated earplugs for general industrial use are available in three to five different sizes. They are made from soft, flexible material that will conform readily to the many different ear canal shapes, thus providing a snug, airtight, and comfortable fit.

Earplugs must be nontoxic and should have smooth surfaces that may easily be cleaned with soap and water. They should be made of a material that retains its shape and flexibility over extended periods of use and is not affected by the presence of earwax. There are a large number of different types of prefabricated earplug available. One of the most versatile and efficient types is the V-51R. This is asymmetrically shaped and carries a single flexible flange that will adapt to a large number of differently shaped ear canals. It provides reasonable protection when fitted correctly, combined with a certain amount of comfort. It is available in five sizes from most manufacturers.

There are also a number of symmetrically shaped earplugs available which provide sufficient protection and comfort under certain circumstances. However, the round and straight types of earplugs generally do not adapt well to sharply bending or slit-shaped ear canals.

Disposable and Malleable Earplugs. These types of earplugs are usually fashioned from low-cost materials such as cotton, wax, glass wool, and mixtures of these and other substances. In general, malleable and disposable earplugs made of nonporous and easily formed materials are reasonably comfortable and capable of providing attenuation values similar to those afforded by prefabricated plugs if made correctly.

The protection provided depends on the material used and how firmly the plug is seated. Ordinary cotton wool by itself is extremely porous and provides very little attenuation. Indeed it is not recommended at all because of its inefficiency and the false sense of security which its use engenders. Glass wool on the other hand, is probably one of the most practical and efficient forms of disposable hearing protection. It is made from extremely fine glass fibers about 1 micron in thickness and provides reasonable attenuation when inserted according to the manufacturer's instructions.

Individually Moulded Earplugs. These are usually made from some form of silicone rubber and are actually moulded in a permanent form within the ear canal. Usually the earplug material is supplied with a curing agent and the two are mixed to a puttylike consistency before being inserted into the ear canal of the person to be fitted. Having cured, the earplugs are in a permanent form and may be removed and reinserted any number of times without affecting their performance. However, the degree of protection provided by individually moulded earplugs is dependent to a certain extent on the expertise of the person making them.

The main advantage of individually moulded earplugs is their greater appeal to the wearer. In situations where difficulty is encountered in persuading men to wear hearing protection, the provision of a personal and individually moulded device that will only fit the person for whom they are intended (like spectacles and false teeth) is a psychological advantage.

Semi-Insert Protectors. These are also called concha-seated hearing protectors or canal caps. They usually consist of two conical soft rubber caps attached to a narrow headband which presses them against the entrance to the external ear canal.

Semi-insert protectors have the advantage that one size will fit the majority of ears, unlike prefabricated plugs. As they are captive and may be reinserted hygienically at any time, they are suitable for industries where the loss of an earplug must be avoided (e.g., the food industry) and for people who must frequently enter noisy environments for short periods, or remain in hot environments for long periods. However, this type of plug is often not as comfortable as other forms of hearing protection, as they must be pressed firmly against the ear canal entrance to be effective.

Earmuffs. Most types of earmuffs are of similar design and are made from rigid cups specially designed to cover the external ear completely. They are held against the sides of the head by a spring-loaded adjustable band and are sealed to the head with soft circumaural cushion seals.

Earmuff seals may be liquid filled or plastic-foam filled. Liquid-filled seals usually provide marginally better protection with less headband tension, all things being equal, but suffer from the problem of leakage if treated roughly. Modern foam-filled seals are almost as good as the liquid seals and have the additional advantage of robustness. In any case, earmuffs should be provided with seals that are easily and separately replaceable in the factory environment.

The attenuation provided by earmuffs is related to the force with which they are pressed against the sides of the head. Maintenance of the correct headband pressure is therefore important and care must be taken that this is not reduced by bending the headband. Furthermore, earmuffs will only provide maximum protection when placed on a relatively smooth surface. Therefore, less protection should be expected when muffs are worn over long hair or spectacle frames and with safety equipment such as goggles and helmets.

Some earmuffs are asymmetrical and thus can only be worn one way, that is, only one cup will fit the left ear and only one the right. In these cases the correct way of wearing the muffs should be prominently indicated.

Earmuffs have the advantage over other hearing protectors of usually providing the greatest protection. Also, one size usually fits most people, and they are easily removed and replaced in a hygienic fashion. This makes them eminently suitable for dirty and high-level noise areas and for people who frequently move in and out of noisy environments. They can also be worn by people who may suffer from minor diseases of the external ear canal or in other circumstances when earplugs cannot be worn.

The disadvantages of earmuffs lie in their bulkiness and cost and the fact that they tend to make the ears hot and exacerbate perspiration. (However, their bulk has the advantage that it can easily be seen that they are being worn correctly.) They are also usually more susceptible to damage than other forms of hearing protection.

Special Types of Hearing Protector. There are a number of earplugs and earmuffs designed for special purposes such as improved communication and the selective attenuation of high-level transient noises.

a. Frequency-selective devices. These are usually fitted with an acoustic low-pass filter which ensures that the attenuation below about 2 kHz is relatively small. This filter enables the lower speech frequencies to be passed and thus permits slightly easier speech communication between wearers. However, improved speech communication will only result if all the external noise is at a higher frequency. This is not the case in the majority of industrial situations and consequently this type of hearing protector is usually unsuitable for use on the factory floor.

b. Amplitude-sensitive devices. This type of hearing protector is designed to attenuate loud sounds more than quiet ones. One type of earplug, which is a modified version of the V-51R plug, is designed so that normal speech and other sounds may be heard but high-level transient noises, such as gunfire, will be attenuated. These plugs are only useful for protection against gunfire and explosive types of industrial noise, such as that generated by cartridge operated tools. They are not suitable for most industrial noises. There are also earmuffs available which incorporate mechanical valves which close when high-level gunfire noise is incident upon them. The main advantage of this type lies in the military aspects of noise.

However, earmuffs are available which incorporate an electronic peak-limiting device. These can be extremely valuable in industrial situations where people are exposed to impulse noise, or any high-level intermittent noise, but wish to communicate easily during the quiet periods between noise bursts. The disadvantages of this type of earmuff are that they are relatively expensive, require batteries, and must be handled with greater care than ordinary earmuffs.

11-8.2 Noise Reduction by Hearing Protectors

The prime function of hearing protectors is to reduce the noise level at the wearers' ears to within safe limits. Information on the ability and consistency of hearing protectors to attenuate sound should be examined in considering which type is most suitable for a particular noise environment.

Acoustic Attenuation. The acoustic attenuation of hearing protectors is usually expressed in decibels attenuation at various test frequencies and is usually described in graphical form (see Figs. 11-10 and 11-11) or in tabular form (Table 11-5).

A second, but equally important, measure associated with acoustic attenuation

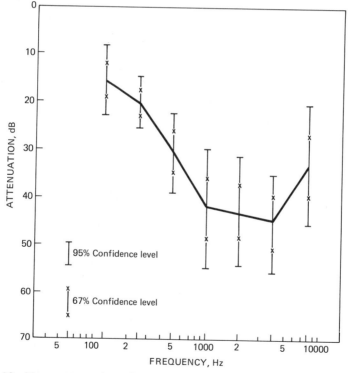

Fig. 11-10 Mean attenuation characteristics of an earmuff plotted with one and two standard deviations.

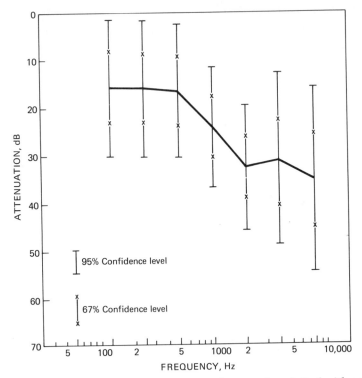

Fig. 11-11 Mean attenuation characteristics of an earplug plotted with one and two standard deviations.

is the degree of scatter of the attenuation as measured on different subjects. This is usually expressed as the standard deviation about the grand mean, or as the interquartile range about the median. This figure should accompany each attenuation datum point when expressing the attenuation, as is shown in Table 11-5. It provides a measure of the hearing protector's ability to fit different individuals and a measure of the accuracy with which the attenuation determinations were carried out. Figure 11-10 shows an example of the mean attenuation characteristics of an earmuff plotted with one and two standard deviations, representing a spread of 67% and 95% of the attenuation data, respectively. Figure 11-11 shows similar attenuation characteristics for an earplug.

The use of a good hearing protector can provide sufficient protection in a large majority of work environments where engineering control measures cannot be used successfully. For those relatively few persons exposed over long periods of time to noise levels in excess of about 115 dB(A), special care should be taken to ensure that the best protectors are used correctly and hearing thresholds are monitored regularly. For higher exposure levels over extended periods of time, it

TABLE 11-5. Typical Mean and Standard Deviation Attenuation Characteristics, of Different Types of Hearing Protection

Type of Protection	Mean (Standard Deviation) Attenuation, dB, at Given Test Frequency						
	125 Hz	250 Hz	500 Hz	1000 Hz	2000 Hz	4000 Hz	8000 Hz
Dry cotton wool plugs	2	3	4	8	12	12	9
	(2)	(2)	(3)	(3)	(6)	(4)	(5)
Waxed cotton wool plugs	6	10	12	16	27	32	26
	(7)	(9)	(9)	(8)	(11)	(9)	(9)
Glass wool plugs	7	11	13	17	29	35	31
	(4)	(5)	(4)	(7)	(6)	(7)	(8)
Personalized ear-mould plugs	15	15	16	17	30	41	28
	(7)	(8)	(5)	(5)	(5)	(5)	(7)
V-51R type plugs	21	21	22	37	32	32	33
	(7)	(9)	(9)	(7)	(5)	(8)	(9)
Foam-seal muffs	8	14	24	34	36	43	31
	(6)	(5)	(6)	(8)	(7)	(8)	(8)
Fluid-seal muffs	13	20	33	35	38	47	41
	(6)	(6)	(6)	(6)	(7)	(8)	(8)
Flying helmet	14	17	29	32	48	59	54
	(4)	(5)	(4)	(5)	(7)	(9)	(9)

may be necessary to use a combination of insert- and muff-type protectors and also limit the time of exposure. Hearing protectors will provide adequate protection for only a very small percentage of wearers when worn in sound levels greater than about 125 dB(A) over long periods of time.

Methods for the Evaluation of Hearing Protectors. When attenuation data are used as guidelines for deciding upon the type and make of hearing protector required, the method by which this information is obtained and the laboratory carrying out the measurements should be carefully noted. It should be determined that attenuation data published by manufacturers are measured following a recognized standard procedure.

11-8.3 Practical Considerations

Hygiene. Earplugs are the type most likely to cause problems of hygiene. Before any form of earplug is issued, the user should be asked about and examined for any ear troubles such as irritation of the ear canal, earache, and discharging ears or whether he is under treatment for any ear disease. In such cases medical opinion should be sought to ascertain whether earplugs can be worn with safety. Obviously, earplugs should be kept absolutely clean and free from chemicals, oil, or grease when being inserted into the ear canal. Earmuffs rarely cause infection or sensitization of the ear canal and are a good alternative where this occurs with earplugs.

Discomfort. The majority of earplugs are a little uncomfortable when worn correctly for long periods of time. However, the wearer should become used to them after a few days. In certain environments earmuffs may tend to feel hot and sweaty, although absorbent seal covers may somewhat alleviate this problem.

It should be noted that there is only a fine dividing line between those types of hearing protectors that are effective and acceptable and those that are effective and unacceptable to the wearer. Consequently those properties of hearing protectors that affect comfort should be considered as important as those that provide protection.

Cost. Where hearing protectors must be supplied in large quantities, cost becomes an important factor. The total cost of providing hearing protection to a group of personnel may be divided into three basic parts: expenditure due to initial purchase, cost of supplying spare parts and replacements, and cost of time spent in administering the hearing protection. It should be noted that the initial cost may not necessarily be the greatest figure; replacements and administration costs are often the largest factors involved.

Effects of Hearing Protection on Communication. Hearing protectors interfere with conversational speech when worn in quiet environments. However, wearing conventional earplugs or earmuffs in noise levels above about 85 dB(A) should not interfere with and indeed may improve speech intelligibility and the reception of warning signals for normally hearing ears. Encouragement to persevere with the new auditory environment is needed so that the initial feelings of insecurity in communication are forgotten.

Existing Noise-Induced Hearing Loss. Unfortunately, persons with noise-induced hearing losses may incur an added impairment when wearing hearing protection. It is probable that speech communication may be slightly impaired by the presence of protectors in such cases. Nevertheless, there is a greater need to protect the hearing of a person who already has a hearing loss. He has, so to speak, less hearing to lose than normally hearing persons and therefore requires greater protection. Although a worker may have experienced many years in a noisy environment without hearing protection and thus have adapted gradually to the implications of hearing loss on his work and social life, his hearing should still be protected to prevent further increase in any social handicap he may have.

Selection of Hearing Protection. It is doubtful whether earplugs or earmuffs alone can satisfy all the needs of a hearing protection program in any one organization. The obvious advantages of each should be utilized wherever possible, so that a hearing protection program may be made as acceptable as possible to the potential wearers. Furthermore, it can be advantageous to give workers a personal choice from a range of different types of potentially adequate protectors.

11-9. GENERAL REQUIREMENTS OF A HEARING CONSERVATION PROGRAM

The basic procedures involved in the instigation of a hearing conservation program are as follows.

a. *Measure the noise.* A detailed noise survey should be carried out in those areas thought to represent a possible hazard to hearing.
b. *Evaluate the hazard.* Measured noise levels should be compared with the appropriate current criterion for noise assessment and all machines, workshops, and noisy areas where this level is exceeded should be designated "noise hazardous areas."
c. *Noise reduction.* All hazardous noise sources should be reduced to a level below the criterion. If reduction of the noise at source or by the use of acoustic barriers is neither economic nor practical, then, as a last and temporary resort, a hearing protection program should be instigated.

11-9.1 Hearing Protection Program

To establish an efficient hearing protection program, the noise exposure patterns, both in terms of sound level and duration of exposure, should be determined for all persons, so that those areas where hearing protection is necessary may be pinpointed.

Monitoring Audiometry. Because hearing protection is seldom used to its best advantage by exposed persons and because current damage risk criteria cover only a certain percentage of a group of people rather than each individual, it is desirable to monitor the hearing of all personnel habitually exposed to noise. Although this type of measurement is often imprecise, regular monitoring audiometry acts as a valuable check to ensure that hearing protection is being used efficiently and that noise exposures have not increased significantly.

A good hearing conservation audiometric program might consist of the following.

a. *Pre-employment audiogram*, preferably measured over a wide range of audiometric frequencies, e.g., 250, 500, 1000, 2000, 3000, 4000, 6000 and 8000 Hz. A detailed pre-employment noise history and description of present work and noise environment. Individual education by the audiometrician or nurse about the hazards of noise and the need to wear hearing protection and individual fitting of the appropriate protectors.

b. *Monitoring audiograms* after 6 months and 1 year and then every year, together with further education and individual discussion about the noise environment and the use of hearing protection.

The initial pre-employment hearing tests must be carried out at the beginning of a working day, preferably after a weekend's rest away from the noise. Otherwise the hearing levels recorded are liable to include temporary changes in threshold due to noise exposure. A practical compromise for monitoring retests is to ask the worker being tested to wear well-fitting, good-quality earmuffs during *all* noise exposure in the 24 hours preceding the retest. If no significant deterioration relative to the initial audiogram is detected, the test result is accepted; if an apparent deterioration is found, he is tested again at the beginning of another working day. Because of the relatively large sources of variation often associated with individual audiometry, an apparent change in the measured hearing level at any one test frequency of the order of 15–20 dB should be recorded before a significant deterioration in hearing acuity can be presumed. If hearing losses are found to have developed or advanced by this amount since the *initial* test, stricter enforcement of the wearing of hearing protection or a change of working environment is necessary.

Conclusion. A hearing conservation program, if carried out with enthusiasm and perseverance and given full backing by management, labor, and medical staff, should drastically reduce the incidence of occupational noise-induced hearing loss within an industry. Perseverance and propaganda are often required to overcome employee reluctance to wear hearing protection properly.

It should be apparent that the cost and effort required to ensure successful hearing conservation by hearing protection may be equal to or even greater than that involved in other engineering noise control measures. A hearing protection program should not be considered as an easy or cheap alternative to the more desirable elimination of the noise from the working environment.

REFERENCES

1. "Specifications for Audiometers." American National Standards Institute, S3.6–1969, 1969.

2. "Specification for a Reference Zero for the Calibration of Pure-Tone Audiometers," Part 2: "Data for Certain Earphones Used in Commercial Practice." British Standards Institution, BS 2497, 1969.

3. "Pure-Tone Audiometers." British Standards Institution, BS 2980, 1958.

4. "Standard Reference Zero for the Calibration of Pure Tone Audiometers." International Organization for Standardization Recommendation ISO/R 389, 1964.

5. W. Burns, *Noise and Man*. Second edition. John Murray, London, 1973.

6. "Criteria for Background Noise in Audiometer Rooms." American National Standards Institute S3.1-1960, 1960.

7. "Information on Levels of Environmental Noise Requisite to Protect the Public Health and Welfare with an Adequate Margin of Safety." Environmental Protection Agency Report No. 550/9-74-004, 1974.

8. W. Burns and D. W. Robinson, "Hearing and Noise in Industry." HMSO, London, 1970.

9. G. R. C. Atherley and A. M. Martin, "Equivalent Continuous Noise Level as a Measure of Injury from Impact and Impulse Noise." *Annals of Occupational Hygiene*, **14**, 11–23, 1971.

10. C. G. Rice and A. M. Martin, "Impulse Noise Damage Risk Criteria." *J. Sound Vib.*, **28**, 359–367, 1973.

11. J. G. Walker and A. M. Martin, "Hearing Conservation." In, Petrusewicz and Longmore, eds., *Noise and Vibration Control in Industry*, Paul Elek, London, 1973.

12. Department of Employment, Code of Practice for Reducing the Exposure of Employed Persons to Noise. HMSO, London, 1972.

13. British Occupational Hygiene Society, "Hygiene Standard for Wide-Band Noise." Pergamon Press, Oxford, 1971.

14. "Assessment of Occupational Noise Exposure for Hearing Conservation Purposes." International Organization for Standardization Recommendation ISO/R 1999, 1971.

15. Safety and Health Standards for Federal Supply Contracts (Walsh-Healey Public Contracts Act). *Federal Register*, 34(96), Part 50–204, 1969.

16. Occupational Safety and Health Act, United States, 1970.

17. F. A. Van Atta, "Federal Regulation of Occupational Noise Exposure." *Sound and Vibration*, 28–31, May, 1972.

18. A. L. Cudsworth, "Noisy Equipment and Compensation Claims for Hearing Loss." *Engineering Digest*, February, 1971.

19. American Medical Association: Committee on Conservation of Hearing's Subcommittee on Noise, "Guide for the Evaluation of Hearing Impairment." *Transactions Amer. Soc. Opthalmol. Otolaryngol.*, March–April, 1959.

20. R. R. A. Coles and A. M. Martin, "Medico-Legal Aspects of Occupational Hearing Loss." *J. Sound Vib.*, **28**, 369–373, 1973.

21. Health and Safety at Work Act, Chapter 37. HMSO, London, 1974.

22. Factories Act, United Kingdom, 1961.

23. Department of Health and Social Security, "National Insurance (Industrial Injuries) Act, 1965" and "Occupational Deafness." Cmnd 4561, HMSO, London, 1973.

24. Ministry of Labour, "Noise and the Worker." HMSO, London, 1963. Republished in 1968 for the Department of Employment and Productivity, and in 1971 for the Department of Employment.

25. A. M. Martin, "The Assessment of Occupational Noise Exposure" *Annals of Occupational Hygiene*, 1974.

26. J. G. Walker and A. Behar, "Problems Associated with the Reproduction of Impulse Noise for TTS Studies and Impulse Noise Measurements." *J. Sound Vib.*, **19**, 349–354, 1971.

27. A. M. Martin and G. R. C. Atherley, "A Method for the Assessment of Impact and Impulse Noise with Respect to Injury to Hearing." *Annals of Occupational Hygiene*, **16**, 19–26, 1973.

12

The nonauditory effects of noise on health

DAFYDD STEPHENS*
and
GRAHAM ROOD†

12-1. INTRODUCTION

Noise has direct and specific effects on the cochlea and indirect disturbance effects on sleep. In addition it has a large number of nonspecific effects on performance and on many aspects of bodily function. In these respects it acts as a general stress, and as such can act and interact with other stressing factors.

Because of the nonspecific nature of the effect, and also because human beings are rarely exposed to an acoustical stress in isolation from other stresses, the exact importance of noise

*Royal National Throat, Nose and Ear Hospital, Gray's Inn Road, London WCIX 8DA, England.

†Royal Aircraft Establishment, Farnborough, England.

This work was performed while the authors were at the Institute of Sound and Vibration Research, Southampton, England.

and its effects in this context have proven more difficult to delineate and to specify than its direct effects on the inner ear.

As with its effects on psychological performance and on sleep disturbance, its short-term physiological effects may be subdivided into two main categories: the startle effect due to its sudden onset, and overall sustained effects of prolonged noise. These have been classified in terms of what Sokolov has described as the Orienting Response to the sudden onset, which is replaced by or modified to the Defense Response with sustained or repetitive stimuli of sufficiently high intensity. It has been argued that the defense response has a higher threshold than the orienting response in terms of the stimulus intensity necessary to elicit it, and that its effect is considerably more sustained.

Many studies have examined different aspects of the nonauditory physiological effects of noise. The approach of these human studies has followed two main lines: the short-term experimental effects of noise exposure measured in experimental subjects, and the long-term retrospective effects on disorders of bodily function. From a medical viewpoint the latter are the most important, but in view of the nonspecific effects of noise, it is very doubtful whether the effects of the noise can be separated from the effects of such other, related factors as the conditions of work, vibration exposure, and the self-selected nature of the labor force studied.

Further approaches have used animals in experiments to evaluate the effects of noise. There again much caution must be applied in any consideration of the applicability of the results to man, because of considerable interspecific differences. Thus, the well-known phenomenon whereby audiogenic seizures may be induced in certain laboratory animals by moderate levels of acoustical stimulation has never been found in man.

A further point here concerns the meaningfulness of the stimuli. This has been shown to influence certain physiological responses to sound partly by a direct effect and partly by the differential amounts of annoyance engendered by stimuli of different meaningfulness or emotive values.

The overall findings in both animal and short-term human experiments have shown very marked and consistent effects. However, the question of the long-term effects of noise on different aspects of the physical and psychological health of those exposed remains much more of an open question. It may be that these permanent effects occur only in the more susceptible individuals, and this will be discussed further later.

The level at which the physiological effects of noise begin to occur varies according to the physiological function measured, but there are few changes of any significance induced by sounds of less than 70 dB(A). Jansen [1] has proposed a series of damage risk criteria (Fig. 12-1) ranging from a level of possible physiological reaction at 60 dB SPL to a possible beginning of injury at 95 dB. He has further shown some frequency dependence of the physiological effects

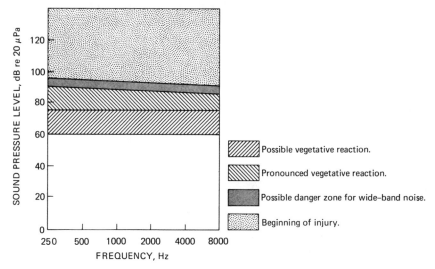

Fig. 12-1 **Damage risk criteria, after Jansen [1].**

of the noise, with greatest effects produced by stimuli in the frequency range of 2000-3000 Hz. The concept of damage risk criteria in the context of the nonauditory effects of audio-frequency sound is less relevant than that of ultrasound and infrasound, as the inner ear is the part of man most susceptible to damage by stimuli in the audio-frequency range. Thus the levels will be more severely limited by the damage risk criteria derived from extensive studies on hearing damage than by anything derived from other physiological effects. The auditory damage risk criteria are extensively discussed in Chapter 11.

From a physiological point of view the nonauditory effects may be regarded as occurring in three stages, each mediated by different physiological mechanisms. There is, first, the rapid tensing of the muscles at the sudden onset of a noise. This is mediated by the main motor nerves. This tensing is followed by slightly slower effects mediated by the autonomic nervous system, comprising changes in the heart rate, respiratory volume, blood vessel diameter, secretions, etc. Finally, there are the effects mediated by hormonal activity, which are controlled largely via the pituitary–adrenal axis, including both parts of these two glands: the neuro- and adenohypophyseal parts of the pituitary, and the adrenal cortex and medulla. The actions of these glands are largely controlled by the part of the brain known as the hypothalamus.

The approach of this chapter is to consider the different effects of audio-frequency noise on the various bodily systems, individually, and to follow this with a consideration of the more specific effects of very low and high frequency sound.

12-2. EFFECTS OF AUDIO-FREQUENCY NOISE

12-2.1 Cardiovascular Effects

The earliest and best documented effect of noise on the cardiovascular system is that it produces a peripheral vasoconstriction. That is to say, it causes a constriction of the small blood vessels in the limbs, particularly in the skin, which results in a reduced blood volume and blood flow in these parts of the body. At the same time, with moderate levels of acoustical stimulation there is a vasodilation with an increase in the blood flow to the head. From an evolutionary point of view this mechanism is considered to be important in preparing the body for avoiding action against whatever threat may be causing the noise. In this context, it is interesting to note that some of the most recent work has shown that this change may occur in response to any change in noise level, and not merely to an increase beyond a certain critical level. Such a change may be a decrease in sound level, as well as an increase.

However, at really high levels of noise there is a vasoconstriction or reduction in the blood flow to the head as well as that to other parts of the body, and this can result in impaired performance. The sort of sustained, high level noise which could cause this would be inconceivable in man's early evolution and is produced only by the benefits of modern technology. It has been agreed that at high noise levels, the change which occurs is from what Sokolov has described as the Orienting Response (OR), a kind of alerting reaction, which precedes what has been defined as the Defense Response (DR). Apart from the fact that these two responses may affect cardiovascular and other bodily functions in different ways, they also have different time courses, with the OR habituating (i.e., its effect becoming smaller) quickly, whereas the DR is a much more sustained reaction.

A particularly critical aspect of this restriction in the blood flow to the head at these high levels is that it affects the head as a whole, including the blood supply to the inner ear. Thus, at the time when the sensory cells of the inner ear are in most need of a good blood supply to provide energy sources and remove metabolites resulting from this high level of auditory metabolic activity, they are in fact receiving an impaired blood supply which enhances their susceptibility to the damaging effects of the noise. The general effects of damage to the inner ear caused by noise are discussed in Chapter 11.

Many different factors influence the blood flow to the limbs, and the role of noise is relatively minor. The changes in blood flow which may be measured with exercise, for example, are considerably greater than those induced by acoustical stimulation. The effects of noise, may, however, be superimposed on the effects of these other factors. Thus, the increase in blood flow produced by exercise is reduced when loud noise is introduced during the exercise, although the reduction is less than that necessary to return the blood flow to its resting level.

Recent studies on the relationship between the peripheral vasoconstriction and the temporary threshold shift or TTS (the reversible impairment of hearing following exposure to an intense noise) have indicated an inverse relationship between these two factors: people who show a large fall in the blood flow to their fingers when exposed to loud noise show relatively little change in their hearing, whereas those showing little change in their finger blood flow show large shifts in their thresholds of hearing. It has been reported that the vasoconstrictive effect is more pronounced in the non-noise exposed Mabaan people, who had been previously reported as having unusually sensitive thresholds of hearing. The Mabaans also show more rapid recovery from the effect once the noise is stopped.

It is also interesting to consider a related study which indicated that the increase in the amplitude of the pulse induced by noise exposure is negatively correlated to the TTS in normally hearing subjects but positively correlated with it in subjects with hearing loss. (It might also be noted that in patients with noise-induced hearing loss, the amount of TTS induced by a given noise exposure is negatively correlated with their degree of permanent hearing loss.)

The vasoconstrictive effect of noise is probably the most extensively studied of its nonauditory effects, and in fact seems to persist longer than other nonauditory effects. It may also be affected by the personality of the subject, being related to his anxiety. This is perhaps not surprising as blood flow measures have been widely used as measures of anxiety, the patient's forearm blood flow being positively correlated with his anxiety level.

In addition to the effects on the peripheral blood vessels, noise has been reported as having effects on other aspects of the cardiovascular system. Changes have been reported in blood pressure, heart rate, cardiac output, and pulse volume. All these components are interrelated, and a change in one may itself induce a change in one of the others. Thus, in response to noise exposure, in addition to the peripheral vasoconstriction, there may be an increase in the diastolic blood pressure, which habituates rapidly, together with minor adjustments to the heart rate. These responses are somewhat variable both in man and in experimental animals, as are the changes in cardiac output (the volume of blood pumped out by the heart per unit time), some subjects showing an increase and others a decrease in cardiac output with noise exposure. An interesting factor in relation to the effects of noise on the blood pressure is that, although the overall effect on normal subjects is negligible, noise exposure will cause a small but definite further increase in pressure in patients whose blood pressure is already high.

From the long-term point of view, there is some evidence from a study in Germany that workers in noisy environments suffer more from cardiovascular disease than those doing similar work in a quieter environment. Perhaps relevant to this are some recent animal studies showing increased levels of cholesterol in the blood and increased atheromatous deposits in the arteries of animals exposed to high levels of noise over a long period of time.

12-2.2 Effects on the Digestive System

It is well known that stressful circumstances can cause an increase in the activity of the digestive tract in the form of increased motility, with waves of contraction passing along the intestines, causing the contents to pass through more quickly and producing either diarrhoea or frequent loose motions in susceptible individuals. In other individuals, stress may cause increased secretion of hydrochloric acid by the stomach, which, coupled with abnormal contractions of the stomach, may lead to dyspepsia and ultimately to peptic ulceration.

The direct effects of noise within this framework are mainly to increase gastrointestinal motility, which increases with increasing noise level, and decreases with decreasing noise level. These changes are reduced with repeated noise stimulation. An interesting fact in human experimental studies is that the noise-induced increase in gastrointestinal motility is greater in subjects who are able to stop the noise but continue to subject themselves to it nevertheless, compared with those who do not have this option. This implies, as in most effects of noise, a complex interaction between noise and psychological effects, both of which are mediated through the parasympathetic part of the autonomic nervous system.

From the point of view of long-term effects of noise on the gastrointestinal tract, there have been two studies which, while not specifically excluding the interactive effects of the monotony of the work and possible vibratory effects, have shown a very high incidence of gastrointestinal disorders (65% of subjects) on X-ray examination of workers exposed to high levels of occupational noise over a period of 15 years or more.

12-2.3 Effects on the Respiratory System

The respiratory system plays a vital role in the metabolic function of the body; perhaps because of this, noise has relatively little effect on aspects of its function. The only definite effect of noise stimulation which has been convincingly demonstrated is to induce slow, deep breathing, which ensures optimal ventilation of the lungs and hence efficient oxygenation of the blood.

This increase in depth of ventilation is related to the level of the acoustical stimulation, high levels (about 120 dB) producing a greater increase than lower levels, although with repeated stimulation the relative effect of the high level stimulation tends to decrease and that of lower levels of stimulation to increase.

No long-term effects of noise stimulation on the respiratory system have been shown.

12-2.4 Effects on the Central Nervous System

The effects of noise on the central nervous system may be subdivided into the psychological effects and the more specific neurological effects. In addition, all other nonauditory effects of noise must be mediated through the central nervous system either by its control of the voluntary nervous system, the autonomic nervous system, or its control of the endocrine system via the hypothalamus.

Those psychological effects of noise are largely mediated by its effect on the level of arousal of the person and its distraction effects, which are extensively discussed in Chapter 14. The psychopathological effects tend to be rather more nebulous, difficult to measure, and complicated by the fact that noise acts in this context as one of a series of often interrelated stresses which may affect the susceptible individual.

The best known study on the psychopathological effects of noise exposure compared the mental hospital admissions from two sections of the London Borough of Hounslow situated immediately to the east of London (Heathrow) Airport. One of these sections was a maximum noise area, in which the noise levels from the aircraft exceed 100 PNdB and where the Noise and Number Index (NNI) exceeded 55. In the other, the level was lower than this.

The study found that mental hospital admissions were significantly higher among people living in the high noise area, and particularly among single women. Subsequent work, however, has shown no more than a trend, albeit consistent, for an increased incidence of hospital admissions from a high noise area.

Other workers have reported a "mild neurotic depressive reaction" on sustained noise exposure and more increases in neurotic symptoms in older individuals working in a high noise environment than in those working under quieter conditions. Further work has suggested that already neurotic individuals tend to be more susceptible to noise stress, developing additional neurotic symptoms under such conditions.

The relationships between these and various psychosomatic conditions remains unclear. One thing which is certain, however, is that all the nonauditory medical conditions which have been reported as being precipitated by noise exposure, have also been considered as having a psychosomatic etiology, so whether in these circumstances noise is acting as a specific factor or as a general, nonspecific stress is debatable. In the context of mental illness, it has been argued that noise can be the "straw that breaks the camel's back" in individuals already under stress because of their work, interpersonal relationships, and a plethora of other factors.

From a neurological point of view, most interest has centered around the audiogenic seizures which may be provoked by relatively low level acoustical stimulation in certain experimental animals. These seizures have been induced in mice, rats, rabbits, chickens, and dogs, and even in cats and monkeys previously treated by certain drugs such as methamine sulfoxine. These same drugs also lower the threshold for seizures in susceptible animals.

Details of the audiogenic seizures in animals vary from species to species, but in the most studied species, mice and rats, the seizures may be divided into four distinct stages. A few seconds after the onset of the stimulus the animal starts to run wildly about its cage, its motion changing gradually to a series of stiff-legged bounds; this is the first stage. The second, or myoclonic, stage consists of a series of convulsive, rapid jerks of all four limbs, ending with the hind limbs stiffening and being drawn right into the body. In the third, or myotonic, stage the hind limbs descend gradually to a fully extended position, with most of the muscles in tonic contraction. The fourth stage is the death of the animal, caused by respiratory failure brought about by the tonic contraction of the respiratory muscles.

Different species of animals are sensitive to different stimulation frequencies. The most effective frequency band varies from 4 kHz to 60 kHz according to species, and the level required varies between 90 and 134 dB. Different strains of mice show different susceptibilities, and this has been shown to be transmitted from generation to generation on a normal genetic basis.

One of the most interesting and specific aspects of audiogenic seizures is the fact that the degree of suceptibility of a mouse may be greatly enhanced by priming the animal. This consists of exposing the animal at 16–19 days after birth to 30 seconds of intense sound (e.g., 103 dB electric bell). It has been argued that the priming has some effect on the developing neural structures at this period.

Fortunately, such a phenomenon does not normally occur in man. In the very few patients in whom epileptic fits have been induced by acoustical stimulation, the evidence seems to be that these responses are to very specific and meaningful acoustical stimuli (e.g., Strauss waltzes), and that the majority of such rare patients may be cured or improved by the use of a behavioral conditioning technique.

12-2.5 Effects on the Special Senses

Apart from its effects on the function of the inner ear, noise has been shown to have effects on the function of two of the special senses, vision and balance (as sensed by the vestibular labyrinth).

That the vestibular labyrinth may be affected by auditory stimulation is not surprising, considering its embryological and evolutionary development from the same source as the inner ear, and its proximity and contiguity with that organ. The vestibular labyrinth on each side consists of five principal components: three semicircular canals, which are sensitive to angular acceleration, and the utricle and saccule, which are sensitive to the position of the head and to gravitational forces in general.

In the 18th century Erasmus Darwin reported that certain patients with vestibular disorders could be made dizzy by the sound of waterfalls. In the early part of this century a number of experimental studies showed that in animals the vestibular system could be stimulated by high intensity sounds. More recently it has been found that when humans are stimulated with very high intensity sound they become dizzy, and the eye movements known as nystagmus may be observed and recorded. The horizontal nystagmus normally considered is evoked by stimulating the lateral semicircular canal on one side, and consists of the eyes moving relatively slowly towards the opposite side and then flicking back quickly. This pattern of eye movements is repeated.

Although in normal individuals very high intensity acoustical stimulation is necessary to evoke such a response (about 130 dB), in patients with certain disorders involving the inner ear and vestibular labyrinth, there may be a hypersensitivity to acoustical stimulation, so that considerably lower intensities (90-100 dB) may provoke vertigo and nystagmus. This is known as the Tullio phenomenon. The interesting aspect of this is that a number of these individuals are sensitive to only a narrow band of test frequencies and are considerably less sensitive to stimulation by other frequencies. Although it was originally considered necessary for there to be two mobile windows in the scala vestibulae of the labyrinth for this phenomenon to occur (e.g., after a fenestration operation for otosclerosis) it has recently been shown that about a quarter of those patients with Ménière's disorder who do not have such an abnormality are abnormally sensitive to acoustical stimulation in this way.

The effects of noise on vision are less direct than those on the vestibular labyrinth, but as with stimulation of that organ, they are temporary in nature and there is no definite evidence for any long-term damaging effect. Many of the effects which have been demonstrated in the past have been somewhat variable, noise stimulation producing a consistent effect in some subjects and little or no effect in others.

The effects which have been described include a narrowing of the visual field, which has been reported by some workers as possibly becoming a permanent effect in people working in a high noise level environment over a long period of time. Other work has suggested that noise may temporarily impair visual acuity, color vision, and the critical flicker fusion frequency (CFF), the frequency of flicker above which a flickering light may be perceived as steady. There is evidence to suggest that noise may cause a reduction in the CFF, but not all studies have substantiated this finding.

More recently, noise has been shown to prolong the integration time of the eye and also to impair the brain's ability to select relevant from irrelevant visual information. Peripherally in the eye itself, noise stimulation has been shown to induce dilation of the pupil, this dilation increasing with the intensity of the stimulus.

12-2.6 Effects on the Endocrine System

Figure 12-2 schematically illustrates the major part of the endocrine system. (The parathyroid glands, which produce parathormone and regulate the calcium

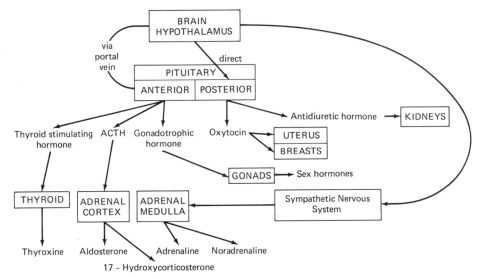

Fig. 12-2   Schematic diagram of endocrine glands and hormones discussed in text.

and phosphate balance of the body, and the islets of Langerhans, which produce insulin and control the sugar balance, have been excluded.)

It may be seen from this figure that all the components of the endocrine system which are affected by noise are directly or indirectly under the control of the brain. This is perhaps inevitable, as noise effects must be mediated initially by nervous pathways arising from the inner ear.

It may also be seen that a vital link between the brain and the rest of the system is the pituitary gland, situated in a bony cavity in the skull beneath the brain. The gland consists of two main parts, the adenohypophysis or anterior pituitary and the neurohypophysis or posterior pituitary. The adenohypophysis is connected to the part of the brain known as the hypothalamus by a blood supply which carries chemical messenger substances which control the activity of this part of the pituitary. The neurohypophysis, on the other hand is linked directly by nerve fibers to the hypothalamus.

The adenohypophysis produces a number of hormones which pass into the blood stream and control various aspects of bodily function. Among the hormones produced here are growth hormone, thyroid stimulating hormone (TSH), adrenocorticotropic hormone (ACTH), and gonadotropic hormone. In experimental animals, noise, like many other stresses, causes a fall in the output of TSH which will result in a fall in the output of the thyroid hormone, thyroxine. This in turn can cause an elevation of the cholesterol level in the blood and may be related to the deposition of atheromatous plaques in blood vessels. However, this effect of noise has only been conclusively demonstrated in experimental animals and not in man.

The effects of gonadotropic hormone will be considered in a later section. What is a more significant effect of stress on an animal is the increased output of ACTH, resulting in an increased output of steroid hormones from the adrenal cortex. In man the hormone principally involved is 17-hydroxycorticosterone. Aldosterone, which is concerned particularly with the control of blood levels of sodium and blood volume, and has its effect on the kidneys, is also released in increased levels into the bloodstream, although this release is not directly triggered by ACTH. The other corticosteroids regulate the general metabolism of carbohydrates, proteins, and fats on a long-term basis, and reduce the inflammatory response in the body and also the allergic response. The level of ACTH is increased by noise exposure, which therefore has widespread effects throughout the body.

It has been shown that, in experimental animals exposed to moderate levels of noise, there is a rise in the level of the corticosteroid hormones, followed by a fall. With higher noise levels, the corticosteroid levels remain elevated over a long period of time and so have more pronounced effects on bodily function. This may be shown by a reduced resistance to infection, which will be considered in a later section, and by changes in different components of the blood. Thus there may be a decrease in the eosinophils (white blood cells involved in allergic reactions), increases in certain plasma proteins (e.g., albumin and alpha-globulin), and decreases in others (e.g., gamma-globulin and fibrinogen). The significance of and the degree of these effects in man remains somewhat uncertain.

The neurohypophysis secretes two hormones, oxytocin and vasopressin (antidiuretic hormone). Although oxytocin's main effects are on the uterus and the lactating breast, it can also have an effect on the fluid balance of the body, influencing the excretion of sodium by the kidneys. In rats exposed to loud noises, the level of oxytocin in the blood has been shown to be increased, together with the excretion of sodium. Similar effects have not so far been demonstrated in humans.

The final endocrine gland to be considered in this section is the adrenal medulla, which developmentally and functionally has little to do with the adrenal cortex. It is controlled directly by a nerve supply from the sympathetic nervous system, and produces two hormones, adrenaline and noradrenaline. Adrenaline is the hormone generally associated with the "fight–flight" response to a stressful situation, causing selective peripheral vasoconstriction and increase in the heart rate, dilation of the pupils, etc., whereas noradrenaline has a more general vasoconstrictive function. The blood levels of both of these rise with acoustical stimulation, and in animal studies the increased level of adrenaline is sustained over a long period of time, whereas the change in noradrenaline levels adapts more rapidly. The extent to which this occurs in man and its possible significance is unclear, although it is reassuring to note that at least one experiment has shown that noise has less effect on the blood levels of these hormones than the subject's attitude towards the experiment.

12-2.7   Effects on the Reproductive System

Noise exposure has been reported as having a variety of effects on reproductive function, the mechanisms for which can be understood in light of the foregoing sections.

In conjunction with the general vasoconstrictive effects of the noise, it is not surprising that it has been shown to cause a reduction in the blood flow to the placenta, which is concerned with the nourishment of the fetus within the uterus. Related to this have been reports of low birth weight both in animals and in one population of humans exposed to high levels of noise, together with a reduced litter size in experimental animals. Various developmental abnormalities and impaired bone formation have also been reported in the fetuses of animals exposed to high levels of noise, and the scientists concerned have related these effects to the alterations in the hormone balance in the mother induced by the noise. There is no evidence for noise causing developmental abnormalities in man.

The second effect comes via the influences of the neurohypophysis on the oestrous rhythm, an effect excited by the gonadotropic hormones. In this context noise has been shown in experimental animals to cause changes in the oestrous rhythm, with reduced fertility but no effect on the reproductive behavior of the animal. The effect of discotheque noise on the latter does not appear to have been scientifically studied in man.

12-2.8   Effects on the Skin and Musculo-Skeletal System

Very little in the way of effects of noise has been demonstrated in this context. The onset of a noise, or a change in its level, causes a brief change in skeletal muscular tension, and this may be recorded using electromyography (EMG). This effect adapts very quickly and is not specific to acoustical stimulation.

Changes in the skin consist of a fall in the blood flow caused by the general vasoconstrictive effects of the noise, coupled with a transitory galvanic skin response. This is an increase in the conductivity of the skin as measured between two electrodes, and is caused by an increased secretion of the sweat glands in the skin. Like the EMG response, this shows marked habituation on repeated stimulation.

The threshold of pain in the ear may also be regarded as a "skin" response as this is a function of the motion of the tympanic membrane (eardrum), which is derived in part from ectodermal structures. Figure 12-3 shows the threshold of pain as a function of stimulus frequency, and this is considered further in a

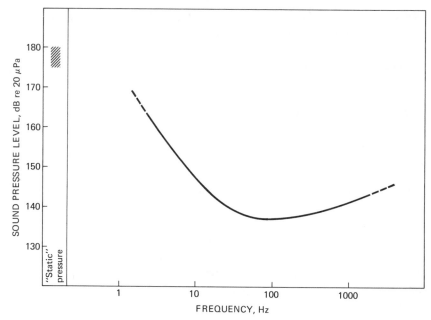

Fig. 12-3   Pain thresholds as a function of frequency.

later section. Studies showing that the pain is mediated by the tympanic membrane have essentially been based on the fact that patients with a complete sensorineural hearing loss, but intact tympanic membrane, have a normal pain threshold, whereas those with a large perforation of the tympanic membrane and only moderately impaired hearing, have pain thresholds which are not measurable.

12-2.9   Possible Permanent Effects of Noise on General Health

Studies of the permanent effects of noise on the general health of human beings up to the present time have been inconclusive, but are suggestive of deleterious effects on the health incurred by long-term exposure. As one of a series of non-specific stresses, noise almost certainly has effects on a number of psychopathological and psychosomatic complaints.

The main problem arises from the fact that studies on this problem are either in the form of animal experiments, which cannot necessarily be extrapolated to human beings, or in the form of human retrospective investigations, in which it is impossible to control all the variables which might have some bearing on the outcome of any study.

In animal studies, noise exposure has been shown to cause an increased susceptibility to viral infection and even to viral leukemia. The mechanisms here are probably related to the increased ACTH levels in the exposed animals, which leads to reduced inflammation and interferon responses in the animals concerned. However, at the time of writing, viruses have not yet been incontrovertibly implicated in any form of malignant neoplastic disease (cancer) in man.

Studies on humans working in noisy environments have suggested an increased incidence of heart disease, problems with peripheral circulation, vestibular problems, absenteeism, psychological problems, and accidents at work. However, it is almost impossible in such studies to control completely factors such as working conditions, the exact nature of the work, the workers' attitude toward it, vibratory effects, and chemical pollution, all of which could have some bearing on any effects found.

What does seem probable is that noise will have its greatest effects on those individuals who, by the nature of their constitution or because of a preexisting disorder, have increased sensitivity to disturbance and deterioration of their condition caused by the noise. Thus, noise seems to have its most marked effect on blood pressure in those individuals whose blood pressure is already elevated, may produce vertigo in those with already damaged vestibular systems, and precipitate a mental disorder in those already suffering from minor psychological problems.

12-3. INFRASOUND

Infrasound is arbitrarily defined as low frequency non-audible sound, and normally encompasses all frequencies below 20 Hz. The term "non-audible" is used loosely; recent research has shown that hearing thresholds extend to frequencies considerably lower than 20 Hz, with thresholds measurable down to 1.6 Hz, albeit at relatively high sound pressure levels (approximately 130 dB at 1.6 Hz).

Infrasound occurs both naturally and artificially in many guises. Naturally occurring infrasound may be produced by winds, thunderstorms, earthquakes, volcanic eruptions, sea waves, auroras, and many other sources, although the sound pressure levels of this type of infrasound are generally low. Where artificial infrasound occurs (i.e., from man-made devices), levels are generally considerably higher than those found naturally, and they are both growing rapidly in intensity and becoming more common. Infrasound at these higher levels may be found in offices (due to air conditioning and ventilating systems), and in the vicinity of large reciprocating engines (e.g., diesel motors in ships), oil burning furnaces, pumps, motor vehicles, helicopters, and gas turbine and rocket motors. Since these levels greatly exceed those generally caused by natural sources, man-made infrasound probably produces the greater potential threat of adverse effects on humans.

To date, there is no really strong evidence that exposure of humans to low frequency sound energy has any significant effects below 130 dB. There are, of course, exceptions to this generalization, as is shown by the effects of infrasonic noise on damaged vestibular labyrinths, but in general, in normal humans, no significant effects are noted until the quoted level is exceeded.

Threshold levels at which detrimental whole-body effects appear are in the region of 150 dB, at least for infrasonic frequencies. At frequencies higher than 20 Hz, human tolerance limits are rapidly reached, Mohr and co-workers [2] showing that levels of 153 dB at 50 Hz, 154 dB at 60 Hz, 150 dB at 73 Hz, and 153 dB at 100 Hz are probable human tolerance limits. At these levels, the effects on the vestibular system for normal humans are noticeable, and corresponding effects on respiration are just detectable above 140 dB, and definitely noticeable at levels around 166 dB.

In the field of performance effects, evidence showing a decline in performance due to infrasonic exposure is not strong enough at present to show that infrasonic exposure causes any particular hazard, at least at levels currently found in the natural and the man-made environments.

12-3.1   The Research of Mohr, Cole, Guild, and Von Gierke [2]

The research of Mohr and co-workers in 1965 provided the first set of limits of human tolerance to intense, low frequency, airborne sound. This study remains the most extensive in this field, and as such will be described in some detail. Whole-body exposure was used in order to attempt to establish human tolerance levels to noise, and examples of the test spectra and tones are shown in Fig. 12-4.

The major conclusions were that human subjects wearing ear protection were able to tolerate broadband and discrete frequency noise in the 1-100 Hz range at 150 dB for short periods of time. Above 150 dB, however, it was felt that the subjective tolerance limit was being reached and that reliable performance was liable to be impaired, at least for the frequency range above 40 Hz. In addition, the individual results are of importance, and some of the main nonauditory conclusions are as follows.

In broadband noise with a peak spectrum level of 128 dB at 40 Hz, but including energy at higher frequencies, heart rate was increased to 10-40 percent above the resting level. Two subjects reported mild nasal cavity vibration (one of these noted perceptible throat fullness), and two others reported mild chest wall vibration. In further tests in broadband noise, but with less high frequency energy, speech signals were completely masked and the speech of one subject was modulated. All subjects reported mild to moderate chest wall vibration, while three subjects noted perceptible, though tolerable, interference with the normal respiratory rhythm; two of these latter subjects noted throat pressure. During all broadband noise exposures, every subject considered the exposures as tolerable for the durations listed, and no changes from normal were experienced in visual acuity, hand coordination, or spatial orientation.

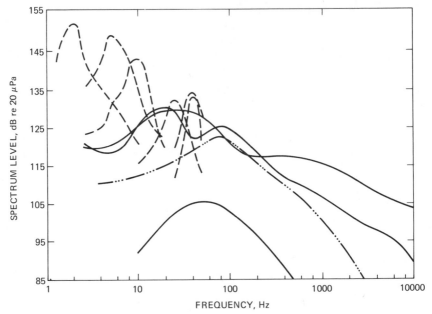

Fig. 12-4 Spectra of acoustic stimuli used by Mohr and co-workers [2].

During the narrow-band noise experiments, where SPLs of 142–150 dB (peak-ing at 2 Hz, 5 Hz, 12 Hz, and 40 Hz) were used, the most conspicuous effects occurred during exposure without hearing protection and at the lowest exposure frequencies (<12 Hz). Uncomfortable sensations, suggesting pressure build-up in the middle ear, were experienced requiring frequent Valsalva manoeuvres to relieve. Mohr and co-workers noted that this phenomenon had been reported earlier in German submarines. Substantial reduction of the effect was possible by the use of insert earplugs, and earmuffs alone helped to prevent this sensation. Without hearing protection, three subjects described an occasional tympanic membrane tickle sensation and one subject experienced noticeable nostril vibra-tion. During exposure to infrasonic noise peaking at 5 Hz, mild abdominal wall vibration occurred in one subject. No shifts were found in the threshold of hear-ing as measured one hour after exposure. Subjects considered that, if hearing protection was worn to prevent the outer ear pressure changes, all of these infrasonic exposures were within tolerance limits.

During tests where the frequency spectrum peaked at higher frequencies (40 Hz), all subjects found perceptible vibration of the visual field, moderate chest wall vibration, and a sensation of hypopharyngeal fullness (gagging). During brief periods without hearing protection, two subjects out of three experienced eardrum pain, the third experiencing no sensation of tickle or pain. While intelligibility scores exhibited no change from control values, recorded speech sounds were audibly modulated. Postexposure fatigue was present, and although subjects did not consider the group of exposures as pleasant, they were considered tolerable.

The remaining tests were carried out at discrete frequencies between 40 Hz and 100 Hz, and while by definition these do not occur in the infrasonic region, the nonauditory effects are substantial and should be noted. Voluntary tolerance was reached at 153 dB (50 Hz), 154 dB (60 Hz), 150 dB (73 Hz), and 153 dB (100 Hz). The subjective effects were often alarming. At 100 Hz they included giddiness, mild nausea, subcostal discomfort, cutaneous flushing and tingling. Coughing, severe substernal pressure, choking respiration, salivation, pain on swallowing, giddiness, and hypopharyngeal discomfort were reported at 60 and 73 Hz. One subject reported a headache at 50 Hz, and another developed both headache and testicular aching at exposure to the 73 Hz discrete tone. During the 43, 50 and 73 Hz exposures all subjects suffered from significant decrements of visual acuity. All subjects complained of marked postexposure fatigue.

As would be expected, considerable individual variation occurred in the responses. Mohr and co-workers, however, concluded that "noise-experienced human subjects, wearing ear protection, can safely tolerate broadband and discrete frequency noise in the 1-100 Hz range for short durations at sound pressure levels as high as 150 dB. At least for the frequency range above 40 Hz, however, such exposures are undoubtedly approaching the limiting range of subjective voluntary tolerance and of reliable performance." [2]

12-3.2 Cardiovascular Effects of Infrasound

The limited research directed toward cardiovascular changes associated with infrasonic exposure, has produced little evidence that infrasonic noise causes cardiovascular symptoms.

In a general sense, many researchers have noted that physiological effects of noise include constriction of the peripheral blood vessels of the skin as a characteristic reaction of the sympathetic nervous system. The degree of reaction appears to be directly proportional to noise level, independent of frequency, but proportional to bandwidth. Thus it would be reasonable to assume that some degree of vasoconstriction may be expected during exposure to infrasonic noise.

During tests with unanesthetised human subjects, levels of up to 144 dB have been used, with each subject exposed to the noise for an 8 minute period, and no significant changes in ECG patterns were noted. No significant variations in heart rate were found when young subjects were exposed for 50 minutes to a 7.5 Hz pure tone at 130 dB. The only change was a discrete, but significant increase in the diastolic blood pressure. Perhaps surprisingly, the systolic blood pressure did not vary significantly.

12-3.3 Endocrine and Digestive Effects of Infrasound

Since both endocrine and metabolic changes occur during exposure to audible noise, similar effects might be expected with infrasonic stimulation. Few data have, however, been presented. Digestive disorders have been reported as occurring with 16 Hz stimulation. A positive correlation has been found between high levels at stress-inducing frequencies in the region of 50 Hz, and eosinophil levels in the blood. High intensity sounds of 120–140 dB at low frequencies induce increased ascorbic acid content in the adrenals of certain strains of rats.

12-3.4 Respiratory Effects

Sound pressure levels of a high enough intensity affect the respiratory system, generally as a result of the respiratory cage or chest being physically compressed, as reported by many exposed subjects. Subjects exposed during Mohr's experiment [2] to broadband noise with a peak spectrum level of 128 dB at approximately 40 Hz, reported mild to moderate chest wall vibration, while three subjects reported perceptible, though tolerable, interference with the normal respiratory rhythm, as noted in Section 12-3.1.

Experiments using infrasonic levels greater than 120 dB with one ear covered and the other free have shown physiological effects which include difficulty in breathing (the breathing frequency of half of the subjects increased at 140 dB) and changes in systolic rhythm.

Modelling of the human body has shown that chest resonance for airborne excitation occurs at between 50 and 60 Hz, and the resonance frequency is proportional to approximately the cube root of body weight. The human thorax-abdomen resonance frequency for airborne infrasonic excitation has also been calculated from the same model and verified, to some extent, from later experiments.

Later stuides of the effects of infrasonic acoustic frequencies upon animal respiration using a small baboon, a monkey, and six dogs in both anesthetized and unanesthetized conditions, showed that with anesthetized animals a reduction of the respiration rate occurred at levels above 166 dB for all animals, while above 172 dB, at frequencies up to 8 Hz, the respiration of the larger dogs ceased. In unanesthetized animals, however, no obvious change in respiration rate was observed.

This and other research suggested that the effect on respiration could be dangerous if it caused excessive hyperventilation of the lungs, as may occur at levels much over 172 dB.

One of the effects consistently noted in experiments with humans during whole-body exposure is voice modulation when the subject speaks into a microphone. This is probably due to the pumping of the chest wall, causing air movement in the trachea. The tickling or choking noted by the subjects during the research of Mohr and co-workers [2] is probably also due to the airflow through the trachea, in this case causing drying of the mucous membrane in this area, leading to tickling or cough reactions.

12-3.5 Effects of Infrasound on the Central Nervous System

The psychological effects of infrasound are somewhat indirect and the evidence for them is largely in the disturbed behavioral patterns of exposed individuals. It has been suggested for some time that intense weather phenomena, such as thunderstorms or high winds, influence human behavior in subtle ways. The effects of infrasonic exposure in the Chicago area of Illinois on automobile accidents and absenteeism among school children has been studied. The infrasound in the Chicago area was produced by severe storms occurring in other regions of North America, up to 1500 miles away, while the weather conditions in Chicago were mild.

Results suggested that there was a positive correlation between the presence of strong natural infrasonic disturbances and the frequency of automobile accidents, and that a similar relationship existed between similar strongly occurring natural infrasound and the absenteeism rates of students in elementary schools.

Previously, work had shown that automobile accident rates were statistically significantly higher during the 4 hour period preceding the Föhn wind in the city of Zurich, and suggested that the pre-Föhn weather conditions had biological effects. Various attempts have been made to explain this phenomenon, including speculation that the effects of combining climatological and man-made infrasound in automobiles may produce the noted spatial clustering of traffic accidents.

At a more physiological level, exposures to frequencies between 245 and 700 Hz at 137 dB in one ear have been reported as producing "a striking and exceedingly prompt abolition and desynchronisation of the parieto-occipital alpha rhythms of the brain." Since the median frequency of the alpha rhythm is around 12 Hz, there have been suggestions that intense infrasound at the relevant frequency may prove particularly hazardous. It has also been suggested that infrasound at lower intensities, but synchronized with the alpha waves of the subject's brain, might enhance intellectual capacity, and that it may also trigger disturbances in very unstable personalities.

As a further step to the suggestion that brain waves may be controlled or driven, an attempt has been made to drive the alpha rhythm of the brain with infrasonic acoustic stimuli, where pure tones were interrupted rhythmically at rates of 1.5–25 per second, and EEG changes (in two out of eight cases) showed acoustic driving in the temporal lobe areas of the brain.

12-3.6 Effects on the Special Senses

As in the case of audio-frequency sound, infrasound has been found to have effects on the vestibular and visual systems.

Mild balance disturbances have been induced by infrasound, at levels of 140 dB and higher, applied monaurally to normal subjects and to patients with mild vestibular disorders. In cases where a subject with an abnormal labyrinthine system was exposed to levels of 105 dB at 2 Hz, severe loss of balance, nausea, and extreme subjective discomfort were experienced. This sound pressure level at 2 Hz is, however, sub-threshold for normally hearing subjects. In another subject with a labyrinthine disorder, levels of 140 dB at 1 Hz induced severe discomfort.

It has been suggested that, at levels of 140 dB, the lower level of the stimulation threshold for balance disturbances in normal subjects is being reached, while it is exceeded in patients with certain vestibular disorders.

In the course of a study on the acoustical stimulation of the vestibular system of guinea pigs by low frequency pressure changes or applications of static pressure, correlations have been found between a variety of eye movements, including nystagmus, and the pressure changes. Interpretation of the results suggested that vestibular stimulation by the acoustic stimulus was a prime factor in the eye movements. Tests on deaf guinea pigs, using those with the vestibular nerve sectioned as controls, showed that the responses originated in the vestibular labyrinth. It has also been suggested that a subject exposed to a rapid pressure change, infrasound, or high intensity sound, could be subject to orientation difficulties, and the orientation disturbances from rapid pressure change were more likely to occur if pressure equalization in the middle ear was delayed, which could occur if the Eustachian tube were blocked. In later research, however, when sinusoidal pressure fluctuations of 1-4 Hz and levels of up to 172 dB were used, no consistent eye movements were observed in either monkeys or guinea pigs.

Chinchillas have been exposed to infrasonic levels of 172 dB, the conditions being chosen such that the total number of fluctuations of the tympanic membrane remained constant (i.e., 2 Hz for a duration of 32 minutes was equivalent to 4 Hz for 16 minutes). During the exposures the chinchilla stood for periods of 5-15 seconds, balanced on their hind legs on the edge of the metal pan, and during this period they exhibited perfect balance, the exposure not appearing to affect their vestibular system.

The discrepant results in several of these cases would appear to be due to the fact that vertigo and nystagmus are induced when the vestibular organ on one side is stimulated disproportionately to that on the other side. In cases like those of the chinchilla in which the animals are exposed to whole-body exposures, both lateral semicircular canals would be stimulated equally, thus counteracting each other's effects.

In the field of performance decrement caused by infrasonic exposure, it has been reported that a band of noise, from 2 to 15 Hz wide at 105 dB, or a 7 Hz pure tone at the same level, produces a 10% increase in visual reaction time in half the subjects of a test group. Further experiments indicated a 10% increase in tracking error in a visual pointer-following experiment for levels of around 95 dB.

As noted previously, in the experiments of Mohr and co-workers [2], during exposure to discrete frequencies, a significant visual acuity decrement occurred for all subjects with frequencies of 43, 50 and 73 Hz at 147-148 dB. For the subjective visual acuity experiments, the subject's ability to read print and instrument dials was tested, while for objective experiments a modified Snellen E test was used.

12-3.7 Musculoskeletal Effects

Very limited research has been carried out in this field, but of the few experiments, those of Mohr and co-workers [2], whose fine finger dexterity tests were carried out in a narrow-band and a discrete tone noise environment, showed no statistically significant changes in objective tests, which in these cases were circle tracing tasks to assess hand coordination and steadiness.

Similar studies on sonic boom transients, where the predominant energy is at infrasonic frequencies, showed no significant effects during finger dexterity tasks.

12-3.8 Effects on the Middle Ear

One of the more common nonauditory effects in exposure to high level infrasonic noise or discrete tones, is a pressure buildup or a fullness sensation in the middle ear. Subjects in the experiments of Mohr and co-workers [2] and of several other workers experienced this phenomenon.

This buildup of negative middle ear pressure has been reported at levels of 132 dB and higher, probably as a consequence of the pumping motion of the tympanic membrane, since the volume changes in the middle ear cavity are at the same frequency as the infrasound. It has been suggested that, since the Eustachian tube is closed for most of the time, the pressure induced in the cavity by the infrasound may result in the partial valvular opening of the tube and expression of some of the air from the middle ear. This pressure may be relieved by swallowing.

Other workers, however, have questioned this explanation, and suggest that this negative pressure could as well be a muscle effect, where the middle ear muscles, in particular the tensor tympani, tire and create an apparent pressure.

The sensation of aural pain is generally produced by the tympanic membrane being displaced beyond its limits of normal operation, and is not produced from the cochlea. It is unrelated to sensitivity, since both normal and subnormal hearing groups have the same mean aural pain thresholds. At infrasonic frequencies, levels which cause pain may not subject the human to damage risk.

Pain thresholds are given in Fig. 12-3, which shows that from 15 Hz to 2 kHz the threshold operates at approximately 140 dB. Below 15 Hz, however, there is a rapid rise in pain threshold; at 10 Hz the threshold level is approximately 147 dB, at 3 Hz it is 162 dB, and it is 175-180 dB at static or quasi-static pressures.

It is interesting to note that for positive static pressure (i.e., inward movement of the tympanic membrane) the pain thresholds appear to be 2-5 dB higher than the negative static pressure (i.e., outward movement of the tympanic membrane). This phenomenon remains unexplained.

12-3.9   Limiting Levels for Infrasonic Exposure

Nixon and Johnson [3] note that the "limiting levels of infrasound exposure effects on the auditory system must consider in addition to potential hearing loss, mechanical effects upon the middle ear system—including pain, speech reception and discomfort. The available knowledge from which limiting levels may be formulated comes from experience in intense infrasound and from laboratory investigations."

In general terms, for discrete frequencies or octave band noise centered on the frequencies noted, the following levels are considered as acceptable: 150 dB at 1-7 Hz, 145 dB at 8-11 Hz, and 140 dB at 12-20 Hz. Maximum exposure duration at these levels and frequencies is 8 minutes, with at least 16 hours' rest between exposures. The authors recommended that levels greater than 150 dB should be avoided, even with the best hearing protection, until more data have been acquired.

TABLE 12-1.   Limiting Values for Infrasound Exposure as a Function of Frequency and Duration.

Duration, minutes	Maximum Permissible Sound Pressure Level, dB[a]								
	0.5 Hz	1 Hz	2 Hz	4 Hz	8 Hz	10 Hz	12 Hz	16 Hz	20 Hz
0.5	169	166	163	160	159	156	155	154	153
1	166	163	160	157	154	153	152	151	150
2	163	160	157	154	151	150	149	148	147
4	160	157	154	151	148	147	146	145	144
8	157	154	151	148	145	144	143	142	141
10	156	153	150	147	144	143	142	141	140
20	153	150	147	144	141	140	139	138	137
30	151	148	145	142	139	138	137	136	135
(1 hour) 60	148	145	142	139	136	135	134	133	132
120	145	142	139	136	133	132	131	130	129
480	139	136	133	130	127	126	125	124	123
(1 day) 1440	134	131	128	125	122	121	120	119	118

[a]As derived from the equation due to Johnson and Nixon [3] given in the text. These researchers advise that no whole-body exposure can be recommended above 150 dB until more data has been amassed. The shaded area is an extrapolation and should be viewed with care.

It was suggested that the use of good insert earplugs tended to reduce the aural contribution to the gross response, and with earplugs permissible levels may be increased by 5 dB.

Table 12-1 is taken from Nixon and Johnson [3] and shows proposed limiting levels as a function of frequency and duration which, if adhered to, will allow the human to exhibit no symptoms of overexposure or abuse to the auditory system. The levels in this table disagree, to some extent, with the general levels quoted previously, but come from the equation

$$SPL_{max} = 10 \log \frac{t}{8} + 10 \log \frac{f}{10} + 144$$

where SPL_{max} is in dB, t is exposure time in minutes, and f is frequency in Hz. This is a semi-empirical equation based on a number of experiments, and appears to be a good approximation for representing exposure to frequencies in the range 1–20 Hz. These levels, however, are based on physiological limits, and the Meeting on Infrasound in Paris in 1973 [4] recommended that, where industrial or transportation conditions prevail, levels of 120 dB at 18 Hz should not be exceeded. The level of 120 dB may be exceeded by 3 dB for each octave below 18 Hz (i.e., 123 dB for 9 Hz).

12-4. ULTRASOUND

Ultrasound is normally referred to as airborne acoustic energy in the frequency region above 20 kHz.

Naturally occuring ultrasound at any appreciable level is rare and the majority of sources which produce levels high enough to affect man are found in production industries in a variety of processes including cleaning, welding of plastics, soldering, drilling, de-icing, emulsification, mixing of liquids, etc. The majority of these processes work in the frequency range 20–40 kHz, and measured sound pressure levels at the operators' working positions rarely exceed 110–120 dB, since atmospheric attenuation is pronounced for acoustic energy at these frequencies.

A further increase in the noise energy may be found in the harmonics and subharmonics of the operating frequency. In addition, in processes where liquids are involved, the SPLs may be increased by cavitation. It has been suggested that the oscillations of the gas bubbles may be responsible for tonal noise, while the collapse of the same bubbles may generate high levels of random noise at frequencies of 3 kHz and above.

The frequencies used for medical purposes for cell destruction and diagnostic techniques lie in the ranges of 1–3 MHz and 1–20 MHz, respectively. The diagnostic side of the medical use is not considered potentially harmful.

Considerably less research into the psychological and physiological effects of ultrasound has been accomplished than for audible sound and infrasound;

consequently, the following paragraphs on the nonauditory effects of airborne ultrasound are not subdivided into discrete problem areas, but only into the broad areas of nonauditory effects and exposure limits.

12-4.1 Nonauditory Effects of Ultrasound

Early work indicated that laboratory workers, having been exposed to ultrasound, showed symptoms of unusual fatigue, loss of equilibrium, nausea, and headaches persisting after the termination of exposure to the noise. These symptoms, excluding loss of equilibrium, but with the addition of tinnitus (see Chapter 11), are often found by operators of industrial ultrasound devices, but levels are not intense enough to result in loss of balance. It has been noted that these effects are probably due to high levels of high frequency audible noise produced as a byproduct of ultrasonic noise.

Previously it had been shown that these effects were reported by young women more often than men, and by younger men more often than older men. Since young women and young men have greater auditory acuity at higher frequencies than men in general and old men, respectively, the effects were probably a function of auditory threshold and not of sex and age per se. It has also been shown that unless auditory acuity extends to approximately 17 kHz, and sound pressure levels are above 70 dB, subjective effects like tinnitus, fatigue, and headaches are unlikely to be experienced.

While these symptoms of "ultrasonic sickness" may well be psychosomatic and may be also due to the high levels of audible content, most authors are of the opinion that the pathway of ultrasonic acoustic energy that is liable to affect man is probably through the inner ear.

There are many reports of research into physiological/biological effects on animals when exposed to ultrasound. The results of such experiments, however, cannot be extrapolated directly to humans, since there are many differences. Generally, in man, ultrasound must be at a high level to cause physiological effects; this is in part due to the poor impedance matching between the skin and airborne sound, resulting in a very poor energy transmission. The effect on animals is quite different, since the fur of small animals acts as an impedance matching device allowing greater amounts of energy to be transmitted. The absorption coefficient of small furred animals at frequencies above 20 kHz is in the order of 21%, compared with between 0.1% and 1% for human skin. Further differences are in the body-mass to surface-area ratios, and thus the total body mass available to dissipate the generated heat. Auditory acuity differences are apparent, and most small animals have thresholds which extend further into the high frequency spectrum than humans, so that some of the high frequencies used, often at high SPLs, are in the audible range of the animals.

Minor biological changes, due to prolonged noise exposure over the range 95–130 dB, with frequencies from 10–54 kHz, have been observed by many

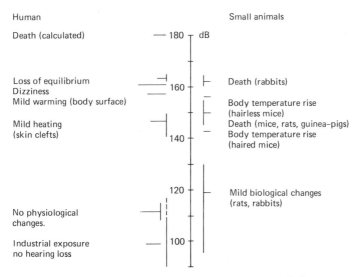

Fig. 12-5 Physiological effects of ultrasound [6].

authors in rats and rabbits. These changes, however, are general changes induced by stress, probably induced by the high level audible content of the noise.

Death of small animals has been found to occur at 150–155 dB at 30 kHz in mice and guinea pigs, and at 144–157 dB at 1–18.5 kHz in rats and guinea pigs.

In humans minor heating of skin clefts at ultrasound exposure levels of 140–150 dB has been found, and slight warming of the body surface has been reported at 159 dB.

Little research has been carried out on the physiological effects of ultrasound on man. Stimulation at 20 kHz at 160–165 dB results in loss of balance and dizziness, while exposure to 20 kHz at 110–118 dB for a period of 1 hour has been reported as producing no physiological changes. These latter levels are, however, fairly low.

An indication of the physiological effects of ultrasound in both humans and small, furry animals is shown in Fig. 12-5 which is taken from the work of Acton [6].

12-4.2 Limits for Ultrasonic Exposure

Two authors have proposed frequency-dependent exposure criteria for ultrasound. The first of these criteria, by Grigor'eva [5] is based upon TTS experimental results at 16 kHz and below. Grigor'eva also proposed a level of 110 dB in the 20–100 kHz frequency range.

Acton [6] based his results on both auditory and nonauditory effects over a long working period. His criterion finishes at the 31.5 kHz one-third octave band,

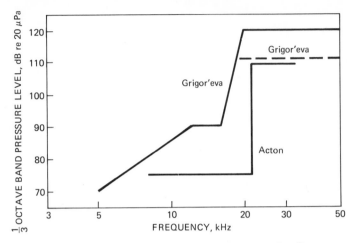

Fig. 12-6   Exposure criteria for ultrasound [5,6].

since experimental work had not gone beyond this frequency when the study was conducted, and extrapolation was not thought advisable until further evidence was available.

Figure 12-6 shows both Grigor'eva's and Acton's exposure criteria for ultrasound; if we assume that 110 dB is acceptable at frequencies of 20 kHz and above, the differences in the lower frequencies (i.e., below 15–20 kHz) could well be due to different percentages of population protected.

12-5. CONCLUSIONS

Any consideration of the physiological effects of noise on man must of necessity be qualified with a number of uncertainties. This is obviously due to the difficulty in practice of separating what are largely nonspecific stress responses over a period of time into the components caused by noise and those caused by a plethora of other factors. The second problem is the ethical consideration involved in exposing individuals to potentially damaging stimuli over a period of time in experimental situations. In the absence of such exposure, only tentative extrapolations can be made on the basis of animal and short-term human studies, the limitations on these approaches being self-evident.

What then can be said about the nonauditory effects of noise on man? One generalization is that it exerts its effect via the ears, the auditory pathway, and the central nervous system. The only significant exceptions to this rule tend to be, first, the effects of certain lower frequency sounds when they coincide with the resonant frequencies of various parts of the body, when they may interfere with such aspects of bodily function as swallowing and respiration; and, second, the potential heating effects of certain high frequency sound on skin clefts.

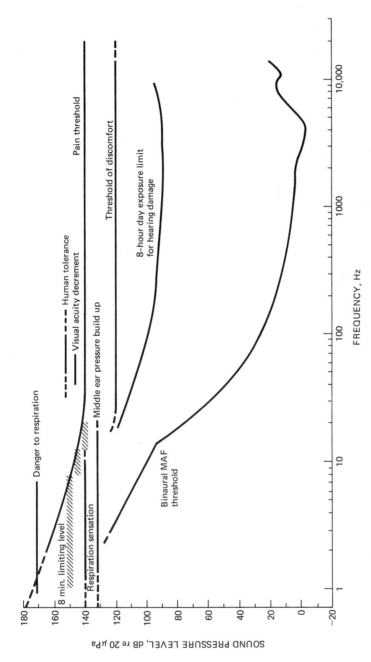

Fig. 12-7 A summary of physiological effects of sound in humans at various frequencies.

However, these few exceptions aside, the fact remains that the vast majority of noise effects, particularly at relatively modest levels, involve hair cell stimulation as an intermediary stage. Those hair cells constitute the most delicate part of the end organ, and hence that most sensitive to damage. Thus, any potentially damaging long-term effect of noise stimulation is likely to result in hair cell damage and an ensuing hearing loss, which may be taken as a warning sign of other potentially damaging effects.

A summary of the overall levels of noise at which the subject is exposed to risk at various levels is given in Fig. 12-7.

REFERENCES

1. G. Jansen, "Effects of Noise on Physiological State." In W. D. Ward and J. E. Fricke, eds., "Noise as a Public Health Hazard," American Speech and Hearing Association Report No. 4, 1969.

2. G. C. Mohr, J. N. Cole, E. Guild, and H. E. Von Gierke, "Effects of Low Frequency and Infrasonic Noise on Man." *Aerospace Medicine*, **36**, 817, 1965.

3. C. W. Nixon and D. L. Johnson, "Infrasound and Hearing." Paper in W. D. Ward, ed., Proceedings of the International Congress on Noise as a Public Health Problem, U.S. Environmental Protection Agency Report 550/9-73-008, Washington D.C., 1974.

4. L. Pimonow, ed., *Colloque international sur les infrasons*, Paris, 1973.

5. V. M. Grigor'eva, "Effect of Ultrasonic Vibrations on Personnel Working with Ultrasonic Equipment." *Soviet Physics—Acoustics*, **11**, 426, 1966.

6. W. L. Acton, "The Effects of Industrial Airborne Ultrasound on Humans." *Ultrasonics*, **12**, 124–128, 1974.

GENERAL READING

7. W. Burns, *Noise and Man,* 2nd edition; John Murray, London, 1973.

8. K. D. Kryter, *The Effects of Noise on Man.* Academic Press, New York, 1970.

9. W. D. Ward, ed., "Proceedings of the International Congress on Noise as a Public Health Problem." U.S. Environmental Protection Agency Report 550/9-73-008, Washington, D.C., 1974.

10. W. D. Ward and J. E. Fricke, eds., "Noise as a Public Health Hazard." American Speech and Hearing Association Report No. 4, 1969.

11. B. L. Welch and A. S. Welch, eds., *Physiological Effects of Noise.* Plenum Press, New York, 1970.

13

Noise and sleep: A literature review and a proposed criterion for assessing effect

---

JEROME S. LUKAS*

13-1. INTRODUCTION

Criteria for acceptable noise effect are required before limits for noise exposure can be established. Whether or not one

*Office of Noise Control, State Department of Health, 2151 Berkeley Way, Berkeley, California 94704.

Work performed while the author was at Stanford Research Institute. This chapter is based on the author's paper of the same title published in *J. Acoust. Soc. Amer.*, 58(6) 1232–1242, 1975, reproduced here by kind permission of the Acoustical Society of America.

agrees with the rationale underlying a specific criterion, it must be agreed that some criterion, albeit a relative one, exists. The criterion for daytime noise might be to limit hearing loss for speech (as in the well-known Occupational Safety and Health Act) or to limit annoyance in some percentage of the exposed population. For example, restrictions on the time, number, and direction of air-craft operations are techniques for limiting annoyance. In the case of the effects of noise on sleep, however, a similar criterion may not exist. Some data and a rationale for a first approximation of a criterion, or standard, to assess the effects of noise on sleep are discussed below.

13-2. WHAT IS SLEEP?

Sleep is typically classified into five stages based on patterns observed in the electroencephalogram (EEG), electromyogram (EMG), and certain eye movements: Stages 1, 2, Delta (a combination of Stages 3 and 4), and rapid eye movement (REM). Details of these patterns have been illustrated by Rechtschaffen and Kales [1].

Stage 1 is the brief, hypnogogic period between awake and Stage 2. Stage 2 is the most prevalent stage, occupying about 50 percent of the usual sleep period. Stage Delta may be thought of as deep sleep, while stage REM is usually associated with dreaming. These stages occur periodically and sequentially throughout the night, as is illustrated in Fig. 13-1. This figure also illustrates how Stage Delta tends to occur in the early third of the night, while Stage REM tends to occur

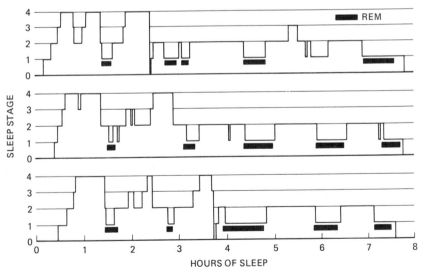

Fig. 13-1. Progression of sleep stages across three nights for one subject [2].

more frequently and with longer duration in approximately the last half of the night's sleep. Interested readers may consult authorities such as Kleitman [3] or others cited in the references for further descriptions of sleep. The functions of sleep and its stages are a matter of current dispute, and beyond the scope of this paper. An introduction to the controversy may be found, for example, in Webb [4], Hartmann [5], and Johnson [6].

Notwithstanding a lack of clear, experimental demonstrations of the need for sleep, of certain of its stages, or of the next-day effects of sleep disruption, it is clear that all people sleep. Many surveys have found that one of the more annoying attributes of noise is its disturbing effects on sleep [7, 8, 9]. It may be valuable to describe research results that help in understanding those complaints.

The studies discussed will be primarily those in which recordings or simulations of meaningful noises were used. The importance of the meaning of noise on the response was demonstrated by Oswald and co-workers [10], Williams and co-workers [11, 12], Zung and Wilson [13], and Langford and co-workers [14]. These studies show that responsiveness will be enhanced by stimuli that are inherently meaningful (such as one's name) or that are made meaningful by instructions or some conditioning procedure. Responsiveness may be defined in terms of frequency of an awakening response, the stimulus intensity required to cause a shift to a lighter stage of sleep, or some other electroencephalographic response. The assumption here is that environmental noises have a meaning that will affect the responses made to those noises, regardless of the instructions given the subjects. A corollary assumption is that studies using only tone bursts, clicks, bursts of white or pink noise, or other uncommon stimuli may underestimate the sleep-disrupting effects of those stimuli.

We must first consider what the significant response to noise may be during sleep. Some controversy exists regarding this question because of the lack of a clear demonstration of the functional significance of sleep. Some investigators [15-17] have emphasized behavioral awakening, as indicated by use of a switch mounted on the headboard of the bed, but they also score and evaluate the EEG responses. Williams and co-workers [18], Keefe and co-workers [19], Thiessen [20], and Berry and Thiessen [21] used scoring procedures similar to those of Lukas, but they do not emphasize EEG responses nor behavioral awakening, as measured by a microswitch taped to a hand. Some, such as Morgan and Rice [22] and Ludlow and Morgan [23], rely on activation of a switch—i.e., behavioral awakening. Others [24-28] score only the EEG responses. LeVere and co-workers [29, 30] perform computer analyses to obtain measures of "cortical desynchronization," and Rylander and co-workers [31] relied on movement of the bed, as well as the use of a switch to indicate awakening.

Other than the studies by Rylander, LeVere, Morgan and their co-workers, scoring of the electroencephalographic response is the common denominator in most of these studies. Within the range of those responses, however, some investigators emphasize awakening while others emphasize significant changes in the EEG such as the frequency of noise-induced changes in sleep stage [32, 33] or disruptions of total nightly sleep patterns [27]. The major emphasis in this presentation is on those studies using EEG changes as measures of response to noise.

13-3. INDIVIDUAL CHARACTERISTICS AS DETERMINANTS OF RESPONSE

13-3.1 Age

Typically, the older the individual is, the more likely he is to be awakened or to change sleep stage as a result of environmental noise. Table 13-1 presents a selected sample of data taken from Refs. 16 and 26. Note that, in general, the frequency of arousal or behavioral awakenings and of sleep stage changes increases with age. An arousal is defined as an EEG response that has all the characteristics of that of an awake individual (Stage 0 in Fig. 13-1), while behavioral awakening is confined to a verbal or motor response indicating the subject is awake. Typically, arousal is coincidental with behavioral awakening.

The generally lower incidence of arousal responses in the Collins and Iampietro

TABLE 13-1. Response Frequencies to Simulated Sonic Booms: Different Age Groups While Asleep.

Study by number, sex, and age of subjects	Response, %		
	No Change[a]	Stage Change to "Shallower" sleep	Arousal or Behavioral Awakening
Lukas et al.[b] 2 males and 2 females, 5–8 years	94.1	5.9	0.0
Collins et al.[c] 8 males, 21–36 years	87.7	10.6	1.7
Collins et al. 8 males, 40–45 years	80.2	13.7	6.1
Lukas et al. 4 males, 45–57 years	80.7	6.3	14.0
Collins et al. 8 males, 60–72 years	72.7	18.6	8.7
Lukas et al. 4 males, 69–75 years	61.7	23.4	14.9

[a]This category includes K complexes, bursts of alpha, eye movements, and other minor changes in the EEG according to Lukas and co-workers [16] and responses 0–4, inclusively, as scored by Collins and co-workers [26].
[b]Lukas and co-workers [16]. Response to a single level (out of four tested) of simulated sonic booms is included: it had an estimated outdoor intensity of 0.65 psf and indoor intensity of 68 dB(A) near the subject's ears.
[c]Collins and co-workers [26]. A single level of boom was studied: 1 psf estimated out of doors. Intensity near the subject's ears was 0.13 psf and 68 dB(A) measured with the "slow" setting of an impulse sound level meter.

study [26] may be attributed to several factors: Even though the booms were presented hourly beginning at 11 pm, the subjects may not have "heard" the booms just as a person does not "hear" a clock striking the hours throughout the night. The subjects in the Lukas studies were told to use the awake switch if they awakened for any reason. Thus the instructions may have made the subjects particularly responsive. That the inherent significance of signals or training may change response patterns was referred to above. The older two groups of subjects used by Collins and Iampietro [26] were younger than those of Lukas and co-workers [16] ; thus the response differences may simply reflect the differences in age of the subjects in the two studies.

13-3.2 Sex

That sleep arousal thresholds are lower in women than in men has been demonstrated by Wilson and Zung [34], Steinicke [35], and Lukas and Dobbs [36]. Table 13-2 gives the results from two studies by Lukas and co-workers [16, 36] in which stimuli were impulse noises (simulated sonic booms with durations of

TABLE 13-2. Response Frequencies to Two Types of Aircraft Noise: Middle-Aged Men and Women During Three Sleep Stages.

			Response, %			
Stimuli	Sleep Stage	Sex	No Change	Stage Change to Shallower Sleep	Arousal or Behavioral Awakening	Chi Square (2 df)[a]
Simulated sonic booms[d]	2	Male[b]	60.1	14.2	25.7	2.27, NS
		Female[c]	54.1	14.1	31.7	
	3 and 4	Male	57.8	32.8	9.5	10.35, S
		Female	34.6	47.4	17.9	
	REM	Male	63.7	11.8	24.5	4.86, NS
		Female	57.7	25.8	16.5	
Jet aircraft fly-over noise.[e]	2	Male	51.8	20.7	27.5	30.07, S
		Female	36.4	10.9	52.7	
	3 and 4	Male	41.8	30.6	27.6	12.19, S
		Female	17.9	37.2	44.9	
	REM	Male	61.9	13.3	24.8	23.34, S
		Female	29.5	16.8	53.7	

[a]NS: not significant; S: significant at least at 0.05 levels.
[b]See Ref. 16.
[c]See Ref. 36.
[d]For the women, three intensities of sonic boom were studied. Near the subject's ears the levels were 68, 79, and 84 dB(A): these levels overlapped the additional intensity studied in men: 68, 73, 79, and 84 dB(A). Corresponding levels outdoors may be estimated at about 25 dB(A) higher than the indoor levels.
[e]Three flyover noises were studied with the women and four with the men. Indoor levels (near the subject's ears) were 68, 74, 80, and 86 dB(A) for the men; the 74 dB(A) was not used with the women. Estimated outdoor levels are about 20 dB(A) higher than the indoor levels shown.

TABLE 13-3.   Response Frequencies to Two Jet Takeoff Noises.[a] College-Age
Subjects (19–24 Years) During Three Sleep Stages.

Sleep Stage	Sex	Response, %			Chi-Square (2 df)
		No Change[b]	Stage Change to "Shallower" Sleep	Arousal	
2	Male	23.1	48.1	28.7	6.84, S
	Female	29.9	55.6	14.5	
3 and 4	Male	31.6	54.4	13.9	3.17, NS
	Female	29.8	63.8	6.4	
REM	Male	76.1	11.9	11.8	2.22, NS
	Female	85.7	6.5	7.8	

[a]Muzet and co-workers [28] studied two jet takeoff noises differing in frequency composition and duration: one was 30 sec long and the other 90 sec. Each was tested at two intensities: the 30 sec noise at 95 and 112 PNdB, and the 90 sec noise at 93 and 100 PNdB.
[b]For purposes of illustration, Muzet's Type I response ("phase of transient activation" [37]) that includes K complexes, bursts of alpha, and brief increases in EMG level and eye movements, are included in the no-change category.

about 275 msec) and aircraft flyover noises with durations of about 10 sec between the 20 dB downpoints. These data indicate that, regardless of sleep stage or the type of stimulus, women were more responsive (hereafter defined more narrowly as a lower frequency of zero responses—i.e., no significant change in the EEG) than were men.

In contrast to the results of Lukas and co-workers [16, 36] (Table 13-2), those of Muzet and co-workers [28] (Table 13-3) suggest that men may be more responsive than women, though the differences are not statistically significant. The difference in results may be due to the fact that Muzet's group [28] stimulated their subjects on only one of four nights of sleep, and hence, their results may reflect some "first night" effect [38]. In contrast, the results of Lukas and co-workers [16, 36] are based on nine test (stimulus) nights in the case of women and twenty nights in the case of men. It is also possible that differences in subjects' ages may explain the inconsistent results: Whereas Muzet's subjects were 19–24 years of age, the men studied by Lukas's group were 45–57 years of age, and the women were 29–49 years of age.

These data suggest that at certain ages the sexes respond differently to noise during sleep. While of college age, women may be less responsive to nighttime noises than are men. Middle-aged (35 years and older) women, however, are more responsive to nighttime noise than are middle-aged men.

13-4. SLEEP STAGE, TIME OF NIGHT, AND ACCUMULATED SLEEP TIME

The data presented in Tables 13-2 and 13-3 are combined in Table 13-4 for a comparison of the effects during the different sleep stages in the two age groups

TABLE 13-4.   Response Frequencies to Jet Aircraft Noises: Men in Three
Sleep Stages.

Study and Subject Age	Sleep Stage	Response, %			
		No Change	Stage Change to "Shallower" Sleep	Arousal or Behavioral Awakening	Chi-Square (4 df)
Muzet et al. [28]	2[a]	23.1	48.1	28.7	
19–24 years	3 and 4[b]	31.6	54.4	13.9	57.73, S
	REM[c]	76.1	11.9	11.9	
Lukas et al. [16]	2	51.8	20.7	27.5	
45–57 years	3 and 4	41.8	30.6	27.8	11.90, S
	REM	61.9	13.3	24.8	

[a]For sleep Stage 2, Muzet versus Lukas, χ^2 = 36.49, 2 df, significant.
[b]For sleep stages 3 and 4, Muzet versus Lukas, χ^2 = 10.41, 2 df, significant.
[c]For sleep stage REM, Muzet versus Lukas, χ^2 = 4.78, 2 df, not significant.

studied. Although subjects respond differently to noise during the three sleep stages, the specific distribution of responses apparently is a function of the age group.

Thus, in the middle-aged group, only small differences were observed in the frequency of awake or arousal responses to the aircraft noises occurring in Stages 2, 3, 4, and REM, but relatively wide differences were observed in the frequency of no-change and sleep-stage changes. In contrast, the younger men showed relatively wide differences in response distributions in each of the three stages. Statistical comparisons of the response frequencies presented by Muzet [28] and Lukas [16] for each of the sleep stages (see the bottom of Table 13-4) show them to be different in the two age groups, particularly during sleep Stages 2 and Delta. These age differences appear to be particularly significant, since Muzet's stimuli had an average intensity some 10 dB(A) greater than those used by Lukas [87 dB(A) vs. 76 dB(A)].

Because of the cyclical occurrence of the sleep stages throughout the night and because Stage Delta occurs infrequently during the later hours of sleep (as shown in Fig. 13-1), relative responsiveness to noise during the different stages is confounded with the time of night and number of hours of accumulated sleep. Williams [39] summarized: "the likelihood of behavioral responding to neutral stimuli is a decreasing function of the amplitude and period of the background EEG rhythms" (p. 504). Note that Williams specifically refers to neutral (i.e., meaningless) stimuli, but that the generalization seems to hold with meaningful stimuli is illustrated in Table 13-5. If we confine the response of interest to behavioral awakening, we can see that typically fewer of those responses to the noise occur during sleep Stage Delta, than during Stages 2 or REM. But note also that the frequency of no change in the EEG was lowest or nearly the lowest

TABLE 13-5. Response Frequencies to Jet Flyover Noise: Women (29–49 Years of Age) in Three Sleep Stages.[a]

Jet Flyover Intensity, dB(A)	Sleep Stage	Response, %				
		No Change	Stage Change to Shallower Sleep	Arousal	Behavioral Awakening	Chi-Square (6 df)
	2	55.0	11.7	8.3	25.0	
68	3 and 4	25.0	56.3	6.2	12.5	17.2, S
	REM	55.9	14.7	5.9	23.5	
	2	24.1	13.0	11.1	51.8	
80	3 and 4	21.1	31.6	10.5	36.8	6.9, NS
	REM	20.8	33.3	4.2	41.7	
	2	27.5	7.8	7.8	56.9	
86	3 and 4	8.3	33.3	20.8	37.5	21.5, S
	REM	10.8	8.1	5.4	75.7	

[a]Based on Ref. 36.

during Stage Delta, regardless of stimulus intensity. This latter observation supports Williams' suggestion [39] that the relative infrequency of behavioral awakening in Stage Delta results from an incompatibility of those stages with "selection and execution of certain motor responses" (p. 508). The frequency of arousal responses in Stage Delta is typically higher than that found during Stages 2 or REM (see Table 13-5). Therefore, when stimuli occur during Stage Delta they are apparently processed and responded to centrally, but the motor command ("push the switch") is executed less frequently than it is during Stages 2 and REM.

In discussing the general finding that thresholds of awakening decrease as time asleep increases, Williams [39] tentatively concluded that the major factor controlling that effect is a circadian biological rhythm independent of the accumulated sleep time. He based his conclusion on the similarity between the curve of behavioral responsiveness over the night and the circadian curve for body temperature. He states, however, that the studies reported have confounded accumulated sleep time and chronological time.

13-5. STIMULUS CHARACTERISTICS AS DETERMINANTS OF THE RESPONSE

As shown in Table 13-5, increases in stimulus intensity generally result in in-increased frequencies of behavioral awakening and arousals, and reductions in the frequency of no-change in the EEG. Lukas [32] has demonstrated that the prediction of frequency of some disruption of sleep (a change in the EEG of *at least* one stage) is likely to be more accurate on the average if the descriptor of the noise includes a term that accounts for intensity and a term that takes stimulus duration into account. In Table 13-6 we see that, when data from a variety of investigators are used, the highest correlations hold if the duration (E in the term EPNdB) [41] and frequency-band weighted intensity (PNdB) are accounted for. Also note that only small differences in coefficient magnitude are obtained if behavioral awakening is treated separately from the arousal response, or if they are combined. That these generalizations hold for the unit EdB(A) follows from the generally high correlation between PNdB and dB(A) units. Kryter [7], for example, has suggested that a constant of 13 be added to dB(A) to estimate PNdB. [PNdB is explained in Chapter 1, and EPNdB in Chapter 3. EdB(A) is the same as EPNdB, except that the dB(A) rather than the PNdB is used in its calculation. The phon unit referred to below is also explained in Chapter 1.]

Note also in Table 13-6 that analogous coefficients in the maximum dB(A) and maximum PNdB columns are of almost identical magnitude. Of possibly greater importance, however, is the apparent improvement in the coefficient if temporary changes (such as *K* complexes, bursts of alpha, or changes of less than one sleep stage) in the EEG are disregarded. It has been noted elsewhere [18]

TABLE 13-6. Coefficients of Correlation Between Responses to Noise During Sleep and Various Measures of the Noise Intensity [32].

Response	Intensity			Number of Data Points
	Max dB(A)	Max PNdB	EPNdB	
None	−0.90	−0.90	−0.84	6[a]
Awakening	0.83	0.84	0.81	6
Awakening and arousal	0.85	0.85	0.79	6
None	−0.79		−0.81	20[b]
Awakening	0.64		0.77	20
Awakening and arousal	0.61		0.75	20
None	−0.64		−0.67	37[c]
None[d]	−0.62		−0.78	37
Awakening	0.53		0.67	39

[a]Includes data only from Lukas, Peeler, and Dobbs [40].
[b]Data from Lukas, Peeler, and Dobbs [40] plus data on middle-aged men and women [16, 36]. Stimuli were other aircraft noises and simulated sonic booms at several intensities each.
[c]Data from Thiessen [20], Berry and Thiessen [21], Morgan and Rise [22], Ludlow and Morgan [23], Collins and Iampietro [26] added to b above.
[d]Responses 0 and 1 combined from Thiessen [20] and Berry and Thiessen [21], and response 0-4 combined from Collins and Iampietro [26], since by their definitions the 0-4 responses are identical to the no response of Lukas, Peeler, and Dobbs [40].

that these patterns occur normally in the EEG and with greater frequency than stage changes or arousals. Thus scoring them as responses biases the results.

Attention must be drawn to an implication of these data. Some investigators [24, 35] have used the technique of increasing stimulus intensity in steps until some desired response is obtained. Obviously, that technique is modeled after one commonly used to obtain thresholds in the awake subject. The probability, however, of obtaining some response in the sleeping subject is related not only to the intensity of stimulus but also to its duration. Therefore, in these studies, the values reported are likely to be biased downward—i.e., the subjects will appear to be more responsive than they actually are. Steinicke's results [35] clearly are a biased estimate of the noise effects. For example, since the stimuli were presented 2 to 3 hours before the subjects usually awoke, a reasonable assumption is that the stimuli occurred primarily in sleep Stages 2 or REM (see Fig. 13-1) in which the frequency of behavioral awakening tends to be the highest regardless of stimulus intensity (see Tables 13-2 to 13-5); and since the noise level apparently was increased every 3 minutes, the probability of obtaining a response because of the noise duration alone was increased. Furthermore, Thiessen [personal communication] has suggested, "Since [Steinicke's] signal strength increased with time and a subject's probability of awakening increases with time, even without any noise, there is bound to be a correlation because of the common factor of time." Considering the test conditions, it is remarkable that about 8% of his subjects never awakened.

The discrepancy between Steinicke's results and those of other investigators is illustrated in Fig. 13-2. About 50% of Steinicke's subjects were awakened when the stimuli attained a level of about 72 EPNdB; whereas the other studies suggest that for stimuli occurring at random times throughout the night, 50 percent of the subjects might be awakened at a level of about 90 EPNdB. To maintain consistency with Steinicke's technique of cumulating responses, the right-hand curve of Fig. 13-2 was based on the assumption that all points shown on the right constitute a population and that the response probability at any intensity is equal to the percentage of points (with respect to the total number of points) at or within 5 dB of that intensity. Thus, four of the 25 points, or about 16% of the subjects, were awakened by noise at a level of about 80 EPNdB.

Before studying Fig. 13-2, the reader should note that since Steinicke increased the noise level without a period of silence between any two levels, the effect on the subject was assumed to be similar to that of a continuous noise. Hence, the higher the level at which a response occurred, the longer the stimulus was present and the higher the EPNdB level. Thus, 33 was added to the highest level (69 PNdB for Steinicke's 70 phon noise [42-44]) to obtain the EPNdB level rather than the 25.5 that would be added if a single 3 minute burst of noise were present. In other words, the longest stimulus was present for 21 minutes before the subject responded. We assumed also that the level was constant throughout any 3 minute presentation.

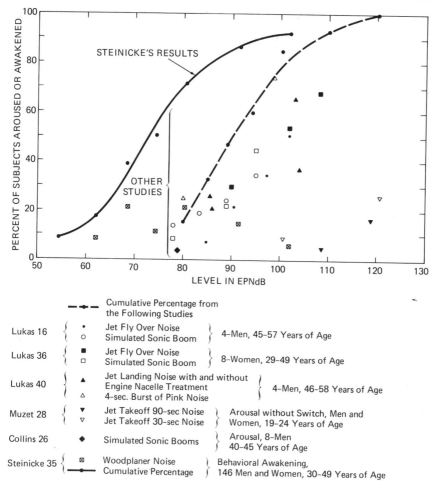

Fig. 13-2. Percentage of subjects behaviorally awakened or aroused by nighttime noise.

In Fig. 13-2 it is worthwhile to observe that, if we ignore Steinicke's and Muzet's data and the cumulative distributions, the frequency of behavioral awakening and arousal appear to be a positively accelerating function of stimulus intensity.

13-6. THE AFTEREFFECTS OF SLEEP DISTURBANCE

Personal experiences and commonsense lead one to expect that disturbance of sleep would result in measurable behavioral effects during the day following sleep disturbance. Demonstrations of such effects, however, have proved to be most difficult. For example, Webb [45] and Lubin [46] have indicated that, in some of their studies, subjects were fully awakened (to the point of being able to complete an immediate recall task correctly) 100 times or more a night. On arising, the subjects remembered being awakened only a few times, believed they had slept reasonably well, and felt rested. In addition, no decreases in various performances were observed the following day.

Two studies with possibly contrary results were conducted by LeVere and co-workers [29] and Herbert and Wilkinson [47]. LeVere's group found an increase of about 0.03 seconds in a disjunctive reaction time task that included a varying memory component. Although the difference observed was of statistical signifi-cance, the practical implications of the result are questionable because of the magnitude of the effect and because the average response to the stimuli (jet flyover noise of 80 dB(A) and 20 second duration) was a change of less than one sleep stage that lasted, perhaps, less than 5 minutes, depending on the type of ongoing EEG activity. Herbert and Wilkinson, who presented about 1200 pairs of clicks randomly at four intensities [65, 75, 80, and 90 dB(A)] during one night, found significant increases in time spent awake and in sleep Stage 1, but concluded that "the effects upon performance . . . were relatively small and con-fined to the early part of the day" (p. 527). Their tasks, measured periodically throughout the day, were the Wilkinson Vigilance and Addition tasks each lasting 1 hour and a short-term memory test lasting about 35 minutes. In contrast, LeVere and co-workers administered their 10 minute task before the subject retired and when he awakened.

Another study [24] of nine consecutive nights of continuous, impulse, or a combination of continuous and impulse noises, with two males in each of three age groups (20, 50, and 70 years) provided inconclusive evidence of an effect on any of six performance tasks. A verbal task, however, suggested some "cognitive impairment" and a decrease in "human relations."

Although brief disturbances of sleep as may be caused by environmental noise have not resulted in large performance decrements, there is no scientific informa-tion suggesting that equivalent results would be obtained if the sleep disturbance continued for long periods of time. Complete deprivation of all sleep (see, e.g., Johnson and co-workers [48]) for two days resulted in poor performance on a number of behavioral measures, but differential deprivation of slow wave (Stages 3 and 4) sleep vs. REM sleep did not. A study of Wilkinson [49] indicated that only after subjects were completely deprived of sleep for about 5 hours out of a "normal" 8 hours (i.e., they were allowed 3 hours of undisturbed sleep) did they show a decrement in performance on a relatively complex task the following day. On the second night, deprivation of about 3 hours was sufficient to result in a performance decrement. Subsequently, Hamilton, Wilkinson, and Edwards [50] found that in subjects deprived of 1.5 or 3.5 hours of sleep (compared with a normal 7.5 hours) for four consecutive days, the degree of performance decrement depended on the type of task. For some tasks, such as calculation of sums, the number of hours and days of sleep loss produced monotonic decrements in performance, but for a vigilance task or a running digit span task the results were not as clear-cut. With these latter tasks, sleep-deprived subjects performed better, at times, than those in the normal sleeping control group. These results appear to be generally consistent with those of Webb and Agnew [51] who restricted the normal sleep regime (7.5–8 hours) of 15 males to a regime of about 5.5 hours for a period of eight weeks; they found some alterations in sleep patterns relative to baseline, but limited objective and subjective behavioral effects.

13-7. IMPLICATIONS REGARDING A CRITERION MEASURE

The data presented above suggest that two criteria are reasonable. The more conservative criterion is to limit sleep disruption to less than a change in sleep stage (a response of 0): the alternative criterion is to limit the frequency of arousal or behavioral awakening. Since the functions of all of sleep or its stages have not been specified unequivocally and the explanations provided heretofore are unverified (see, e.g., Hartmann [5] and Webb [4] or the earlier sections of this chapter), the safer policy is to limit noise levels indoors so that the responses to noise are limited to electroencephalographic changes of less than one sleep stage. A rationale for this conclusion is discussed in the following paragraphs.

Although the available data are equivocal regarding the possible short-term or next-day effects of frequent behavioral awakenings or arousals, no data are available regarding their actual or possible long-term effects. In addition, arousals or behavioral awakenings have been assumed to be correlated with the degree of annoyance of noise. Whether annoyance is correlated with induced changes in sleep pattern, arousals or awakenings, or some combination of these responses has yet to be demonstrated. Some data suggest a iow correlation between annoyance and frequency of behavioral awakening: After 100 or so awakenings nightly, the subjects of Webb [45] and Lubin [46] reported no drastic untoward effects. It may be argued that their subjects were of college age and possibly unique or atypical; however, their results differ little from those reported by others who have studied older subjects. For example, Lukas, Dobbs, and Kryter [16] reported that after eight males ranging in age from 45 to 75 years were exposed on two nonconsecutive nights to 16 noises nightly, an average of 57% of their responses to a questionnaire administered after each night of sleep with noise indicated that they arose fully rested. In contrast, after nights without noise, an average of 87.5% of the responses indicated that the subjects felt fully rested. It is important to note, however, that the old (at least 69 years of age) subjects more frequently reported they arose fully rested, despite being behaviorally awakened more frequently, than did the middle-aged males. In addition, five of the subjects reported no deficit in daytime performance at home or at work, while the remaining three reported some fatigue only. Eight women, 29–49 years of age, after three consecutive nights of exposure to a noise environment similar to that used with men, showed little correlation between the number of reported awakenings and the reported severity of sleep disruption. There was also a discrepancy between the reported number of awakenings by simulated sonic booms and jet aircraft noise and the type of noise reported as being most disturbing. These results are presented in Table 13-7.

In the laboratory, after behavioral awakening, middle-aged males returned to sleep Stage 2 in about 4 minutes on the average. In contrast, after electroencephalographic arousal, the subjects needed about 2 minutes to return to sleep Stage 2. In general, the time spent awake after both behavioral awakening and electroencephalographic arousal was reduced as additional nights were spent in the noise environment [33]. Because of this rapid return to sleep after arousal by noise, reasonable doubt arises about the ability of people to judge how frequently they were awakened or what the disturbing stimulus might be. This conclusion is similar to that reported to Frankel and co-workers [52] regarding the tendency of insomniacs and normal controls to underestimate the frequency of spontaneous awakenings and total sleep time.

Despite the lack of a high correlation between the frequency of awakening or arousal and subjective estimates of the severity of sleep disruption, the lack of knowledge about the long-term effects of these presumably extreme disruptions of sleep indicates that a conservative criterion should be used for limiting disruption caused by noise. This conservative criterion appears to be particularly necessary in light of Hartmann's 1973 review [5], theorizing that during sleep Stage Delta (or slow-wave sleep in Hartmann's terminology), certain biochemicals are synthesized that appear to be necessary during sleep Stage REM for synthesizing other catecholamines needed during wakefulness. It appears possible, therefore, that disrupting the usual pattern of sleep, particularly for a long period of time, may result in potentially significant or hazardous biochemical changes. However, that the no-response category means the biochemistry of sleep is undisrupted must be demonstrated.

TABLE 13-7. Perceived Frequency of Awakening and Disturbance Attributed to Flyover Noise and Simulated Sonic Booms [36].

The Subjects	Flyover Noises	Booms
Thought they were awakened by	56 times (77.8%)	16 times (22.2%)
Thought the stimulus indicated was the most disturbing	11 times (15.0%)	61 times (85.0%)

Furthermore, the proposed criterion has an additional advantage. On the basis of the data shown in Table 13-6, we are better able to predict the frequency of no significant changes in the EEG than the frequency of arousals or behavioral awakenings. Correlation coefficients were higher between the frequency of 0 responses (no significant change in EEG) and the various measures of stimulus intensity than for the frequency of either arousals or behavioral awakenings or the combination of these responses. For example, a coefficient of 0.52 was obtained when the frequency of arousals and behavioral awakenings was correlated with the stimulus intensities shown in Fig. 13-3. In contrast, a coefficient of about -0.80 was obtained when a frequency of no EEG change was correlated with the same stimulus intensities. It is worthwhile to recall that statistical significance can be obtained with only small differences in the magnitude of the coefficients (of the order of 0.1-0.2 units) because of the high intercorrelations between the various response measures and between the various intensity measures [33]. Therefore, to judge the coefficients on the basis of their predictive power rather than on the basis of such statistically significant but small differences in coefficient magnitudes is perhaps more reasonable.

Using no significant change in sleep EEG patterns as the criterion of noise effect will permit some predictions on the basis of data now in the scientific literature. Plotted in Fig. 13-3 are curves illustrating the relationship between the frequency of no significant change in the sleep EEG and the intensity (calculated in units of EPNdB) of a variety of noises. As shown in the legend of the figure, the data were obtained with men and women ranging from about 20 to 55 years of age and with a variety of stimuli that occurred several times nightly. The curves were plotted using the coefficient of correlation and other data shown on the left, but without the data of Muzet and co-workers [28]. Three of their data points appeared to be out of line—possibly for the reasons referred to earlier. The coefficient of correlation obtained if their data [53] are included is shown on the right. The inclusion of data from Muzet's group reduced the coefficient only slightly (about 0.09 units), and both coefficients are of statistically significant magnitude.

Using the curves shown in Fig. 13-3, we can make some tentative predictions. Clearly, if noise is limited so that there is little or no probability of disturbing sleep, levels indoors must be maintained below 70 EPNdB. By referring back to Fig. 13-2 and ignoring Steinicke's [35] results, we see that the probability is small that people will be awakened or aroused by noise of this level. If we are willing to have a 50–50 chance that someone will be disturbed, then any noise should be limited to about 90 EPNdB. On the other hand, assume that 10,000 people live in the vicinity of some freeway and indoor noise levels at night occasionally peak at 100 EPNdB. The data suggest that each time traffic noise reaches this level it will evoke at least a change in sleep stage in about 75 percent (or 7500) of the people exposed.

Obviously, other assumptions (or measurements) can be made about the fre-

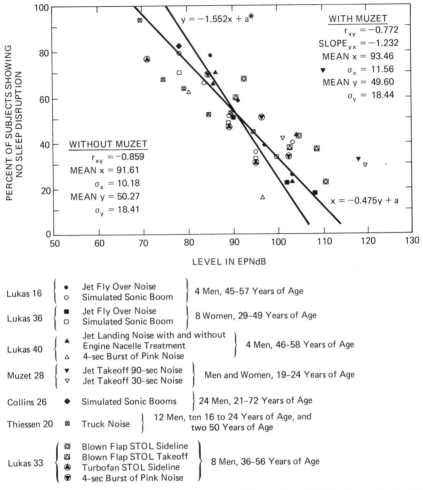

Fig. 13-3. **Frequency of no significant disruption of sleep by a variety of noise types at various levels.**

quency at which the noise reaches different levels, as can extrapolations about the extent of sleep disruption during given nights. Perhaps such extrapolations may be used in predicting complaint behavior among residents of some localities. However, the limited amount of data currently available precludes drawing too many extrapolations and conclusions. The brief exercise in extrapolation presented here primarily illustrates the potential of how the available data may be used to establish noise exposure limits during the hours of sleep.

13-8. GAPS IN SLEEP-NOISE RESEARCH

This article suggests that, despite differences in nighttime responsiveness to noise attributed to age, sex, type of noise, and certain other variables, it may be possible to make some important predictions of noise effects on the basis of available data. Nonetheless, there are gaps in our grasp of the problem. Some of the more important of these gaps (as we see them) are addressed below.

Reference was made earlier to the theoretical nature of Hartmann's analysis [5] of the biochemistry of sleep and to the need to demonstrate that disruption of sleep, as defined herein, changes those biochemical processes. However, even if sleep disruption is shown to cause biochemical changes, we have only a small amount of information about adaptation to noise, how adaptation may affect the biochemical changes, or over what period of time the biochemistry must be altered before the physiological or psychological welfare or health of the individual is jeopardized.

How do patterns or the density of noise during the night affect sleep? For example, Schieber and co-workers [55] (see also Muzet [56]) observed that disturbance from low-density street traffic (1.8 vehicles/minute) was greater than that from high-density traffic (4.3 vehicles/minute), but the opposite result was obtained with aircraft noises at two densities. This study should clearly be replicated.

The existing research data are applicable to generally healthy people. Does it apply to insomniacs, the ill, or the very old?

What is the correlation between studies conducted in the laboratory and conditions in the home? In other words, to what extent or how accurately can we predict sleep disruption at home from data gathered in the laboratory? The data reported herein are from laboratory studies, but whether similar results would be obtained with noise in the home is unknown; presumably they would.

Reference was made (Section 13-2) to the importance of the meaning of significance of a noise in determining relative sensitivity to that noise. It seems possible that an individual who expresses annoyance with daytime noise perhaps may inadvertently be sensitizing himself to noise disturbance during sleep; this same individual may, therefore, be more likely to complain about the sleep-disturbing effects of noise. To our knowledge, little, if any, study has been directed to problems of this type and their corollaries.

13-9. CONCLUSIONS

Although some rather serious deficiencies remain in our understanding of both the functions of sleep and the effects of environmental noise thereon, the conservative criterion of preventing disruption of normal sleep patterns is warranted in light of the evidence available.

Given the acceptance of this criterion, available data suggest that reasonably accurate predictions of the frequency of sleep disruption can be made if the noise is described in units of EPNdB [or EdB(A)] that account for its spectral characteristics and its duration.

To limit the probability of sleep disruption, single-event noise levels should not exceed 70 EPNdB or about 57 EdB(A).

ACKNOWLEDGMENTS

My sincere thanks to Dr. Karl D. Kryter, Prof. Wilse B. Webb, and Dr. George J. Thiessen for their sage comments about drafts of this chapter.

Preparation of this chapter was supported, in part, by the Environmental Protection Agency under contract No. 68-01-3120. The final report of that work is Reference 57, to which the reader is also referred for recent information on this subject. Studies by the author described herein were also supported by a number of contracts with the National Aeronautics and Space Administration.

REFERENCES

1. A. Rechtschaffen and A. Kales, eds., "A Manual of Standardized Terminology, Techniques and Scoring System for Sleep Stages of Human Subjects." Natl. Inst. Health Publ. No. 204, 1968.

2. W. B. Webb, "Sleep as a Biorhythm." In W. P. Colquhoun, ed., *Biological Rhythms and Human Performance*, Academic, London, 1971, pp. 149–177.

3. N. Kleitman, *Sleep and Wakefulness*. University of Chicago Press, Chicago, 1963, rev. ed.

4. W. B. Webb, "Sleep as an Adaptive Response." *Percept. Mot. Skills*, **38**, 1023–1027, 1974.

5. E. L. Hartmann, *The Functions of Sleep*, Yale University Press, New Haven, 1973.

6. L. C. Johnson, "Are the Stages of Sleep Related to Waking Behavior?" *American Scientist*, **61**, 326–338, 1973.

7. K. D. Kryter, *The Effects of Noise on Man*, Academic, New York, 1970.

8. E. Grandjean et al., "A Survey of Aircraft Noise in Switzerland." In W. D. Ward, ed., Proc. Int. Congr. Noise Public Health Probl., EPA No. 550/9-73-008, 1973, pp. 645–659.

9. S. Sörensen, K. Berglund, and R. Rylander, "Reaction Patterns in Annoyance Response to Aircraft Noise." In W. D. Ward, ed., Proc. Int. Congr. Noise Public Health Probl., 1973, pp. 669–677.

10. I. Oswald, A. M. Taylor, A. M. Triesman, "Discriminative Responses to Stimulation During Human Sleep." *Brain*, 83, 440–453, 1960.

11. H. L. Williams, H. C. Morlock, and J. J. Morlock, "Instrumental Behavior During Sleep." *Psychophysiology*, 2, 208–215, 1966.

12. H. L. Williams, "The Problem of Defining Depth of Sleep." In S. S. Kety, E. V. Evarts, and H. L. Williams, eds., *Sleep and Altered States of Consciousness*, Williams and Wilkins, Baltimore, 1966, pp. 277–287.

13. W. W. K. Zung and W. P. Wilson, "Response to Auditory Stimulation During Sleep." *Arch. Gen. Psychiat.*, 4, 548–552, 1961.

14. G. W. Langford, R. Meddis, and A. J. D. Pearson, "Awakening Latency from Sleep for Meaningful and Non-meaningful Stimuli." *Psychophysiology*, 11, 1–5, 1974.

15. J. S. Lukas, "Awakening Effects of Simulated Sonic Booms and Aircraft Noise on Men and Women." *J. Sound Vib.*, 20, 457–466, 1972.

16. J. S. Lukas, M. E. Dobbs, and K. D. Kryter, "Disturbance of Human Sleep by Subsonic Jet Aircraft Noise and Simulated Sonic Booms." NASA Report CR-1780, July, 1971.

17. J. S. Lukas and K. D. Kryter, "Awakening Effects of Simulated Sonic Booms and Subsonic Aircraft Noise." In, B. L. Welch and A. S. Welch, eds., *Physiological Effects of Noise*, Plenum, New York, 1970, pp. 283–293.

18. H. L. Williams, J. T. Hammock, R. L. Daly, W. C. Dement, and A. Lubin, "Responses to Auditory Stimulation, Sleep Loss, and the EEG Stages of Sleep." *Electroencephalog. Clin. Neurophysiol.*, 16, 269–279, 1964.

19. F. B. Keefe, L. C. Johnson, and E. J. Hunter, "EEG and Autonomic Response Pattern During Waking and Sleep Stages." *Psychophysiology*, 8, 198–212, 1971.

20. G. J. Thiessen, "Effect of Noise During Sleep." In B. L. Welch and A. S. Welch, eds., *Physiological Effects of Noise*, Plenum, New York, 1970, pp. 271–275. The information about Thiessen's studies was gleaned also from Ref. 21, from G. J. Thiessen, "Noise Interference with Sleep," Natl. Res. Counc. Canada (undated), and from personal communications.

21. D. Berry and G. J. Thiessen, "The Effects of Impulsive Noise on Sleep." Natl. Res. Counc. Canada, Report No. NRC 11597, 1970.

22. P. A. Morgan and C. G. Rice, "Behavioral Awakening in Response to Indoor Sonic Booms." Tech. Report No. 41, Institute for Sound and Vibration, Univ. Southampton, England, December, 1970.

23. J. E. Ludlow and P. A. Morgan, "Behavioral Awakening and Subjective Reactions to Indoor Sonic Booms." *J. Sound Vib.*, 25, 479–495, 1972.

24. M. Kramer, T. Roth, J. Trindar, and A. Cohen, "Noise Disturbance and Sleep." FAA No. 70-16, 1971.

25. L. C. Johnson, R. E. Townsend, P. Naitoh, and A. G. Muzet, "Prolonged Exposure to Noise as a Sleep Problem." In W. D. Ward, ed., Proc. Int. Congr. Noise Health Probl., EPA No. 550/9/73-008, 1973, pp. 559–574.

26. W. E. Collins and P. F. Iampietro, "Effects on Sleep of Hourly Presentations of Simulated Sonic Booms (50 N/m^2)." In W. D. Ward, ed., Proc. Int. Congr. Noise Health Probl., EPA No. 550/9-73-008, 1973, pp. 541–558.

27. G. Globus, J. Friedmann, H. Cohen, K. S. Pearsons, and S. Fidell, "The Effects of Aircraft Noise on Sleep Electrophysiology as Recorded in the Home." In, W. D. Ward, ed., Proc. Int. Congr. Noise Health Probl., EPA No. 550/9-73-008, 1973, pp. 587–591.

28. A. Muzet, J. P. Schieber, N. Olivier-Martin, J. Ehrhart, and B. Metz, "Relationship Between Subjective and Physiological Assessments of Noise-Disturbed Sleep." In W. D. Ward, ed., Proc. Int. Congr. Noise Health Probl., EPA No. 550/9-73-008, 1973, pp. 575–586.

29. T. E. LeVere, R. T. Bartus, and F. D. Hart, "Electroencephalographic and Behavioral Effects of Nocturnally Occurring Jet Aircraft Sounds." *Aerospace Medicine*, 43, 384–389, 1972.

30. T. E. LeVere, G. W. Morlock, L. P. Thomas, and F. D. Hart, "Arousal from Sleep: The Differential Effect of Frequencies Equated for Loudness." *Physiol. Behav.*, 12, 573–582, 1974.

31. R. Rylander, S. Sörensen, and K. Berglund, "Sonic Boom Effects on Sleep: A Field Experiment on Military and Civilian Populations." *J. Sound Vib.*, 24, 41–50, 1972.

32. J. S. Lukas, "Predicting the Response to Noise During Sleep." In W. D. Ward, ed., Proc. Int. Congr. Noise Public Health Probl., EPA No. 550/9/73-008, 1973, pp. 513–525.

33. J. S. Lukas, D. J. Peeler, and J. E. Davis, "Effects on Sleep of Noise from Two Proposed STOL Aircraft." NASA Report No. CR-132564, 1975.

34. W. P. Wilson and W. W. K. Zung, "Attention, Discrimination and Arousal During Sleep." *Arch. Gen. Psychiat.*, 15, 523–528, 1966.

35. G. Steinicke, "Die Wirkungen von Lärm auf den Schlaf des Menschen." In *Forschungsberichte des Wertschaft- und Verkehrsministeriums Nordheim Westfallen*, No. 416, Westdeutscher, Köln and Opladen, 1957.

36. J. S. Lukas and M. E. Dobbs, "Effects of Aircraft Noise on the Sleep of Women." NASA Report No. CR-2041, 1972.

37. J. P. Schieber, A. Muzet, and P. J. R. Ferrière, "Les Phases d'activation transitoire spontanées au cours du sommeil normal chez l'homme." *Arch. Sci. Physiol.*, 25, 443–465, 1971.

38. H. W. Agnew, W. B. Webb, and R. L. Williams, "The First Night Effect: An EEG Study of Sleep." *Psychophysiology*, 2, 263–266, 1966.

39. H. L. Williams, "Effects of Noise on Sleep: A Review." In W. D. Ward, ed., Proc. Int. Congr. Noise Health Probl., EPA No. 550/9-008, 1973, pp. 501–511.

40. J. S. Lukas, D. J. Peeler, and M. E. Dobbs, "Arousal from Sleep by Noises from Aircraft With and Without Acoustically Treated Nacelles." NASA Report No. CR-2279, 1973.

41. In calculating EPNdB, each successive doubling (or halving) of stimulus duration beyond the reference duration of 0.5 sec added (or subtracted) 3 dB to the PNdB level estimated or reported for the reference duration. Stimulus duration is the time between the points in time when the stimulus rises and falls 10 dB below its peak level. As far as possible, frequency spectra were used to calculate the PNdB level of each stimulus. If spectra were not available but dB(A) or other measurements were, the maximum PNdB levels were estimated from data for similar stimuli provided by Kryter [7]. These procedures are consistent with those recommended by Kryter (pp. 472 and 483), but his pure-tone, onset, and other corrections were not used.

42. On the basis of the publication dates of Steinicke's paper [35] and Zwicker's 1960 paper on calculating loudness [43], Stevens's method [44] probably was used to calculate phons.

43. E. Zwicker, "Ein Verfahren zur Berechnung der Lautstärke." *Acustica*, 10, 304–308, 1960.

44. S. S. Stevens, "The Calculations of the Loudness of Complex Noise." *J. Acoust. Soc. Amer.*, 28, 807–832, 1956.

45. W. B. Webb, personal communication, 1973.

46. A. Lubin, personal communication, 1973.

47. M. Herbert and R. T. Wilkinson, "The Effects of Noise-Disturbed Sleep on Subsequent Performance." In W. D. Ward, ed., Int. Congr. Noise Public Health Probl., EPA No. 550/9-73-008, 1973, pp. 527–539.

48. L. Johnson, P. Naitoh, A. Lubin, and J. Moses, "Sleep Stages and Performance." In W. P. Colquhoun, ed., *Aspects of Human Efficiency*, English Universities Press, London, 1972, pp. 81–100.

49. R. T. Wilkinson, "Sleep Deprivation: Performance Tests for Partial and Selective Sleep Deprivation." In L. A. Abt and B. F. Riess, eds., *Progress in Clinical Psychology*, Grune, New York, 1969, Vol. 7, pp. 28–43.

50. P. Hamilton, R. T. Wilkinson, and R. S. Edwards, "A Study of Four Days Partial Sleep Deprivation." In W. P. Colquhoun, ed., *Aspects of Human Efficiency*, English Universities Press, London, 1972, pp. 101–113.

51. W. B. Webb and H. W. Agnew Jr., "The Effects of a Chronic Limitation of Sleep Length." *Psychophysiology*, 11, 265–274, 1974.

52. B. L. Frankel, R. Buchbinder, and F. Snyder, "Sleep Patterns and Psychological Test Characteristics of Chronic Primary Insomniacs." Abstr., 13th Ann. Meeting, Assoc. Psychophysiol. Study Sleep, 1973, p. 24.

53. Dr. Muzet kindly provided time histories of his stimuli; from these, EPNdB levels were estimated by using the 10 dB downpoints as the significant duration. His 90 sec stimuli had durations of 22–30 sec, while his 30 sec stimuli had durations of 2 and 3 sec according to our technique of calculation.

54. W. L. Hays, *Statistics for Psychologists*. Holt, Rinehart and Winston, New York, 1963, pp. 503–505.

55. J. P. Schieber, J. Mery, and A. Muzet, "Etude Analytique en Laboratoire de l'Influence du Bruit sur le Sommeil." Centre d'Etudes Bioclimatiques du C. N. R. S., Strasbourg, France, April, 1968.

56. A. Muzet, "Evaluation Experimentale de la Gêne et de la Nuisance des Perturbations du Sommeil par Divers Bruits de Circulation Automobile et Aérienne." Centre d'Etudes Bioclimatiques du C. N. R. S., Strasbourg, France, undated.

57. J. S. Lukas, "Measures of Noise Level: Their Relative Accuracy in Predicting Objective and Subjective Responses to Noise During Sleep." Prepared by Stanford Research Institute for the Environmental Protection Agency under contract 68-01-3120, September, 1975.

14

Effects of noise on human work efficiency

G. R. J. HOCKEY*

14-1. INTRODUCTION

14-1.1 Scope of the Chapter

This chapter is concerned with the effects of noise on the efficiency of ongoing mental work. Most of the evidence comes from laboratory studies, for reasons outlined later, though the findings have general implications for work situations in industry, transport, and the home. We shall confine our discussion to *temporary* effects on efficiency, rather than permanent changes resulting from long-term exposure to noise environments. Effects of long-term exposure are likely to be related to problems of health, sleep, deafness, and annoyance, which are treated elsewhere in this volume.

*Department of Psychology, University of Durham, Durham DH1 3LE, England.

Our main concern is with the performance of tasks which are essentially of a nonverbal nature, since it is with this kind of work that effects of noise are, perhaps, least expected. The commonly observed interference with spoken communication is well known and normally guarded against to some extent. This is a much simpler problem theoretically, since it is relatively easy to explain why the resultant "masking" reduces the discriminability of the speech signals. Even so, effects on communications can occur at noise levels which do not impair intelligibility, and we shall briefly summarize work in this area in Section 14-2. In fact, recent research has indicated that such effects may have much in common with those observed in visual tasks, over and above the separate effects attributable to the sense modality involved. In Section 14-3, we will examine the effects of brief, intermittent noise. Such effects, again, appear to have a quite specific basis, and are to be distinguished from those resulting from exposure to continuous, unchanging noise. Section 14-4, the longest of the chapter, is devoted to effects of this latter kind, and it is here that the greatest uncertainty is found, both in terms of findings and explanations. In Section 14-5 we shall present an overview of what is known about noise and efficiency, and attempt to provide some useful practical recommendations.

14-1.2 Varieties of Noise

Research on noise has tended to be rather unsystematic in its manipulation of the parameters of sound: studies have tended to use almost any kind of noise in an experiment. This probably stems largely from the common interpretation of all noises as "distracting," but ignores other possible effects; different forms of noise may be exciting, irritating, startling, or boring. In the main, three kinds of noise have been used: sudden noises (e.g., sonic bangs), intermittent noise, and continuous (mainly broadband) noise. Other studies have used meaningful sound such as music or speech, though these may not always be regarded as "noise," and their effects are probably more complex than those found with other forms; people are more likely to listen to music or speech for its own sake, rather than simply treating it as an unwanted background event. We shall deal with sudden and intermittent noise together (Section 14-3), and deal most thoroughly with the effects of continuous broadband noise (Section 14-4). This is one of the most common kinds of noise experienced in the work situation and is the one used most often in research. This division is necessarily a rough one, and we will refer to other, more specific, varieties where they are relevant, including music and speech. The majority of studies deal with increases in the sound pressure level (SPL) of a particular kind of noise, though some are concerned with effects of modulation around a mean level or of degree of variety. The relevance of these different procedures will become clear when we outline the theoretical bases for noise research in Section 14-1.4. Lastly, it should be noted that only a small number of studies give information about the weighting scale on which SPLs are measured, so that often only a general dB level is available.

14-1.3   Methodological Problems

Limitations of Field Studies. The principal difficulty in measuring the effects of noise on efficiency is to ensure that the observed effect is indeed attributable to the noise, and not to some other changing aspect of the situation. This is the strongest argument for preferring laboratory studies to field studies, since only in the laboratory can conditions be controlled in such a way that this state of affairs prevails. If the level of noise is reduced in one part of a factory, but not in another part, one ought to be able to measure the relative changes in productivity in the two locations and, hence, assess the effect of the noise treatment. While this is possible in the most careful field studies (such as the Broadbent and Little study reported in Section 14-4.5), the difficulties in interpretation are immense. The knowledge that such research is being carried out in the factory is known to give rise to a generalized improvement in production throughout the plant, which is maintained when the treatment is removed [1-3]. Furthermore, the reorganization associated with any permanent noise treatment may well result in simultaneous changes in other aspects of the environment, such as improved ventilation or better lighting, while, during the period of measurement, there may be changes in the weather or in the state of factory politics which may also systematically distort the findings of the study.

Although it is certainly possible to overcome most or all of these problems, the evidence suggests that this has seldom been done. As a result, only a small proportion of field studies can be used as reliable evidence of effects of noise on efficiency. Rather than report these in a separate section we have discussed them along with the relevant laboratory results. Of course, from the practical standpoint, it is essential that laboratory findings be tested thoroughly under actual working conditions. With the guidance of clear effects obtained from controlled laboratory studies, it may be possible to allow for some of the concomitant changes mentioned above, but, with few exceptions, this rather unsatisfactory state of affairs is the rule rather than the exception.

Control and Design Problems in Laboratory Studies. The advantage of a laboratory study is that all variables other than that (or those) being studied can be controlled. In practice, this apples only to those which are likely to influence performance, but, typically, a noise experiment should control such variables as ambient temperature, time of day, degree of practice at the task, level of motivation (by instruction), and the influence of other factors such as sleep deprivation, alcohol, and tobacco. It is also important that the task be identical for each person taking part in the experiment (the subjects). In a field study, the actual requirements of the job may vary widely from day to day as a function of other ongoing processes in the plant.

In a typical experiment, a number of subjects are tested on a set task under conditions of noise and quiet, and their scores compared in the two conditions. It is essential to test a sufficient number of subjects to allow statistical generalizations to be made. Some subjects will be required to carry out the task in the order quiet–noise, and some in the reverse order (noise–quiet), usually on separate days. This is because the second condition may be better (or worse) simply becuase of the effects of practice (or boredom), and any such effects must be properly counterbalanced. The quiet condition is more exactly described as "less noisy," and may be as high as 70 dB(A) (compared to 100 dB(A) in the noise condition). If this were not so, and no noise at all were present in the quiet condition, one or both of two effects might occur. Subjects may be distracted by extraneous sounds (traffic noise or aircraft), or they may obtain feedback from nearby scoring apparatus. Even when these artifacts are removed (as in the use of an efficient, sound-proofed room for testing), it is desirable to change only one parameter of the noise at a time, e.g., the SPL. The same kinds of consideration apply, however, to studies which manipulate the degree of variety of the noise.

Before examining the evidence from empirical studies, it may be helpful to outline briefly the major theoretical interpretations of noise effects. We shall refer to these theories continually throughout the chapter.

14-1.4 Theories of Noise Effects

Two rather different theories have been put forward to account for the effects of noise on human efficiency. The first of these accords closely with the commonsense view and may be called the distraction theory. Naturally, this theory predicts that the presence of noise will impair ongoing performance, though, as we shall see, there is more than one way in which it could do this. Such a view has existed in experimental psychology for nearly a century, probably because of its commonsense nature, and, as a theory, it is largely inductive, growing up out of cumulative observation. Its influence in this field in recent times is due to Broadbent's filter theory [4].

The second approach considers noise not as an unwanted intruder but as one of the means of maintaining an adequate level of background stimulation necessary for efficient functioning. This view, which has its origins in physiological psychology [5-7], is known as the arousal theory. It, too, has its commonsense proponents (in particular, those who cannot work without the radio on), though it is a comparatively recent influence in research on noise.

It will be helpful to outline the basic assumptions of these two theories, and the broad predictions which they make. This should help the reader to evaluate them continually in the light of the empirical findings reported in the body of this chapter.

Distraction Theory. The modern version of the distraction theory is known as the filter theory and has its origins in Broadbent's book *Perception and Communication* [4]. The human being was regarded by Broadbent as operating as an information-processing device, having a single central processing unit of limited information-handling capacity. Filtering was the mechanism by which sensory inputs were either selected for processing or discarded. The filter, assumed to operate at an early stage in the system, could be thought of as shifting from source to source in accordance with factors such as built-in priorities and momentary changes in the attractiveness of competing inputs. Such shifts could be involuntary, e.g., when an irrelevant but attention-catching stimulus was selected after a long period of processing information from one source. These interruptions, although brief, were regarded as being the cause of, among other things, impaired efficiency with prolonged monitoring, and of the distracting effect of noise.

Noise was considered to increase the probability of filter shifts by acting as a strong competing source of stimulation: it had its effect by capturing attention. This view was, in fact, consistent with all known effects of noise and prolonged work at that time, and may still apply to certain situations (e.g., the effect of sudden noises, discussed in Section 14-3). Certain difficulties with filter theory as an overall explanation of noise effects have recently become apparent, however. The major alternative explanation is that of arousal theory.

Arousal Theory. Arousal refers to the function of generalized activation of the central nervous system associated with various mid-brain structures, notably the reticular activating system (RAS). This function is quite separate from the specific sensory processing carried out by the brain, in that it depends on stimulation from all sources, both external and internal, and provides the central nervous system with the necessary background activity for efficient functioning. Its influence in the understanding of sleep and effects of sensory deprivation has been considerable [6, 7], though it has only recently been used as an explanation of the effects of noise [5, 7].

Noise, as one of the more prominent sources of background stimulation, may be regarded as increasing the level of arousal: this can occur either through an increase in the intensity of the noise, or through an increase in its variety. The principal advantage of arousal theory over distraction theory is that it is a more flexible concept, being able to account for facilitatory effects as well as inhibitory ones. This is because arousal appears to be related to efficiency in a curvilinear manner, as in Fig. 14-1. Optimal efficiency occurs when arousal is

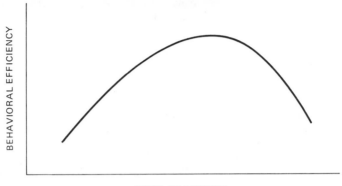

Fig. 14-1. Hypothetical relationship between arousal (or activation) level and behavioral efficiency.

intermediate, and either very high levels (e.g., strong emotional states) or very low levels (e.g., drowsiness) are associated with inefficiency (see Ref. 6 for a fuller discussion of this). This advantage is also the major drawback to the theory; since almost any result can be explained, it cannot easily be disproved. In fact, the best evidence for sticking to arousal theory as an interpretation of the effects of noise comes from studies in which noise occurs in combination with other stresses (especially sleep loss and incentive). These will be discussed later in Section 14-4.5. Fortunately, more recent work provides a more convincing rationale for an arousal interpretation of noise effects, and if we bear in mind the dangers of using this concept in a gratuitous fashion, it still appears to be the most useful way of thinking about such effects.

The Problem of Interpretation. Having outlined the two major explanatory concepts of distraction and arousal, let us now turn to an examination of the actual findings. More startling at first sight than the contrary expectations of the two viewpoints are the apparently inconsistent experimental results. Where distraction has been invoked as the explanation of noise effects, a decrement in efficiency has usually been found. Where arousal has been assumed to underlie the observed effects, performance has usually been shown to improve (in this case the noise is often referred to as auditory stimulation). One of the main tasks of this chapter is to explain these discrepancies in terms of methodological differences between the two groups of findings. These are principally in terms of (a) the levels and characteristics of the noise used, and (b) the demands made on the subjects by different tasks. Our treatment of the results will be somewhat selective, since it is directed toward the resolution of these ambiguities, and, in the main, discusses only those studies which appear to be scientifically sound.

Although there is, necessarily, considerable overlap with previous reviews of noise effects, the problem of determining which kinds of situation are likely to give rise to an impairment of efficiency, and which are not, is one that was last tackled effectively in Broadbent's 1957 survey [2, 4]. Much has happened in this field since then, and a number of new ideas are available. For a complete coverage of the literature, however, the reader is referred to the extensive reviews by Kryter [1, 3] and that of Broadbent. Plutchik [9] provides a useful overview of research on intermittent noise up to 1959.

14-2. PROBLEMS OF COMMUNICATION

14-2.1 Masking of Speech

One of the most obvious aspects of reduced efficiency with high ambient noise levels is the interference with spoken communication. While it may be possible to overcome this problem to a certain extent, e.g., by the use of visual gestures and exaggerated articulation, such operations are only of value in limited situations, such as close, face-to-face conversation. Effects of masking are, of course, even more serious in purely auditory communication systems, such as the telephone: not only is the listener unable to make use of any such visual cues, but he has to cope with an additional source of noise from the transmission channel itself. In general, the degree of interference with communication depends on the complexity of the transmitted message [10]. In the case of digits, 100% intelligibility can be reached with signal-to-noise ratios of up to −3 dB, and 70% (sufficient for most purposes) with −10 dB. However, for more unpredictable messages involving unfamiliar names or detailed instructions, a signal-to-noise ratio of about +20 dB is needed for 70% intelligibility—similar to the critical level for completely unfamiliar items such as nonsense syllables. Figure 14-2 provides a useful graphical representation of these findings. One important point about masking is that it is generally confined to information in the same frequency range as the noise. This means, of course, that since human speech contains a high proportion of lower frequencies, masking is generally a more serious problem with noise having most of its energy in the lower part of the sound spectrum (say, below 2 kHz).

It is now possible to estimate the efficiency of transmission over a communication system by referring to data on articulation tests [3, 8]. Such tests aim at providing an index of critical noise levels for a message to be heard correctly. This assumes that if a listener can hear each word correctly the efficiency of the communication system is not being impaired. However, recent work in this field has demonstrated that this is not the case; perception and comprehension may be more difficult even though speech is not actually being masked.

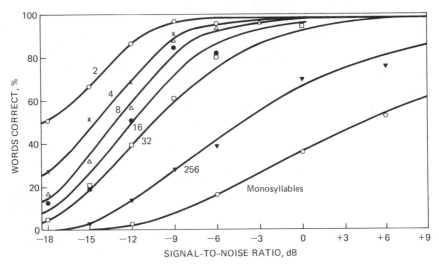

Fig. 14-2. Articulation scores as a function of speech signal-to-noise ratio and number of alternatives. Redrawn from Miller, Heise, and Lichten [10].

14-2.2 Interference with Cognitive Operations Due to Nonmasking Noise

In a typical articulation test a list of words is read over a communication channel (e.g., a loudspeaker) with different levels and frequencies of noise built into the system, and subjects are required to simply write down each word as it is presented. Pollack [11] showed in one study that although the number of errors on the test words were minimal when nonmasking noise was used, the latency of the response (the time taken to recognize and report the word) increased steadily as the level of (nonmasking) noise was raised. In other words, the actual process of speech perception was being made more difficult. The increased attention necessary for efficient recognition in this experiment may be expected to limit the capacity of the listener to carry out any additional task at the same time. This interpretation was supported in a study by Rabbitt [12], who presented listeners with lists of words, some accompanied by nonmasking noise and some with no additional noise present. In a recognition test 10 seconds later, subjects were significantly less accurate in picking out the test words from other (dummy) items when they had been presented in noise. This can be regarded as an effect of reduced capacity for committing the words to memory due to the greater demands made upon listening. Finally, an experiment by Holloway [13] required pairs of subjects who could not see each other to work as a team. One had to read out a list of digits while the other transformed them in some way (e.g., by adding 3 to each) and reported the answer; only then was he provided with the next digit. Holloway found that performance was noticeably slower in the presence of nonmasking noise, even though, once again, there was no problem about hearing the digits.

All these findings indicate that noise may affect communication in a more subtle way than simply masking the speech signal. It may have the effect of slowing down the rate at which information can be taken in and acted upon by the listener. This reduces overall efficiency by causing him to concentrate more on the actual reception of the message, and less on either comprehension or on other relevant sources of information. In addition, he may be less able to remember the message when he come to act upon the information. It would, therefore, be a serious mistake to assume that efficiency is being adequately maintained in an office where the noise level is not sufficient to impair intelligibility of speech. For a fuller coverage of the extensive literature on masking and speech perception the reader is referred to the comprehensive reviews by Kryter [3, 8], while Chapter 1 of this volume explains how noise is analyzed to determine the degree to which it impairs speech communication.

We shall now turn our attention to tasks where the verbal requirement is minimal. Tasks of this sort normally require the subject to process visual information and to produce adequate perceptual–motor responses. It is probably fair to say that effects of noise are less serious here than in the case of jobs where verbal communication is essential, but it is now quite clear that several broad classes of task are susceptible to the effects of noise; furthermore, some of these effects are, like those above, rather subtle and difficult to predict. Only by being fully aware of them can efficiency be maintained at a high level.

14-3. SUDDEN AND INTERMITTENT NOISE

14-3.1 Reaction to Environmental Change

The sudden onset of intense auditory stimuli such as explosions or bangs has been shown to affect behavior in quite well-defined ways. Events of this kind are, of course, distracting in a real sense, in that the noise actually appears to capture attention. A number of early experiments [14–17] showed that subjects could be distracted by loud buzzers or gongs from efficient performance of various mental tasks such as addition or simple reaction. The effects were, however, very slight, and there was no tendency for performance to be disrupted at times other than when the sounds actually occurred. Moreover, there was some evidence that subjects were even able to make up for the distraction, if it occurred, by working harder between noises, since there was no overall difference between the distraction and control conditions.

This kind of experiment cannot be considered as reliable evidence because of the lack of any real experimental control and the arbitrary definition of what is a distracting noise. The noise in these experiments may be characterized less by its acoustical properties (SPL, sound spectrum, etc.) than by its novelty value. Such a distinction becomes meaningful when it is noted that several experiments [15, 16] have found distraction effects when noises were turned off as well as on. This conclusion has been confirmed by the findings of two recent experiments, which have investigated performance disruption as a function of the amount of change in the ambient noise level [18, 19]. In both studies performance decrement was a linear function of the amount of change in SPL, irrespective of the direction. The distraction effects in such experiments cannot, therefore, be regarded strictly as effects of noise: if the normal condition is noisy, as it is in some of the abovementioned experiments, then any impairment in efficiency may more accurately be attributed to an effect of quiet. The general interpretation of these findings is that distraction is a function of the amount of change in the environment.

14-3.2 The Orienting Response

Effects such as the ones we have been describing have a close physiological correlate. An intense noise, as well as distracting attention from a task being carried out, is also found to elicit a complex set of physiological responses (see also Chapter 12) known as the startle pattern [20] or the orienting response [21]. Typically, this reaction includes pupil dilation, lowered skin resistance, various complex changes in cardiovascular and respiratory activity, a general increase in muscular tension, and desynchronization (activation) of the EEG. Gross behavioral changes of orientation and posture usually accompany these effects, and the general pattern of response is an essential feature of adjustment to sudden alterations in energy impingements in both human and animal species.

Such changes are found to be a function of the novelty of the stimulus, being maximal on the first presentation of the stimulus, but becoming less pronounced with repeated exposure. With this habituation the organism comes to be largely unresponsive, and we can assume this to be the result of the stimulus no longer being novel or unfamiliar [22]. Such habituation may, however, be quite specific, a slight change in the frequency of a repeated tone often being sufficient to produce a reinstatement of the complete orienting response [21]. In addition, habituation is less pronounced for stimuli which arrive irregularly in time. The phenomenon of habituation is of central importance in a discussion of distraction effects due to sudden noises. If the degree of interference follows the observed changes in the orienting response with repeated exposure, then subjects ought to be less affected the more the noise is continued. As we have already mentioned, this has been a general finding of the early studies in this field, and was shown to occur by Pollock and Bartlett [23] for a wide range of tasks.

Not all noises are successful in arousing the orienting reaction. The most important consideration is intensity, a very loud sound being more likely to arouse the response than one of moderate loudness. A sound of high frequency is also more successful than one of low frequency [21, 22]. These facts coincide with further behavioral evidence [24, 25] that loud or high-pitched sounds interfere more with ongoing attention processes than soft or low sounds. Broadbent's study provides a clear insight into the nature of the disruption. The kinds of noise associated with the greatest distraction in a visual serial reaction time task were also those which produced the fastest reactions when they themselves served as relevant stimuli in an auditory reaction time task. Thus, the effect is not one of shutting off the operation of the central information processing system, but that of actively capturing attention away from the main task. When these strong attention-getting stimuli are themselves the relevant task features, the resulting orienting reaction may well serve the purpose of preparing the system for efficient response.

This consideration applies, of course, to the general nature of the orienting response. It is an adaptive mechanism for preparing the organism for emergency action. The particular stimuli which arouse this reaction are, to a great extent, species-specific, but loud or high-pitched noises are among the more widespread releasers of the reaction. The irony is that, as far as human beings are concerned, these biologically useful warning signals are largely inappropriate in everyday life—yet they are all too common in the modern acoustic environment. Fortunately, the process of habituation ensures that if an irrelevant stimulus is repeated often enough, the physiological and behavioral response to it is soon diminished. Even very loud sounds, such as sonic bangs, must eventually become ineffective if they occur with sufficient frequency and regularity.

14-3.3 Localized Effects of Sudden Noises

Possibly because of this habituation to distraction, very few experiments using noise of this kind have, in fact, demonstrated any sizeable disruption on visual task performance. Although distraction may have occurred in some of these studies, it is likely to have been confined to the beginning of the task; averaging scores over the whole test would have the effect of concealing any such localized effect. The importance of this consideration is neatly illustrated in some experiments by Woodhead [26-28] on the effects of bursts of rocket noise on visual inspection tasks. The overall effects in her data are rather small, but closer inspection of the performance at the time of the distraction revealed a marked depression in performance in the period following the burst, sometimes lasting up to half a minute. Figure 14-3 shows an example of such an effect. In these experiments little habituation could, in fact, occur, since the bursts were infrequent (about every 3 or 4 minutes) and unpredictable in time. Such findings

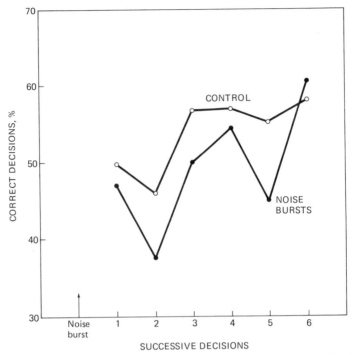

Fig. 14-3. Localized effect of 1 sec noise bursts on decisions following the burst: the time period covered by six decisions varies between 17 and 31 sec. Redrawn from Woodhead [27].

strongly support the idea that the organism takes a considerable time to recover from distraction (much longer than the duration of the distracting event), as the pronounced physiological changes involved in the response are allowed to revert to normal levels. It seems likely, however, that the duration of this distraction state may well depend on the task involved. A more recent study by Fisher [29] has found localized effects of 80 and 100 dB(A) noise bursts on the 5-choice serial reaction task (see Section 14-4.1), but these are confined to the first response after noise onset (i.e., less than 0.5 sec). A similar result was obtained by May and Rice [30] for the effect of pistol shots (peak level 124 dB) on pursuit motor performance. The impairment in this case was quite drastic but lasted only two or three seconds. Clearly, much more information is needed on the precise effects of noise bursts, and their relation to task structure.

14-3.4 Sensory Facilitation Effects

To complicate this picture even further, there are a number of studies which have found performance to be improved with short bursts of noise, across a variety of tasks involving manipulation [31], flicker fusion threshold [32, 33], visual search [34, 35] , and visual detection [34]. One interesting result, which relates to the discussion of Section 14-3.3, is that of Watkins [36] . This experiment examined visual detection performance in a forced-choice situation, where subjects are required to report the observation interval in which a target is presented. Performance was improved when bursts of noise accompanied the target interval, but detection was actually impaired when the burst occurred in some other interval. It is not at all clear why a brief burst of noise should ever facilitate performance. One possibility is that, as we implied earlier, the concomitant orienting response is actually adaptive if it occurs in conjunction with the need to react at that moment; normally, of course, the bursts arrive at moments quite unrelated to the task structure. Some of the results in this area, and the broader area of sensory interaction [37] , suggest that a proper understanding of the mechanism of distraction (and the related phenomenon of facilitation) is still some way off, though some broad generalizations can clearly be made on the basis of the work summarized in this section.

Effects of sudden and intermittent noises, as we have pointed out, cannot be regarded in the same way as those associated with continuous noise. Any change in efficiency tends to diminish with repeated presentation, as expected from an understanding of the habituation process. Changes with continuous noise, rather surprisingly, are found to increase with repeated exposure; it is with effects of this type that we shall be most concerned in this review.

14-4. CONTINUOUS NOISE

Since the second world war many studies have been carried out on the effects of continuous noise in the work period. It is perhaps surprising, therefore, that very few studies have succeeded in demonstrating a reliable effect on efficiency. (This includes a number of investigations which were very well designed, even by the standards of modern research, [e.g., 38, 39]). An understanding of the reasons for this failure, however, have enabled us to examine the problem in a more analytic way. Subjects in these experiments were required to perform tasks such as card-sorting, choice reaction time, and tests of manual skill and perceptual judgment. While these are fairly representative of the sorts of operations used in everyday life, they are all limited in one important respect: they do not continually "stretch" the man's capacity. It was recognized in the late 1940s that man is a very efficient "self-regulating device" [40], able to overcome temporary interruptions in performance and maintain a near-constant output (e.g., by working harder between disruptions). Effects on his performance would therefore be more likely to be detected in terms of more subtle measures of efficiency, such as errors of timing or the production of inappropriate responses. It will be recalled that short-term effects such as these were found by Woodhead in her study of the effects of bursts of rocket noise (see Section 14-3.2). Furthermore, such errors are only likely to become pronounced when the task allows the man no spare time in which he can safely disregard the possibility of the occurrence of relevant task information. Laboratory tasks which meet these requirements are of two general types, both of which are typified by a decline in efficiency with time spent at work, generally thought of as a fatigue effect.

One such task is that of monitoring (sometimes called watchkeeping or vigilance). Generally, this kind of situation requires the subject to continuously watch out for barely detectable signals which might arrive at any time during a prolonged session (as in radar operating or conveyor belt inspection jobs). The second type of situation may be called the serial choice reaction. In this, the subject must react as quickly as he can to a series of possible light signals, each one coming on as soon as the last has been reacted to: this may be thought of as simulating a number of real-life situations calling for rapid organized sequences of actions. In both cases the man's attention is continually needed, and any momentary lapses can be detected in terms of missed signals or errors. In the serial reaction task it is also possible to observe occasional very long reaction times in the response sequence (known as gaps or blocks). Compensatory bursts of fast work can of course occur in these situations, but the use of more analytic measures of efficiency enables any previously occurring lapses of efficiency to be observed. This is not true in all cases; e.g., in card-sorting tasks only an overall measure of output or time is obtained.

In reviewing the literature on noise effects up to 1957, Broadbent [2] summarized the necessary conditions for demonstrating impairment in efficiency:

1. The task should be continuous and of relatively long duration (at least 30 min)
2. Task information should be presented at a high rate and/or with a high degree of temporal and spatial uncertainty
3. The "microstructure" of performance should be examined, if possible, rather than relying totally on gross measures of efficiency
4. In addition, effects were unlikely if the noise level was less than 90 dB.

These generalizations, based on a comparison of studies which had found effects of noise with those which had not, provided a useful guide to research on noise over the last 15 years. More recent work in the area of memory and perceptual selection has in some ways contradicted these criteria, by showing effects on quite brief tasks and with SPLs of less than 90 dB. There are various reasons for this, and we shall discuss such evidence in Section 14-4.3. In general, however, where continuous work is involved, the above principles may still be regarded as a useful summary of the situation.

14-4.2 Prolonged Work Situations

Time at Work. It is well established that the effects of noise increase with time spent on the task, performance in the first 10 minutes or so showing no clear differences between a SPL of, say, 100 dB and one of 70 dB. It is important to note that this generalization applies not only to experiments showing impairment in efficiency, but to those which show an overall facilitation.

Impairment has been demonstrated in a large number of studies using complex monitoring tasks [41-45]. The most extensive investigations are those of Broadbent and Jerison. In their experiments, subjects were required to monitor one or more signal sources for long periods of time (30 minutes to 2 hours). In nearly all these reports the effects on efficiency are greater toward the end of the task, the earlier work periods in fact showing very little change at all. Similar effects have been observed with the 5-choice serial reaction tasks [46], a version of the task outlined in Section 14-4.1. Noise has been shown to produce either an increase in the number of errors or a greater proportion of slow responses (gaps), but again only toward the end of the work period [24, 47, 48].

The same effect of time at work is evident in studies which show an overall facilitation of performance. In continuous work situations this usually takes the form of a reduced decrement with prolonged work [49-53]. It is obviously necessary to ask why noise may have what appear to be opposite effects in situations which are objectively similar. Two possible factors are the characteristics of the noise and those of the task.

Of the abovementioned studies, a number are interesting in that they show better performance resulting from a change not in the intensity of ambient noise but in its quality. McGrath [49] showed that the visual detection efficiency of subjects working in a steady, low level (72 dB) noise was poorer than that of subjects receiving varied, meaningful noise of the same intensity (music, speech, etc.). Similarly, McBain [51] found a reduction of errors in a monotonous printing task when subjects were presented with varied noise (in this case, a tape of speech played backwards). Other studies have supported this general finding of improved efficiency with varied rather than unchanging noise. Again, the advantage of the varied noise condition is greater later in the work session.

In our attempts to reconcile the distraction and arousal viewpoints, considerations such as these may well be important. For example, one could argue that noise only acted as a competing source of stimulation at high intensities (say, above 90 dB). Below this level the noise may have very little effect, except that any variation in its characteristics serves to increase the general variety of the otherwise monotonous task situation, and tends to enhance efficiency. The studies reported above are generally consistent with this argument, showing facilitation in continuous work when low level noise is given some variety. A number of other results, however, complicate this simple picture.

One experiment which is directly contradictory to this argument [54] shows impaired efficiency due to amplitude modulation of a modal 75 dB noise background. It should be noted that the task, in this case, was one of perceptual classification, so that the finding may not be directly relevant in the present context. One study which does show such an effect in monitoring is that of McGrath

[55] , reported in more detail below in the section on signal rate and event rate: he found impairment with varied noise only on a "fast" version of the monitoring task. It certainly appears that varied noise does not always lead to facilitative effects. A more important line of argument, though, is the observation that facilitation is often found with increases in SPL; both Davies and Hockey [52] and Tarrière and Wisner [53] showing quite convincing increases in detection efficiency with an increase in SPL. In the Tarrière and Wisner study the noise level was 90 dB (vs. 35 dB in quiet), though the same level of meaningful noise improved detection even more. Davies and Hockey used 70 and 95 dB(A) as the two conditions, and found an enhancement with the latter, but only for a lower signal rate condition (45 signals per hour).

Since both inhibitory and facilitative effects of noise are a function of time spent at the task, this feature clearly does not differentiate the two effects, though it does suggest that the underlying mechanism may be the same. It is possible that the level or the kind of noise used may be relevant to the direction of the effect but here, again, there are no clear-cut pointers. It is necessary to look in more detail at the characteristics of the situations under which the two kinds of effect are found. There are two principal factors which we may examine— the number of separate components of the display and the rate at which relevant information has to be responded to.

Spatial Complexity. The importance of temporal uncertainty in demonstrating reliable distraction effects has been emphasized by Broadbent [2] , and may well be a necessary condition for finding adverse effects of noise. Yet all Broadbent's positive findings were obtained using tasks which required more of the subject than simply attending to a single source of information. His tasks could all be described as possessing a high degree of spatial complexity, in that a signal could occur in any of a number of possible locations.

One such situation is the 20 dials test [41] , in which subjects were required to watch for critical positions of each of 20 pressure gauges spaced around three sides of a room. The task was carried out for 90 minutes in both noise (100 dB) and quiet (70 dB), with critical signals (dials which remained in the "danger position" until reported) only occurring 15 times during the work period. A large effect of noise was found, signals taking much longer to detect in the noise condition. A second study [42] used an easier, though similar, task in which the dials were replaced by small light bulbs. No overall effect of noise was found with this version of the task, however, although a more detailed analysis did show an increase in the latency of detection for signals in the central part of the display in noise. It is clear from these two findings that the way in which noise affects performance may depend very much on the requirements of the task. In addition, when effects on complex, multi-component tasks are considered, there is an obvious need to examine the microstructure of the performance, rather than relying totally on gross, overall measures of efficiency.

It is not clear from Broadbent's findings whether noise has an effect because of the essential temporal uncertainty of the situation (not knowing when action may be required) or because of the additional spatial uncertainty (not knowing where to look for information). In support of this latter possibility is the task which used lights, which showed a smaller effect of noise, and could be said to require less complex scanning: a signal could be detected by a single sweep of the eyes over a particular section of the display, whereas individual fixations of each source are necessary in the case of the dials task. The question is further complicated by findings of Jerison and his co-workers [56, 57] using Mack-worth's clock test [58]. In this monitoring task the signal is an occasional double jump of a pointer moving around a dial; the situation has been shown to produce considerable decrement in efficiency of detection when subjects have been working for 30 minutes or so. Jerison compared performance on the clock test in noise of 113 and 79 dB over a session lasting 1 hour and 45 minutes, but found no differences either in overall level of detection or in the degree of decrement, as can be seen in Fig. 14-4(a). In a further experiment these workers looked at the effects of noise (114 and 83 dB) on the simultaneous monitoring of three such clocks. The overall level of detections is lower in this condition, but, as Fig. 14-4(b) shows, there is a clear effect of noise at the end of the work period (despite the separation of the curves, the difference is only meaningful in the last half-hour, since all subjects performed for the first half hour in 83 dB noise).

Jerison suggested that noise only affected multi-source tasks, by reducing the "flexibility of attention" [57]. In other words, the implication is that the noise

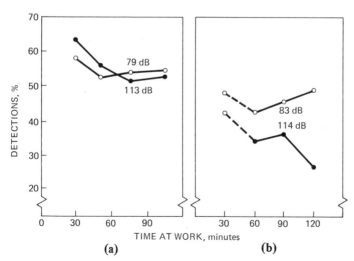

Fig. 14-4. Effects of noise on (a) one-source, and (b) three-source monitoring. Redrawn from Jerison [56, 57].

will only impair performance when subjects have to divide their attention over a number of sources of information. There are, of course, other ways in which tasks may possess a high degree of spatial complexity, and it is not clear whether Jerison's interpretation can be broadened to include these.

One study which can be thought of as relevant to the question of spatial complexity is an experiment by Samuel [59], though the independent variable here is not the number of sources but their degree of separation. Subjects had to add single digits presented at a rate of one per second on two displays, either close together or widely separated. Noise (100 dB vs. 80 dB) had no effect on performance for the adjacent displays, but actually improved performance when they were separated. This is a surprising result, since the separated condition clearly involved a more difficult spatial scanning factor, and would seem to require a more flexible attention pattern. No explanation can be given for the finding (apart from a number of ad hoc suggestions made by Samuel himself), though some of the findings described in Section 14-4.3 (below) do suggest a possible effect of noise on the ease of selecting the relevant information from the background in the difficult separated condition. The conditions for obtaining facilitation or improvement with spatially complex displays are discussed at this more appropriate later stage. For the present, we should regard the Samuel result as a further warning about the dangers of oversimple extrapolation from a very general hypothesis.

The results discussed in the present section do not, then, allow for any firm conclusions to be reached on the importance of spatial complexity, although, in general, impairment does appear to be more likely in situations requiring more complex division of attention. An obvious problem of interpretation is that many factors have been allowed to vary in these studies: temporal uncertainty, signal rate, event rate (task speed), number of sources, the need to make head or eye movements, and so on. A major group of task variables is that concerning the pacing or speed of the situation, signal rate, and event rate. These factors represent a quite different way in which to describe the difficulty of monitoring tasks.

Signal Rate and Event Rate. Signal rate has long been known to be a critical variable in monitoring efficiency, higher detection rates being associated with a more frequent occurrence of signals [60]. Event rate (or task speed) is the rate at which information of any kind (signal or nonsignal) is presented. More recent work in this area has demonstrated an important role for this factor [61]. Indeed, it may well be the probability of a signal occurring (as a function of the total number of events) that is the most important single independent variable in monitoring behavior [5, 7, 62].

Returning now to Jerison's findings reported above, it could be asked whether noise effects are only serious in spatially complex situations, or whether they may be found in all situations in which the overall signal rate is high. These two variables are clearly confounded in the Jerison studies, the three-clock task requiring the subject to look for three times as many signals. An experiment by Broadbent and Gregory [63] attempted to separate the two effects experimentally. This is a complex experiment, and the data are analyzed using statistical decision theory [64], so the findings are not easily summarized. For the present purposes it is sufficient to note that performance appeared to be equally impaired by noise (100 dB) with a high rate on one source or the same number of signals spread over three separate sources. A third condition, in which a low signal rate was employed on one source, showed no effects of the noise. The effects, in this case, were not on the number of signals detected, but on the confidence with which subjects made their decision after each event. Broadly speaking, noise caused subjects to make less use of the "not sure" category, and to increase their tendency either to report "certain signal" or "certain nonsignal" (normally in this kind of task a high proportion of decisions are accompanied by an uncertain response from the subject). Whether overall detection efficiency improves or deteriorates in noise may thus be seen to depend on the extent to which these uncertain responses are converted into detection responses, on the one hand, or denials of signal occurrence on the other. The data do, however, suggest an interesting new effect of noise on the mechanics of the decision-making process itself. We shall return to it later.

Although it is far from clear how this complex result relates to the more conventional findings which we have discussed, it does offer a possible interpretation of Jerison's one- and three-clock studies. Broadbent's data show that it is not necessary to have three separate sources, merely a high signal rate. Furthermore, in a similar study, Broadbent and Gregory [65] demonstrated that a drop rather than an increase in sure detections is expected in noise when the overall level of false reports of signals is high, as it is in Jerison's three-clock task: the reader is referred to Ref. 65 for a fuller discussion of this rather technical argument. From these experiments the importance of signal rate is quite clear, and independent of any effects of spatial complexity. In addition, there is the strong suggestion that, in general, situations which encourage a high incidence of false reporting are more likely to show an overall reduction in efficiency in noise. This might occur either if the costs associated with false reports are low, or in situations where the subject has a greater expectation of signals occurring (high signal rate tasks).

One other experiment which supports this general effect of signal rate is a study of visual checking by Davies and Hockey, mentioned earlier [52]. The overall effect of 95 dB(A) noise in this study [compared with a control level of 70 dB(A)] was one of facilitation, though the improvement was more pronounced for a low signal rate than for a high signal rate. Although the general effect of noise was opposite to that found by Broadbent, the two effects are quite consistent, in both cases being in the direction of increased impairment (or decreased facilitation) with higher signal rates. Lastly, in this section, a similar result has been shown by McGrath for an increase in event rate, rather than signal rate [55], and using varied auditory stimulation (VAS), rather than steady, broadband noise. Subjects had to detect an occasional increase in the brightness of a light which either came on for 1 sec and off for 2 sec (slow task), or on for $\frac{1}{3}$ and off for $\frac{2}{3}$ sec (fast task). The effect of VAS was to facilitate detection in the case of the slow version of the task, but to impair efficiency in the faster task.

Although there are a number of rather obvious differences between these various situations they all show a greater likelihood of performance being adversely affected when the subject has more to do in the task; all these situations may be expected to be associated with guessing or a lack of caution in reporting signals. They demonstrate the sometimes critical nature of certain changes in the structure or information-processing demands made by the task situation, in determining the precise form of the effect of changes in the noise environment. These considerations are taken one step further in the next section, where subjects are required to carry out tasks involving a hierarchy of demands; the job involves the efficient allocation of attention over a number of components.

14-4.3 Priorities of Attention

All that can be inferred from the experiments discussed in Section 14-4.2 is that performance is more likely to be impaired by noise in more complex or high information-load tasks. We have very little idea of how it has its effect. Most of the situations in which the effects of noise have been studied have required the subjects either to attend to a single source of information, or to process information equally from a number of sources. Even in these multi-source situations (such as Jerison's three-clocks and Broadbent's 20-dials tasks), efficient performance depends basically on the subject treating the display as one large source of information, since signals may occur equally often in all locations.

In real-life situations (from process-control monitoring down to the cooking of a complex evening meal) it is clear that people do not approach the performance of multi-component jobs in such a uniform way. Different parts of the task are not treated as of equal importance; significant information is more likely to become available in some places than in others. What this means is that there are normally priorities of attention in multi-source monitoring situations. These priorities determine which parts of the display are examined most, and depend, among other things, on the assessed likelihood of a certain component requiring checking, or on the long-term value of the information from each component— or, probably, a combination of the two. Such considerations are important in discussing the effects of noise, since, in practical terms, jobs which have to be carried out in loud noise are often of this kind. More fundamental, then, than the question of how overall efficiency is affected by noise, is what kinds of changes occur in the pattern of performance. Indeed, it is not possible to make any statements about overall efficiency unless we can compare these changes in pattern with the optimum profile of attention allocation for the job, as we shall see later.

We have been suggesting throughout that noise may generally be considered to produce an increase in the level of arousal. Some earlier research in other areas has indicated that arousal may be an important factor in determining the pattern of attention allocation in multiple-task situations. Broadly speaking, it has been found that procedures which are thought to increase arousal (stress, threat, danger, high incentives, strong emotional states, and so on) are associated with a restriction in the range of attention over the subcomponents of the task. The effect is normally one of maintaining adequate performance on the central (high priority) task at the expense of an impairment of less important components [5, 66, 67]. We shall refer to this restriction of attention deployment as an increase in selectivity, for reasons that will become apparent later: the pattern of attention in states of high arousal is more selective than it is at more normal levels. Since noise can be regarded as arousing, does it also change the pattern of attention in this way? I have recently been looking at this question, using a task meant to simulate the essential components of driving [67].

The basic situation used in these experiments was modeled on a task developed by Bursill [68] to examine effects of high ambient temperature on performance. It consists of a pursuit tracking task, around which are spaced six lamps, as in Fig. 14-5. The instructions to the subject emphasize that the tracking task is of high priority, which means that he should attempt to maximize his performance on this component, even if the secondary task suffers. The tracking task requires

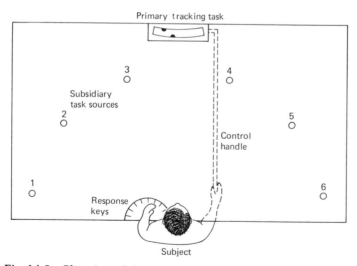

Fig. 14-5. Plan view of the dual-task situation. From Hockey [67].

the subject to follow a moving pointer (analogous to steering the vehicle along a road). While he is doing this the lamps in the secondary task display light up at random intervals at any of the locations (corresponding to secondary visual information in the driving situation). He is required to report any of these signals that he observes by pressing one of a panel of buttons. Each subject performed in both 70 and 100 dB(A) noise for sessions of 40 minutes.

The main interest in the experiments carried out with this task was in the way in which the balance of attention between the tracking task and the secondary monitoring task would be changed by loud noise. The results were rather more complicated than anticipated, but the effects are quite unequivocal. As can be seen from Fig. 14-6(a), tracking efficiency was maintained under the 100 dB(A) condition, whereas it showed a decrement over time under the quieter condition. This is clear evidence that the primary task is being dealt with, if anything, more efficiently in noise than in quiet. The data on the secondary task [Fig. 14-6(b)] show a rather complex effect of noise. Detection of centrally located signals is, like tracking, slightly improved, whereas signals appearing in the more peripheral locations are detected much less efficiently (although the data are not expressed here as a change over time, the same trend of an increasing effect of noise was evident). It appears, then, that noise has the same kind of effect on the selectivity of attention as do other arousal variables. It alters the way in which subjects structure their attention to the environment by increasing existing priorities and biases in behavior. One point which may not be clear, however, is why there should be an improvement in the detection of central lights, while those in the

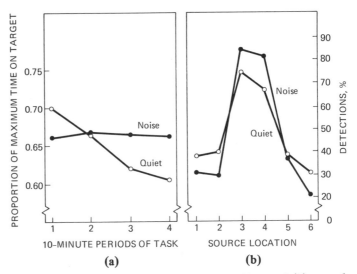

Fig. 14-6. Effects of noise on (a) primary task tracking, and (b) secondary task detection efficiency. From Hockey [67].

periphery are neglected; after all, they were all given the same low priority level. A further experiment [69], which specifically examined this point, has made it clear that it is because these signals are detected more frequently in the first place than those in the periphery. Subjects may thus give them a higher priority rating because they appear to be presented more often. It is not because noise causes a contraction in the visual field, for example, as has been suggested for effects of this type [68].

This is clearly confirmed by a number of other studies. First, when the peripheral signals are themselves given a high priority, the effects of noise are to improve detection of them [70]; a similar finding has been obtained for the effects of another arousal variable, threat of shock [71]. These results may help to clarify the cause of the unexpected facilitation found with separated sources by Samuel (discussed in Section 14-4.2). Spatial complexity by itself is not a sufficient condition for noise to interfere with performance—only in the case of relatively high information loads, when this results in the subject having to neglect some of the sources of information. It is, then, generally the less relevant components that will tend to suffer most. One further situation in which priorities were manipulated is a 3-source monitoring situation in which signals (described as "faults" in this case) were more likely to occur in some places than others (as do the faults in industrial monitoring situations). In this task subjects had to make an active sampling response in order to obtain a brief look at the state of a particular source (whether it contained a fault or not), but they are only allowed to examine one of the sources once every 2 sec; the task is based on one developed by Hamilton [72]. The main effect of noise was to increase the existing tendency for subjects to sample most often the source which was most likely to indicate a fault state. As in the abovementioned task, noise increased the bias toward the high priority activity. This effect could clearly not be attributed to a restriction of the peripheral visual field, since the three sources were arranged in the form of an equilateral triangle [73]. This is why the effect seems best described as an increase in selectivity, rather than a narrowing of attention or a restriction in the range of cue utilization, for example.

It will have been noticed that we have, as far as possible, avoided discussing the effects of noise on efficiency in this section. The results of experiments on attention priorities have made it clear that whether efficiency may even be improved depends on the appropriateness of the underlying increase in selectivity to the actual demands of the task. This observation goes some way toward explaining why noise may have such markedly different effects in monitoring situations. It will be recalled from our discussion of the relevant task variables in Section 14-4.2 that, although there is no clear-cut divison between tasks, those which are spatially complex (involving a number of sources) or which present information at a high rate are more likely to suffer impairment. Facilitative effects, on the other hand, occur generally on simple tasks (a single source of information or a low signal rate). These effects may be understood in terms of the underlying changes in selectivity. Simple tasks will clearly benefit from an increased tendency to attend to high priority aspects, since this applies to the whole task. Where a number of equally important sources have to be monitored, on the other hand, the increase in selectivity is unlikely to benefit efficient performance, since none of the components can be safely disregarded. In the case of high signal rates, the impairment may be more directly attributable to an effect on decision-making policy, as implied in the Broadbent findings mentioned earlier, though it is likely that the two effects are not unrelated. In fact, in the 3-source sampling task described above, both effects were observed—an increase in selectivity and a decrease in the number of cautious responses (in this case, double checks on the presence of a fault). Broadent [5] and Hockey [73] discuss the possible relationship between the two effects.

Recent interest has been focused on the effects of noise in a different kind of task situation, involving more cognitive operations such as memory and verbal information processing. Before attempting to integrate the wide range of findings discussed in this review, we shall examine findings from this area. Tasks of this type tend to be fairly brief, and noise levels are rarely above 90 dB(A). Nevertheless, a number of interesting effects of noise has been found, and there are indications that these may even be accounted for by the same general kind of explanation as that found useful for interpreting effects on continuous work situations, namely changes in selectivity due to an increase in arousal level.

14-4.4 Cognitive Tasks

It should perhaps be said that the main interest in the research reported here has been less in obtaining practical information about the effects of noise than in using noise-induced arousal changes in order to examine the fundamental nature of memory and perceptual processes. Even so, a number of useful facts have emerged which may well have practical value.

One major area is the effects of noise on verbal learning and memory. Berlyne and co-workers [74] examined the efficiency of learning a list of paired associates as a function of noise level (either 58 or 72 dB of white noise, or no noise at all). This task requires subjects to learn which items are paired together in a list, so that they are able to provide the appropriate response when provided with each stimulus word. Immediate learning and recall of the response words was less efficient at 72 than at 58 dB (though this latter condition was better than no noise at all). In another study, however, when recall was tested the next day, subjects who had received noise during training did significantly better on the delayed test. Similar effects (impairment with immediate recall but improvement with delayed recall as a result of receiving noise during the initial presentation session) have been reported also by McClean [75]. As with the findings on selectivity, the same kinds of effects have been observed for other arousing treatments (see Ref. 76 for a review of these), and an explanation has been put forward in terms of the consolidation process involved in learning [77]. On this view, states of high arousal are more effective in laying down the memory trace in the brain, but, at the same time, somehow make the stored information less accessible for immediate use. There are a number of objections to this interpretation, as we shall see later. For the moment it is sufficient to note that, in another of Berlyne's experiments [78], no such effect on long-term memory was observed if noise was present for a period after the training phase, a procedure which would certainly be expected to speed up consolidation of learning. A further point about this study is that no interference with immediate recall was observed, even though the noise level of 75 dB had been found to affect this aspect of performance in Berlyne's experiment.

An effect of noise on a complex cognitive task was demonstrated by Broadbent [79]. Subjects had to carry out a difficult mental arithmetic task with a heavy memory component. Here, 100 dB noise had a slight effect in slowing down the rate at which subjects could carry out the problem, though it did not result in an increase in errors. This result is strongly reminiscent of some of the effects of nonmasking noise on speech communication, which we discussed in Section 14-3.2. It is probable that the two effects have a common basis, in terms of an effect of greater difficulty of taking in the information, irrespective of modality. Some support for this interpretation comes from a similar study of Woodhead's [28]. She found that noise (in this case 1 sec bursts of 110 dB rocket noise) resulted in increased errors if the bursts occurred while subjects were in the process of memorizing the numbers, but actually speeded up performance if they occurred in the calculation stage. The difference in detail between the two sets of findings is accounted for by the fact that in Woodhead's task the rate of presentation was externally controlled, while in Broadbent's it was controlled by the rate of work of the subject. If the detrimental effect of noise is on memorizing and not on speed of processing of the stored information, as we have suggested, it is evident that this will be seen as an increase in errors in the externally paced task and a slowing of work rate in the self-paced task.

Although the picture is not entirely clear there is, then, a fair amount of evidence to suggest that operations involving the immediate recall of material will be impaired by noise. Broadbent's filter theory (see Section 14-1.4) is able to account for this quite well, since memory and attention were regarded as intimately related processes, both susceptible to interference in stimulus intake by competing distraction. The filter theory is, however, less able to explain the apparent enhancement of long-term memory observed in some of these studies. One clue about the kinds of changes that occur with noise is provided by an experiment of Hörmann and Osterkamp [80]. Essentially, they found a correlation between the performance change in noise and the ability to disregard irrelevant information, as measured by the Stroop color word test [81]. The degree of organization in the recall of a list of words was higher in noise for those subjects who could ignore the irrelevant information in the Stroop test, but reduced for those who could not do this easily. This implies a change in the information-structuring process brought about by noise. Information appears to be processed in a more structured way in noise when subjects are already operating selectively. The opposite effect may occur, however, in subjects who are not efficiently attending to the relevant stimuli. Clearly, much more needs to be known about such effects. There are a number of instances in the literature of intriguing individual differences of this sort being found in noise. Unfortunately, they have rarely been followed up and clarified.

Assuming that noise does produce some change in the information structuring process, one can ask whether this effect occurs only at the retrieval stage of long-term recall, or, as the monitoring findings indicate, at an earlier (input) stage in the system. Such an effect may, for example, be simply the result of a bias in perceptual selection, and may be evident in tasks which do not demand new learning or memorizing. There is some evidence for this interpretation.

Broadbent and Gregory [82], using a tachistoscopic presentation technique, showed that certain categories of visual stimuli (e.g., words printed in black) could be recognized more quickly in the presence of 100 dB broadband noise than under control (70 dB) conditions. This occurred only when subjects were instructed to report one of two words presented together (one black and one red), and not when they were told to report either word presented on its own. The facilitation due to noise may thus be seen to be one of increased ability to ignore the irrelevant (red) word, and not simply an overall effect on the speed of recognition (retrieval of appropriate verbal responses from memory). A similar result was obtained by Houston [83] for the effect of variable noise on the Stroop test mentioned earlier. Noise improved the speed of naming the colors in which words were printed when the words were, in fact, conflicting color words ("green" printed in blue ink). Noise had no effect, however, on the other two tasks in this test: naming colors, or reading color words printed in black. The effect, again, appears to occur at the stimulus selection stage, rather than at the response stage.

Their findings are, of course, very similar to those observed in Section 14-4.3. Noise appears to increase perceptual selectivity by biasing attention toward high priority aspects of the stimulus. Two final experiments in this section offer further support to this interpretation, and shed some light on the interaction of the effects of noise on memory and the time at which recall is tested. Hamilton, Hockey, and Quinn [76] used four separate groups of subjects who were required to learn a set of ten paired associates (adjectives) either in noise (80 dB broadband noise) or quiet (60 dB). The list pairs were either kept in a constant order from trial to trial or randomized, as is usual in such studies. Recall scores were better in noise when the order remained the same but were worse when order was changed. The result suggests that subjects were structuring their attention to the list more in noise, by learning the order of the pairs as well as the appropriate responses. This cue is normally useless, of course, when the order changes randomly, and actually impairs recall by providing false information about the correct response in this situation. For both kinds of presentation, however, the noise condition resulted in fewer wrong guesses than occurred in quiet; i.e., subjects tended either to give a correct response or not to give one at all. This may, again, be regarded as a change in the decision-making component of performance, noise seeming to reduce the number of doubtful responses made by subjects. It will be recalled that this kind of effect has also been found in the context of monitoring behavior [63, 65, 73].

One perhaps oversimplified generalization that can be made from this result is that noise produces a more "all or none" strategy for taking in information, by focusing attention more effectively on a limited sample of the task stimuli. These items are thus better learned and less subject to interference or decay. This interpretation helps us to understand the opposite effects of short- and long-term memory, without having to resort to the difficult concept of consolidation. If consolidation alone were responsible for the effect of noise on memory then the same pattern of performance change would be expected to occur in both presentation conditions of the above experiment; the actual findings strongly implicate a change in information selection strategies used by subjects. This view is also upheld by the outcome of a further experiment by Hockey and Hamilton [84]. Subjects were shown a sequence of eight words, one at a time, in the form of slides, and asked to report the words in their correct order by writing them down. The number of words correctly recalled was, in fact, slightly higher in 85 dB noise than in a 60 dB control condition, though the difference was not significant. An incidental feature of the task, however, was that words were presented in different spatial positions on the slides. When subjects were unexpectedly tested on this aspect of the stimulus, performance was significantly worse in the noise condition. The finding has recently been confirmed by Davies and Jones [85]. In other words, noise can, once again, be considered to result in increased selectivity, by focusing attention more onto the central or relevant parts of the task (in this case the word list itself). The effect on doubtful responses was again evident in the data.

14-4.5 Further Environmental and Social Considerations

Adverse working conditions are rarely simple. In particular, a factory which has a high ambient noise level is also likely to be uncomfortably hot, or require men to work for long unbroken sessions. In addition, men are often under additional stress through an inability of the body to adapt to shift work schedules or resultant loss of sleep. It is clearly important not only to know how noise affects working efficiency, but how the effect depends on other factors such as these. It would not matter if one could assume that all stressful conditions had the same effect (i.e., lowering efficiency) since these effects would simply add together. We have seen from the evidence reviewed so far that noise itself does not have a simple effect on efficiency, and it is also true that other stresses differ in the manner in which they affect performance. It is therefore important to examine the changes in efficiency that occur when noise is combined with other working conditions. Most of the relevant work on this problem comes from the Applied Psychology Unit at Cambridge, England, principally using the five-choice serial reaction task discussed in earlier sections [46]. A comprehensive review of such effects is provided in the recent book by Poulton [87], though we will summarize the main findings here.

We have already stated (Section 14-4.1) that the effect of noise alone on this task is to increase errors (or sometimes the number of long response times, or gaps), but not output. These effects increase with time at work. Loss of sleep is known also to impair efficiency on this task, though the effects of this stress are more widespread. One might therefore suppose that the joint effects of the two stresses is to impair efficiency even more, but studies by Corcoran [48] and by Wilkinson [47] have shown this not to be the case. In fact, noise produced fewer errors after subjects had lost a night's sleep than when they had slept normally. The two effects tend to cancel each other out. Even more surprising is the effect of incentives. Normally, efficiency is greater under incentives, on all measures [86], so one would expect this procedure to be effective in removing some of the impairment which occurs with noise. Wilkinson showed, however, that performance on this task in noise was worse in the presence of incentives than under control conditions [47]. This rather surprising result may have important consequences for industrial work, since incentives and bonus schemes are common in jobs where very loud noise is present. Broadbent and Little [88] found a great reduction in breakages in a film processing plant when the noise level was reduced from 98 to 90 dB by acoustic treatment in one of the bays. A major reason for the originally high level of breakages may well have been the high bonus rates operating in this situation. The men were thus under some pressure to maintain a high output. This is, possibly, the only field study of the effects of noise which is sufficiently carefully controlled to be of any general value. Even so, Kryter [3] makes the point that the reduction in noise level may have improved performance by allowing the operators to use the auditory cues of the film threading process more efficiently, rather than by some more fundamental change in efficiency. Ongoing industrial situations do not readily lend themselves to the methodological rigor necessary to demonstrate results of an unequivocal nature.

It is possible to account for these results, up to a point, by referring back to the arousal interpretation of noise (see Section 14-1.4). If noise is arousing and sleep loss de-arousing (since it makes one less alert) then the two stresses in combination will tend to balance these opposing changes in the state of the person. Incentives can also be regarded as arousing, but their effect seems to be to bring the level of arousal up to an optimal point for maximum efficiency (there is evidence that people normally do not work with maximum effort). It is not difficult to see, then, that if performance is already poor because of too high a level of arousal, it is likely to become even worse when incentives are added. The additional pressure to work quickly may have the effect of focusing attention too much onto each component or movement, and result in errors in the integration of task performance. This is more likely, of course, in the case of complex tasks, such as that involved in film processing, where a whole sequence of intricate operations has to be carried out, in this case in darkness.

There are good reasons for doubting that the arousal process does not operate quite as simply as this explanation would suggest [5]. However, the model does provide a convenient way of discussing the complex interactions of stresses observed, and, as far as it goes, is an acceptable simplified version of the underlying mechanism relating excitation to action. What these results do suggest is that incentives and sleep loss affect the same basic process as noise, since the separate effects are not simply additive [86]. This does not appear to be the case with all stresses. Heat, for example, appears to have an effect which is independent of that of noise [86, 87], and, in any case, affects efficiency right at the beginning of the work period as well as later. Alcohol also appears to have an independent effect, though it is in some ways similar in its effect on attention distribution [89]. Much more research is needed in this area before we have a clear picture of how the effects of noise depend on other environmental variables.

One further consideration, of course, is the interaction between the effects of noise and the social environment in which these effects are examined. Azrin [90] has argued that subjects in noise experiments may develop stimulus and response dependencies: either the subjects associate the task with an unpleasant experience, and make less of an effort to perform well, or the noise may induce greater effort toward proficient performance if the subjects believe that this will be instrumental in removing the distraction. Considerations of this sort are not likely to account for the bulk of findings in the areas of research we have been discussing, though they are clearly of some importance in assessing effects of noise reduction in industrial settings (the Hawthorne effect). Kryter [3], rather naively, puts forward this kind of argument in order to dismiss effects of noise on efficiency, but it clearly carries little weight: it is merely an ad hoc suggestion with minimal experimental foundation. It may be true that subjective reactions to the presence of noise, and the social environment in which studies are carried out, are relevant factors in determining the kinds of effect obtained in the study, just as the state of fatigue or intoxication of the subject may be. The value of this approach, however, can only be assessed by experiments designed to examine this class of effects independently. Since Azrin's paper in 1958 very little systematic work of this kind has been done, although a recent monograph by Glass and Singer [91] does attempt to look at the role of variables such as the degree of perceived control over the noise. In general, effects of noise are less pronounced when subjects are told they can terminate the noise at any time, though little information is given about whether noise per se has an effect on performance. It is probable that, under certain conditions, noise will either impair or improve efficiency, and will do so as a direct consequence of its effect on the central arousal process. On the other hand, it is also likely that the way in which the subject reacts to the presence or absence of noise in the task will be a relevant factor in the manner of its execution. It is important to realize that these are two separate classes of independent variable.

14-4.6   Individual Differences

The question of the way in which the subject reacts to the noise is examined more directly in the study of individual differences in the effects of noise on efficiency. From a practical point of view it is apparent that some individuals are better able to withstand noise than others, and a reliable means of identifying such persons would be of considerable value in personnel selection.

It has, in fact, proved difficult to isolate major sources of individual variation. Surprisingly enough, for example, there is no firm evidence that the unaffected individuals are those who are least annoyed by the presence of noise. The most useful evidence has come from studies in which differences in temperament or personality have been assessed. The personality variable of "introversion-extraversion" may be measured quite reliably using an inventory-type test [92, 93]. Extraversion is a measure of the extent to which a person's behavior is outward-oriented and impulsive (extraverted) or inward-looking and careful (introverted), though these are, of course, only the extremes of a continuous distribution of temperament. Generally speaking, extraverted subjects are more likely to show effects of noise, and this appears to be the case whether the overall effect of noise is either an impairment or a facilitation of efficiency [94]. Extraverts tend to suffer a greater decrement with prolonged, monotonous work, and this decrement has been shown to be either increased [47] or decreased [52] in noise. Introverted subjects do not appear to be seriously affected in either direction, and it is possible that these subjects are better able to maintain a stable level of efficiency in the face of changes in stimulation [94].

The state of knowledge in this area is very incomplete, though there is no shortage of theories to explain temperament differences of this kind. Eysenck [95] has suggested that extraverts are in a state of stimulus hunger, and are continually involved in seeking out additional stimulation from their surroundings. This view is supported by the findings of a number of studies in which subjects are allowed some control over the noise environment [94, 96]. Extraverts, in these experiments, regularly select both higher levels and more frequent periods of noise than introverts. This factor is probably quite an important one for determining which individuals will tend to suffer most from a noisy environment, or, indeed, benefit most from the introduction of low-level sound, or music, into an otherwise monotonous task situation. What is needed, here, though, are experiments in which efficiency is examined as a function of increases or decreases in noise level, with respect to that selected by a particular individual as optimal. It cannot be assumed that the selected level is the one that will produce the most efficient performance.

Other personality measures have been largely unsuccessful in accounting for individual differences in the effect of noise. No differences are apparent in the efficiency of older and younger workers, nor between men and women. Anxiety-proneness, or neuroticism, may be expected to result in greater impairment in noise, since feelings of irritation and annoyance are very common in such environments. There is some evidence that this is the case, but other studies have reported that anxious subjects are, in fact, less prone to interference. As Kryter [3] rightly points out, many of these experiments suffer from poor control of other factors, such as differences in the initial ability on the task, and, in general, no clear inferences can be drawn from them. Intelligence level, too, might be thought to be an important factor, but several studies have found no differences in the effects of noise on efficiency between normal and mentally retarded subjects (see Ref. 3 for a critical discussion of these studies).

One of the problems with this area of research is that the tasks used have often been of the kind that do not show reliable effects of noise at all (simple reaction time, problem solving, etc.). Since the identification of individual differences is known to be a difficult problem it is clearly necessary to use tasks which are sensitive in the first place. On the positive side, the variable of extraversion does appear to provide a fairly useful guide to which individuals are likely to suffer most from exposure to noise, even though there are still many gaps in our understanding of the precise nature of the relationship.

14-5 CONCLUSIONS

14-5.1 Summary of the Facts

It will now be apparent that no simple conclusions can be drawn about the effects that noise has on efficiency. On the other hand, we would clearly disagree strongly with Kryter [3], who concludes that "other than as a damaging agent to the ear and as a masker of auditory information, noise will not harm the organism or interfere with mental or motor performance" (p. 587). A careful examination of the literature has revealed a fairly clear pattern of performance change under noise, provided that the task being carried out is sensitive to changes in the basic state of the subject. In this final section we shall attempt to summarize the basic facts which have emerged, and present a tentative guide to noise criteria that may be applied in practical situations. In doing this there is, of course, a risk of overgeneralizing, but the loss that this may entail in theoretical precision is probably offset by the gain in practical utility. At some stage it is always necessary to put theory into practice, and this can only be accomplished easily if the conclusions are sufficiently general.

    a. Noise may improve or impair efficiency, but in either case the effects are more likely to occur later in the work session than immediately.

    b. Adverse effects normally occur with complex, multi-component tasks, or those in which the information load is high.

    c. Improvement can occur in simple, routine operations, normally associated with boredom and loss of attention.

    d. Effects of steady, broadband noise are rare with SPLs of less than 90 dB, though changes in other noise parameters can affect performance at lower levels.

    e. Cognitive tasks, such as those involving memory, may be affected at SPLs as low as 70 or 80 dB, and after only brief periods of work.

    f. The comprehension of spoken communication may be impaired at noise levels which do not result in a reduction in intelligibility.

    g. The effect on cognitive processing is to increase concentration during the reception of information, and to reduce the immediate availability of the material. Immediate recall may suffer, whereas long-term recall may be facilitated.

    h. The most fundamental effect of noise is to bias attention more strongly toward important or subjectively relevant aspects of the task.

    i. Subjects may become more decisive in noise, in that they are subjectively more confident about the adequacy of a decision, even though the evidence may not warrant such confidence.

To these broad conclusions, we may add the suggestion that extraverted individuals are more likely to show effects of noise in their behavior, though the full implications of this have not been explored.

Such a summary of the facts may be useful for two reasons. First, they help in the selection of a suitable theory of noise effects, since the value of a theory will necessarily be judged primarily by its success in taking account of the principal findings of the area. Secondly (and this is more important for the present purposes), they provide a working guide to the kinds of situations which are likely to be at risk in the presence of noise, and enable the plant manager or factory inspector to be more objective in their assessment of the problem. We shall not pursue the search for a theory in this review, though we can regard the arousal–selectivity theory as an adequate account of the underlying process, as far as it goes. Instead, we shall attempt to provide more practical suggestions for the kind of noise control that is required.

14-5.2 Some Practical Implications

The main aim of this chapter has been to present a sufficiently detailed survey of effects of noise on work efficiency to allow genuinely useful and valid practical recommendations to be made. What this amounts to, in essence, is an answer to the question "How much noise is too much?" It is unfortunate that no clear-cut answers can be given to such a question, since research on noise has been anything but systematic with regard to the manipulation of the parameters of sound. Experiments have been almost exclusively concerned with simply showing that noise does affect performance. It is only in recent years, as this review has attempted to make clear, that investigators have even worried about the nature of the task which subjects have been given to do. The suggestions offered below are, then, to a certain extent subjective, though they are generally in accord with the available evidence. The maximum level for efficient work can be seen to depend on the job requirements, on the kind of noise present, and possibly on the temperaments of the people involved. We will limit the rest of this discussion to a consideration of the two factors about which most agreement has been reached: task factors and noise factors.

Task Factors. Two main considerations are important here: noise is more likely to impair complex tasks than simple ones, and tasks involving a high component of cognitive processing are affected at lower levels than operations of a primarily perceptual nature, such as monitoring. There seems little doubt that where a job demands extensive use of memory and decision-making, noise levels should be kept as low as possible, and in any case not higher than, say, 80 dB. For less intellectually demanding jobs, such as inspection work or simple motor skills, levels up to 100 dB should not be harmful, unless a number of components are involved in the operation, or the operator is working at maximum capacity. Examples of such demanding situations are plentiful, but air traffic control, film processing, or certain kinds of process control are all cases where levels of 90 dB or more may well be a serious problem. A more familiar example is that of driving, where high noise levels are common (say, inside the cabs of large trucks). Driving is a complex monitoring and multi-component motor skill situation, with a hierarchy of built-in priorities. Steering is a major priority, since if it is not carried out with near 100% efficiency everything else is irrelevant. Instrument checking, mirror use, and control operations, down to route finding and conversation with passengers, are lower priorities, with the ordering of these depending on the circumstances of the external situation. Effects of high noise levels, from the findings discussed in Section 14-4.3, are likely to leave steering unimpaired, though the speed of response to vehicles emerging unexpectedly from side roads, or children running into the road to collect balls may be impaired. These implications have not been checked in real-life settings (e.g., by careful survey of accident statistics), but, if it turns out that this is the kind of pattern of impairment that can be expected, some allowance should perhaps be made for it in training programs.

Noise Factors. It seems likely that an increase in the level of broadband noise will have less serious consequences for efficiency than an increase in the level of variable or unpredictable noise, though surprisingly little is known about this. Basically, effects of continuous, broadband noise are rare below 90 dB, whereas variable or unpredictable noise may affect behavior at much lower levels. On the other hand, most studies using this latter kind of noise have used fairly simple tasks, and usually found an improvement in performance. What is needed are studies which use complex tasks, and which manipulate the level of variable noise up to, say, 100 dB. It is likely that, for the same task, the maximum level for efficient work is lower for variable than for steady noise, though it is not possible to be more precise. The effects of regular, intermittent noise (e.g., that produced by a machine) will depend on its temporal relationship to the information processing structure of the job. Bursts of noise in conjunction with the arrival of task information are more likely to disrupt performance than noise which is out of phase with the task, though this, again, may well depend on whether the information has to be memorized for immediate use. If not, as in the case of an article delivered for inspection, a possible benefit may even accrue.

A Final Comment. These are fairly general recommendations, based on reasonably firm evidence. There is no doubt that noise criteria must take into account the actual job being done, and the characteristics of the noise in question. What is not clear, since there are virtually no data on the question, is whether these laboratory-based findings, using relatively unpracticed tasks, do, in fact apply to real-life, ongoing work situations. Do people adapt to noise over a long period, so that it no longer affects them? Can they be trained to use different strategies to counteract any effects of noise that do occur? Do they adjust their behavior naturally, as a function of feedback in the situation? These are important practical questions, to which no answers are as yet available.

REFERENCES

1. K. D. Kryter, *J. Speech Hearing Disorders*, Monograph Supplement 1, 1950.

2. D. E. Broadbent, "Effects of Noise on Behavior." In C. M. Harris, ed., *Handbook of Noise Control*, McGraw-Hill, New York, 1957.

3. K. D. Kryter, *The Effects of Noise on Man*. Academic Press, New York, 1970.

4. D. E. Broadbent, *Perception and Communication*. Pergamon Press, London, 1958.

5. D. E. Broadbent, *Decision and Stress*. Academic Press, London, 1971.

6. D. O. Hebb, *Psychol. Rev.*, **62**, 243-253, 1955.

7. D. R. Davies and G. S. Tune, *Human Vigilance Performance*. Staples Press, London, 1970.

8. M. E. Hawley and K. D. Kryter, "Effects of Noise on Speech." In C. M. Harris, ed., *Handbook of Noise Control*, McGraw-Hill, New York, 1957.

9. R. Plutchik, *Psychol. Bull.*, **56**, 133–151, 1959.

10. G. A. Miller, G. A. Heise, and W. Lichten, *J. of Exp. Psychol.*, **41**, 329–335, 1951.

11. I. Pollack and H. Rubenstein, *Language and Speech*, **6**, 57–62, 1963.

12. P. M. A. Rabbitt, *Psychonomic Sci.*, **6**, 380–383, 1966.

13. C. Holloway, Proceedings of the Fifth International Symposium on Human Factors in Telephony, 1970.

14. J. J. B. Morgan, *Amer. J. Psychol.*, **28**, 191–208, 1917.

15. E. E. Cassell and K. M. Dallenbach, *Amer. J. Psychol.*, **29**, 129–142, 1918.

16. A. Ford, *Amer. J. Psychol.*, **41**, 1–32, 1929.

17. H. B. Hovey, *Amer. J. Psychol.*, **40**, 585, 1928.

18. W. H. Teicher, E. Arees, and R. Reilly, *Ergonomics*, **6**, 83–97, 1962.

19. R. W. Shoenberger and C. M. Harris, *J. Engineering Psychol.*, **4**, 108–119, 1965.

20. C. Landis and W. A. Hunt, *The Startle Pattern*. Farrar and Rhinehart, New York, 1939.

21. E. N. Sokolov, *Perception and the Conditioned Reflex*. Pergamon Press, London, 1963.

22. D. E. Berlyne, *Conflict, Arousal and Curiosity*. McGraw-Hill, New York, 1960.

23. K. G. Pollock and F. C. Bartlett, Industrial Health Research Board, Report 65, Part I, HMSO, London, 1932.

24. D. E. Broadbent, *Ergonomics*, **1**, 21–29, 1957.

25. J. V. Grimaldi, *Ergonomics*, **2**, 34–73, 1958.

26. M. M. Woodhead, *Brit. J. Indust. Med.*, **15**, 120–125, 1958.

27. M. M. Woodhead, *J. Acoust. Soc. Amer.*, **31**, 1329–1331, 1959.

28. M. M. Woodhead, *Amer. J. Psychol.*, **77**, 627–633, 1964.

29. S. A. Fisher, Personal communication.

30. D. N. May and C. G. Rice, *J. Sound Vib.*, **15**, 197–202, 1971.

31. A. Weinstein and R. S. Mackenzie, *Percept. Motor Skills*, **22**, 498, 1966.

32. McCroskey, School of Aviation Medicine Research, Farnborough, England, Report No. NM/18/02/99, Sub. 1, no. 7, 1957.

33. C. D. Frith, *Brit. J. Psychol.*, **58**, 127–131, 1967.

34. H. D. Warner, *Human Factors*, **11**, 245–250, 1969.

35. H. D. Warner and N. W. Heimstra, *Human Factors*, **14**, 181–185, 1972.

36. W. H. Watkins, *J. Exp. Psychol.*, **67**, 72–75, 1964.

37. N. Loveless, J. Brebner, and P. Hamilton, *Psychol. Bull.*, **73**, 161–199, 1970.

38. S. S. Stevens, et al. OSRD Report No. 274, Part I, Harvard University, 1941.

39. M. S. Vitles and K. R. Smith, *Trans. Amer. Soc. Heating Ventilating Eng.*, **52** (1291), 167, 1946.

40. K. Craik, *The Nature of Explanation*. Cambridge University Press, Cambridge, 1946.

41. D. E. Broadbent, MRC/APRU Research Report No. 160-51, 1951.

42. D. E. Broadbent, *Quart. J. Exp. Psychol.*, **6**, 1–5, 1954.

43. H. J. Jerison, *J. Appl. Psychol.*, **43**, 96–101, 1959.

44. H. G. Broussard, R. Y. Walker, and E. E. Roberts, AMRI Research Report No. 6-95-20-001, 1952.

45. M. Loeb and G. Jantheau, *J. Appl. Psychol.*, **42**, 47–49, 1958.

46. J. A. Leonard, MRC/APRU Research Report No. 326-59, 1959.

47. R. T. Wilkinson, *J. Exp. Psychol.*, **66**, 332–337, 1963.

48. D. W. J. Corcoran, *Quart. J. Exp. Psychol.*, **14**, 178–182, 1962.

49. J. J. McGrath, Human Factors Research Inc., Los Angeles, Technical Report No. 6, 1960.

50. R. F. Kirk and E. Hecht, *Percept. Motor Skills*, **16**, 553–560, 1963.

51. W. N. McBain, *J. Appl. Psychol.*, **45**, 309–317, 1961.

52. D. R. Davies and G. R. J. Hockey, *Brit. J. Psychol.*, **57**, 381–389, 1966.

53. C. Tarrière and A. Wisner, *Travail Humain*, **25**, 1–28, 1962.

54. A. F. Sanders, *Ergonomics*, **4**, 253–258, 1961.

55. J. J. McGrath, "Irrelevant Stimulation and Vigilance Performance." In D. N. Buckner and J. J. McGrath, eds., *Vigilance: A Symposium*. McGraw-Hill, New York, 1963.

56. H. J. Jerison and S. Wing, WADC Technical Report No. 57-14, 1957.

57. H. J. Jerison and R. Wallis, WADC Technical Report No. 57-206, 1957.

58. N. H. Mackworth, "Researches on the Measurement of Human Performance." HMSO, London, 1950.

59. W. M. S. Samuel, *Quart. J. Exp. Psychol.*, **16**, 264–267, 1964.

60. H. M. Jenkins, *Amer. J. Psychol.*, **71**, 647–661, 1958.

61. J. F. Mackworth, *Vigilance and Attention*. Penguin, London, 1970.

62. W. P. Colquhoun, *Ergonomics*, **4**, 41–52, 1961.

63. D. E. Broadbent and M. Gregory, *Human Factors*, **7**, 155–162, 1965.

64. W. P. Tanner and J. A. Swets, *Psychol. Rev.*, **61**, 401–509, 1954.

65. D. E. Broadbent and M. Gregory, *Brit. J. Psychol.*, **54**, 309–323, 1963.

66. A. D. Baddeley, *Brit. J. Psychol.*, **63**, 537–546, 1972.

67. G. R. J. Hockey, *Quart. J. Exp. Psychol.*, **22**, 28–36, 1970.

68. A. Bursill, *Quart. J. Exp. Psychol.*, **10**, 113–129, 1958.

69. G. R. J. Hockey, *Quart. J. Exp. Psychol.*, **22**, 37–42, 1970.

70. G. R. J. Hockey, Ph.D. Dissertation, University of Cambridge, 1969.

71. D. M. Cornsweet, *J. Exp. Psychol.*, **80**, 110–118, 1969.

72. P. Hamilton, *J. Exp. Psychol.*, **82**, 34–37, 1969.

73. G. R. J. Hockey, *J. Exp. Psychol.*, **101**, 35–42, 1973.

74. D. E. Berlyne, D. M. Borsa, M. H. Craw, R. S. Gelman, and E. E. Mandell, *J. Verbal Learning Verbal Behav.*, **4**, 291–299, 1965.

75. P. D. McClean, *Brit. J. Psychol.*, **60**, 57–62, 1969.

76. P. Hamilton, G. R. J. Hockey, and J. G. Quinn, *Brit. J. Psychol.*, **64**, 112–129, 1973.

77. E. L. Walker and R. D. Tarte, *J. Verbal Learning Verbal Behav.*, **3**, 112–129, 1963.

78. D. E. Berlyne, D. M. Borsa, J. H. Hamacher, and I. D. V. Koenig, *J. Exp. Psychol.*, **72**, 1–6, 1966.

79. D. E. Broadbent, *J. Acoust. Soc. Amer.*, **30**, 824–827, 1958.

80. H. Hörmann and H. Osterkamp, *Z. Exp. Angew. Psychol.*, **13**, 31–38, 1966.

81. A. R. Jenson and W. D. Rohwer, *Acta Psychol.*, **25**, 36–93, 1966.

82. D. E. Broadbent, Personal communication.

83. B. K. Houston, *J. Exp. Psychol.*, **82**, 403–404, 1969.

84. G. R. J. Hockey and P. Hamilton, *Nature*, **226** (5248), 866–867, 1970.

85. D. R. Davies and D. Jones, Personal communication.

86. D. E. Broadbent, *Quart. J. Exp. Psychol.*, **15**, 205–211, 1963.

87. G. C. Poulton, *Environment and Efficiency*. Thomas, Springfield, 1970.

88. D. E. Broadbent and E. A. J. Little, *Occupational Psychol.*, **34**, 133–140, 1960.

89. P. Hamilton and A. Copeman, *Brit. J. Psychol.*, **61**, 149–156, 1970.

90. W. H. Azrin, *J. Exp. Anal. Behav.*, **1**, 183–200, 1958.

91. D. C. Glass and J. E. Singer, *Urban Stress*. Academic Press, New York, 1972.

92. A. Heron, *Brit. J. Psychol.*, **47**, 243–257, 1956.

93. H. J. Eysenck and S. S. G. Eysenck, *Eysenck Personality Inventory*, University of London Press, London, 1963.

94. G. R. J. Hockey, *J. Sound Vib.*, **20**, 299–304, 1972.

95. H. J. Eysenck, *The Biological Basis of Personality*. Thomas, Springfield, 1967.

96. D. R. Davies, G. R. J. Hockey, and A. Taylor, *Brit. J. Psychol.*, **60**, 453–457, 1969.

Appendices

1. BASIC ACOUSTICS

To assess noise, one must know what it is. This appendix is a short introduction to acoustics for readers who have picked up this book without prior acoustics training. Given here are the elements of noise generation, measurement, propagation and control. This information may be supplemented, as the reader advances to noise assessment, by an introduction to subjective acoustics (Chapter 1) and a description of the human hearing mechanism (Chapter 11).

Noise Generation

Sound—and noise is just an expression for "unwanted" sound—is the sensation we experience when we perceive the vibrations of air* particles on our eardrums.† The vibrations of these particles give rise to fluctuations in air pressure above and below the prevailing atmospheric pressure. This fluctuating air pressure is the physical characteristic of sound we most often refer to.

*Vibrations in liquids and solids can also transmit sound—for example, we can hear when underwater.

†We can also hear, though not very efficiently, from bone conduction through the skull.

Sound is generated, therefore, by whatever causes air pressure fluctuations or air particle vibrations in the first place. This may be the vibrating panels of a piece of machinery, causing the air next to the panel to vibrate, which in turn causes adjacent air particles to vibrate until eventually the air particles close to the eardrum vibrate also. It may, on the other hand, be the fluctuating pressure in a vehicle exhaust, or the air disturbance from the successive close passages of fan blades past a fixed object, or the turbulence in air caused by unsteady combustion, or the turbulence that results when a fast-moving airstream mixes with a slower one.

For the air pressure fluctuation to be audible, two conditions must be satisfied in addition to the obvious one that the receiver of the sound should be able to hear. First, the frequency of the fluctuating air pressure—i.e., the rate at which the fluctuations occur—should have components within the audible frequency range of the ear. Second, the amplitude of the pressure fluctuations should be high enough in level to exceed the threshold of hearing at the frequency involved.

Frequency is the physical characteristic of a sound that makes us feel it is low or high in pitch. It is measured in Hertz (abbreviated Hz), which is the number of cycles of pressure fluctuation per second. Most sounds have irregularly fluctuating pressures—i.e., the air pressure does not vary sinusoidally with time—but these irregular fluctuations can be represented as the sum of a number of regular (sinusoidal) fluctuations of different frequencies, the component frequencies mentioned above. Chapter 1 further describes the frequency analysis of sound and its importance to our subjective responses.

The amplitude of the pressure fluctuation is known as sound pressure level when expressed in the right units (see "Noise measurement" below). The ear can perceive sound pressures over an enormous range. The threshold of hearing is, at 1000 Hz, around 20 μPa (or 2×10^{-5} N/m^2 or 2×10^{-4} μ bar), while we can also hear (though pain is near) pressures about a million times higher.

Noise Measurement

The most basic noise measurement is the determination of sound pressure level. A microphone is a device which converts air pressure fluctuations to electrical voltage variation; it is placed at the point where the sound pressure is to be measured and the electrical voltage, after processing (see below), is displayed on a meter calibrated in units of sound pressure level.

The accepted unit of sound pressure level (SPL) is the decibel (dB), defined by

$$SPL = 10 \log \left(\frac{p^2}{p_{ref}^2} \right)$$

where p is the root mean square (rms) of the pressure difference from atmospheric, and p_{ref} is a reference pressure. The fluctuating pressure is processed electrically to yield its rms value, the rms averaging being chosen because it gives a positive value; in contrast, the arithmetic mean of the pressure differences from atmospheric would be an unhelpful zero. The reference pressure p_{ref} can be given any nonzero positive value. By specifying for p_{ref} a value of pressure that people can just hear, 20 μPa, the SPL has the convenient value of 0 dB near the threshold of hearing at the particular reference frequency of 1000 Hz. Sound pressure levels using this reference pressure have the unit dB re 20 μPa though the reference pressure is not always quoted.

An rms sound pressure a thousand times higher than 20 μPa (see the preceding section, "Noise generation") now has the value 60 dB, while an rms pressure a million times higher has the value 120 dB. Use of the decibel scale achieves the purpose of describing sounds throughout the enormous dynamic range of the ear in numbers that are manageable. The decibel scale also bears a relationship to the relative subjective magnitudes of the sound pressures involved (see Chapter 1).

A sound level meter (SLM) is a device which, properly operated, measures sound pressure level. Improper operation includes the use of the meter outside the wind, temperature, and humidity conditions allowed, though surrounding the microphone with a windshield increases the maximum acceptable wind speed.

The rms sound pressure signal can also be processed to give frequency-weighted sound levels having a relationship to the ear's frequency response, or to give the sound pressure levels in a number of frequency bands. Numerous other analyses can also be performed.

Noise measurements have most meaning when performed as part of an accepted test procedure. Appendix 2 contains a list of the more important noise measurement procedures, as well as a list of standards governing sound measurement accuracy.

Noise Propagation

Sound generated by a given source does not have the same sound pressure level at all distances, even for sources whose sound output is equal in all directions. The sound field around a source depends on the directivity of its output and on the extent to which the sound pressure falls off with distance and other attenuating factors, or is increased by reflections.

Sound is propagated when an air particle close to the source disturbs one further away, and so on. A pressure disturbance propagates in this way at approximately the speed of sound, reaching distant points after a time interval. The front of this disturbance is known as a wavefront and is similar to the wavefront which occurs when one drops a stone into water and a wave advances radially from the point of disturbance.

The energy in this sound wave is proportional to the area that the wavefront occupies and the square of the sound pressure. It follows that, even if there is no sound energy dissipated by excess attenuation (see below), the pressure must decrease at increasing distances from the source as the area of the wavefront becomes larger and larger.

The distance attenuation thus occurring has a value of 6 dB per distance doubling from a point source. Thus a source that is small compared with one's distance from it (a primitive but adequate definition of a point source) will have sound pressure levels that reduce by 6 dB every time one doubles one's distance from it. If 50 ft from the source the level is 80 dB, then another 50 ft away, at 100 ft, it will be 74 dB; and a further 100 ft away, at 200 ft, it will be 68 dB.

The figure of 6 dB is a consequence of the spherical wavefront which occurs with a point source, and the figure holds true for a source in free space or near the ground. A line source, on the other hand, which is a source or series of sources spread in a line that is rather longer than one's distance from it, has a distance attenuation of 3 dB per distance doubling. As one moves away from a finite line source it becomes more and more like a point source, and the figure of 3 dB gradually changes to 6 dB.

To the attenuation due to this divergence of the sound wave, one must add the effects of other attenuating or amplifying influences. Such other attenuations are known as excess attenuations, and are due to atmospheric influences on the one hand and contact with solid objects in the other. The atmospheric influences are absorption in the air, with or without fog or precipitation, the effects of wind and temperature gradients, and changes in the product of air density and speed of sound (a product of basic importance in all physical acoustics). To these minor, but often significant, excess attenuations must be added that due to absorption of sound by the ground and by vegetation, that due to obstructions like walls and buildings, and that due to the fact that sound waves may reach a point by more than one path and may, to a degree, cancel one another.

Sound amplification can also occur, the result for example of a reflection from a solid surface adding to the sound pressure from the direct wave.

The complexities of accounting for these effects when trying to assess the noise of a source results in the approach of operating the source in the controlled conditions of special measurement rooms. Such rooms are of two types. One is an anechoic ("no echoes") room, whose surfaces are highly sound absorbent so that, in theory at least, all of the sound is direct, i.e., comes from the source without reflection, and is subject to little atmospheric dissipation in the distances involved. Since there are no reflections, these rooms provide a means of studying the directivity of a source.

The other type of room is a reverberant one, whose surfaces are highly reflective so that the sound at any point in the room has got there from multiple reflections as well as directly from the source. The sound field in the room is therefore diffuse (has approximately the same qualities everywhere), which permits the source's overall output to be studied without needing several microphones to account for its directivity.

These rooms are sometimes specified in the test procedures listed in Appendix 2.

Noise Control

Noise control may be achieved at the source, in the propagational path, or at the receiver.

Noise may be reduced at the source by altering the amount of air vibration generated there. This may involve reducing the amplitude of structural vibrations in a machine by eliminating their cause, isolating them, or damping them—or reducing the area of solid surface that so vibrates. The frequency of the force generating the vibration, or the frequency response of the part that vibrates, may be altered to reduce the amplitude of the resultant vibration or to shift the frequency of vibration into an insensitive region of the audible frequency range. If the noise is caused by air turbulence, ways may be sought of reducing the scale and intensity of the turbulence by reducing airflow velocities, by reducing duct dimensions (though avoiding velocity increases), by the use of turning vanes, by avoiding lift fluctuations, by steadying combustion, and so on. Probably also falling within the term noise control at the source are the orientation of the source so that its maximum noise output occurs in the least sensitive direction, and the siting of the source furthest from sensitive points of reception. Noise can also be reduced at the source by operating controls—hour restrictions, engine or vehicle speed or acceleration limits—and occasionally by improved attention to maintenance.

Noise control may be achieved in the propagational path through the use of partial barriers (e.g., screens and fences) or total barriers (e.g., an airtight sound enclosure round a machine). The first of these is only partially effective, and depending on the situation may be very expensive. The second of these is usually effective, but may not be feasible in the case of sources needing large amounts of air for cooling or combustion, or requiring easy access for maintenance or material feed, or when the source is a moving one that would suffer from carrying the enclosure with it. Airflows can, however, be handled through absorptively lined ducts or the reactive mufflers common in motor vehicles.

Particularly when source and receiver are both in one room, the use of absorptive material on the surfaces of the room and the use of screens may also be effective noise control measures. If the source is inside the room and the receivers are not, the walls of the room may serve as an enclosure if thick enough and reasonably airtight—windows may need to be double-glazed with an appreciable air gap, and be kept closed.

Noise control at the receiver, if the receiver is indoors and the source is not in the same room, may be accomplished in the ways described in the paragraph above. The ultimate noise control at the receiver is the use of ear protection, as described in Chapter 11.

2. ACOUSTICAL STANDARDS

The two international standards organizations of most importance in acoustics are the International Organization for Standardization (known as ISO) and the International Electrotechnical Commission (IEC). A list of those of their standards which are relevant to noise assessment is given in this appendix, bearing in mind that noise measurement is a fundamental part of such a task.

ISO and IEC standards may be ordered from each nation's standards organizations; a list of the more important of these addresses is also given below. These international standards are sometimes identical to (i.e., an international endorsement of) a particular national standard. Sometimes, however, they are different, and within each country the national standard is often the more accepted. Thus in the U.S. particularly the American National Standards Institute (ANSI) has at least as much influence as ISO or the IEC, and so various relevant ANSI standards are listed below too.

In addition to international and national standards organizations, various industrial and professional institutes and associations have their own acoustical standards or test methods. Among these the Society of Automotive Engineers (SAE) is important, so relevant SAE documents are listed. A number of other association addresses are also given for reference, as well as the addresses of U.S. Government agencies that issue acoustical standards as part of their procurement practices.

International Organization for Standardization (ISO)

Address orders to the relevant national standards organization (addresses follow).

Note: ISO documents prefixed "R" are Recommendations; they may later become Standards.

ISO 31/VII-1965	Quantities and units of acoustics.
ISO 16-1975	Standard tuning frequency (standard musical pitch).
ISO/R 131-1959	Expression of the physical and subjective magnitudes of sound or noise.
ISO/R 140-1960	Field and laboratory measurements of airborne and impact sound transmission.
ISO/R 226-1961	Normal equal-loudness contours for pure tones and normal threshold of hearing under free field listening conditions.
ISO 266-1975	Preferred frequencies for acoustical measurements.
ISO/R 354-1963	Measurements of absorption coefficients in a reverberation room.
ISO/R 357-1963	Expression of the power or intensity levels of sound or noise.
ISO/R 362-1964	Measurement of noise emitted by vehicles.
ISO 389-1975	Standard reference zero for the calibration of pure-tone audiometers.
ISO 454-1975	Relation between sound pressure levels of narrow bands of noise in a diffuse field and in a frontally-incident free field for equal loudness.

ISO/R 495-1966	General requirements for the preparation of test codes for measuring the noise emitted by machines.
ISO/R 507-1970	Procedure for describing aircraft noise around an airport.
ISO 532-1975	Method for calculating loudness level.
ISO/R 717-1968	Rating of sound insulation for dwellings.
ISO/R 1680-1970	Test code for the measurement of the airborne noise emitted by rotating electrical machinery.
ISO/R 1761-1970	Monitoring aircraft noise around an airport.
ISO/R 1996-1971	Assessment of noise with respect to community response.
ISO 1999-1975	Assessment of occupational noise exposure for hearing conservation purposes.
ISO 2151-1972	Measurement of airborne noise emitted by compressor (primemover-units intended for outdoor use).
ISO 2204-1973	Guide to the measurement of airborne acoustical noise and evaluation of its effects on man.
ISO 2249-1973	Description and measurement of physical properties of sonic booms.
ISO 2922-1975	Measurement of noise emitted by vessels on inland water-ways and harbors.
ISO 2923-1975	Measurement of noise on board vessels.
ISO 3095-1975	Measurement of noise emitted by railbound vehicles.
ISO/TR 3352-1974	Assessment of noise with respect to its effect on the intelligibility of speech.
ISO 3741-1975	Determination of sound power levels of noise sources—precision methods for broad-band sources in reverberation rooms.
ISO 3742-1975	Determination of sound power levels of noise sources—precision methods for discrete-frequency and narrow-band sources in reverberation rooms.

International Electrotechnical Commission (IEC)

Address orders to the relevant national standards organization (addresses follow).

IEC 50-08 (1960)	International electrotechnical vocabulary—electro-acoustics.
IEC 123 (1961)	Recommendations for sound level meters.
IEC 177 (1965)	Pure tone audiometers for general diagnostic purposes.
IEC 178 (1965)	Pure tone screening audiometers.
IEC 179 (1973)	Precision sound level meters (includes supplement 179A).
IEC 303 (1970)	IEC provisional reference coupler for the calibration of earphones used in audiometry.
IEC 318 (1970)	An IEC artificial ear, of the wide band type, for the calibration of earphones used in audiometry.

U.S.A.: American National Standards Institute

1430 Broadway
New York, NY 10018

ANSI S1.1-1960 (R1971)	Acoustical terminology (including mechanical shock and vibration).
ANSI S1.2-1962 (R1971)	Method for physical measurement of sound.
ANSI S1.4-1971	Specification for sound level meters.
ANSI S1.6-1967 (R1971)	Preferred frequencies and band numbers for acoustical measurements.
ANSI S1.7-1970 (R1975)	Method of test for sound absorption of acoustical materials in reverberation rooms.
ANSI S1.8-1969 (R1974)	Preferred reference quantities for acoustical levels.
ANSI S1.10-1966 (R1971)	Method for calibration of microphones.
ANSI S1.11-1966 (R1971)	Specifications for octave, half-octave and third-octave band filter sets.
ANSI S1.12-1967 (R1972)	Specifications for laboratory standard microphones.
ANSI S1.13-1971	Methods for the measurement of sound pressure levels.
ANSI S1.21-1972	Methods for the determination of sound power levels of small sources in reverberation rooms.
ANSI S1.23-1976	Method for the designation of sound power emitted by machinery and equipment.
ANSI S2.2-1959 (R1971)	Methods for the calibration of shock and vibration pickups.
ANSI S2.10-1971	Methods for analysis and presentation of shock and vibration data.
ANSI S2.11-1969 (R1973)	Selection of calibrations and tests for electrical transducers used for measuring shock and vibration.
ANSI S3.1-1960 (R1971)	Criteria for background noise in audiometer rooms.
ANSI S3.2-1960 (R1971)	Method for measurement of monosyllabic word intelligibility.
ANSI S3.3-1960 (R1971)	Methods for measurement of electroacoustical characteristics of hearing aids.
ANSI S3.4-1968 (R1972)	Procedure for the computation of loudness of noise.
ANSI S3.5-1969 (R1973)	Methods for the calculation of the Articulation Index.
ANSI S3.6-1969 (R1973)	Specifications for audiometers.
ANSI S3.7-1973	Method for coupler calibration of earphones.
ANSI S3.8-1967 (R1971)	Method of expressing hearing aid performance.
ANSI S3.13-1972	Artificial head-bone for the calibration of audiometer bone vibrators.
ANSI S3.17-1975	Method for rating the sound power spectra of small stationary noise sources.
ANSI S3.19-1974	Method for the measurement of real-ear protection of hearing protectors and physical attenuation of earmuffs.

ANSI S3.20-1973	Psychoacoustical terminology.
ANSI S5.1-1971	Test code for the measurement of sound from pneumatic equipment.
ANSI S6.2-1973	Exterior sound level for snowmobiles.
ANSI S6.3-1973	Sound level for passenger cars and light trucks.
ANSI S6.4-1973	Definitions and procedures for computing the Effective Perceived Noise Level for flyover aircraft noise.
ANSI S9.1-1975	Guide for the selection of mechanical devices used in monitoring acceleration induced by shock.

Australia: Standards Association of Australia

Standards House
80–86 Arthur Street
North Sydney, NSW 2060

Canada: Canadian Standards Association

178 Rexdale Blvd
Rexdale, Ontario M9W 1R3

France: Association Française de Normalisation

Tour Europe
Cedex 7
92080 Paris–La Défense

Federal Republic of Germany: Deutscher Normenausschuss

4–7, Burggrafenstrasse
1 Berlin 30

India: Indian Standards Institution

Manak Bhavan
9 Bahadur Shah Zafar Marg
New Delhi 110001

Italy: Ente Nazionale Italiano di Unificazione

Piazza Armando Diaz 2
1 20123 Milano

Japan: Japanese Industrial Standards Committee

Ministry of International Trade and Industry
3-1, Kasumigasekil, Chiyodaku
Tokyo

The Netherlands: Nederlands Normalisatie-Institut

Polakweg 5
Rijswijk (ZH)-2108

Sweden: Sveriges Standardiseringskommission

Box 3295
S-103 66 Stockholm 3

Switzerland: Association Suisse de Normalisation

Kirchenweg 4
Postfach 8032 Zurich

United Kingdom: British Standards Institution

2 Park Street
London W1A 2BS

USSR: Gosudarstvennyj Komitet Standartov

Soveta Ministrov SSSR
Leninsky Prospekt 9b
Moskva 11 70 49

Society of Automotive Engineers (SAE)

400 Commonwealth Drive
Warrendale, PA 15096

SAE J34	Exterior sound level measurement for pleasure motorboats.
SAE J47	Maximum sound level potential for motorcycles.
SAE J57	Sound level of highway truck tires.
SAE J87	Exterior sound level for powered mobile construction equipment.
SAE J88a	Exterior sound level measurement for powered mobile construction equipment.

SAE J184	Qualifying a sound data acquisition system.
SAE J192a	Exterior sound level for snowmobiles.
SAE J247a	Instrumentation for measuring acoustic impulses.
SAE J331a	Sound levels for motorcycles.
SAE J336a	Sound level for truck cab interior.
SAE J366b	Exterior sound level for heavy trucks and buses.
SAE J377	Performance of vehicle traffic horns.
SAE J919a	Sound level measurements at operator station.
SAE J952b	Sound levels for engine powered equipment.
SAE J986a	Sound level for passenger cars and light trucks.
SAE J1046	Exterior sound level measurement procedure for small engine powered equipment.
SAE J1060	Subjective rating scale for evaluation of noise and ride comfort characteristics related to motor vehicle tires.
SAE J1074	Engine sound level measurement procedure.
SAE J1077	Exterior sound level for trucks with auxiliary equipment.
SAE J1105	Performance, test and application criteria of electrically operated forward warning horn for mobile construction machinery.

Air-Conditioning and Refrigeration Institute (ARI)

1815 N. Ft Meyer Drive
Arlington, VA 22209

Air Moving and Conditioning Association (AMCA)

30 West University Drive
Arlington Heights, IL 60004

American Society for Testing and Materials (ASTM)

1916 Race Street
Philadelphia, PA 19103

American Society of Heating, Refrigerating and Air-Conditioning Engineers (ASHRAE)

United Engineering Center
345 East 47th Street
New York, NY 10017

Association of Home Appliance Manufacturers (AHAM)

20 N. Wacker Drive
Chicago, IL 60606

Compressed Air and Gas Institute (CAGI)

122 East 42nd Street
New York, NY 10017

Diesel Engine Manufacturers Association (DEMA)

2130 Keith Building
Cleveland, OH 44115

General Services Administration (GSA)

Criteria and Research Branch
GSA Building
19th and F Street NW
Washington, DC 20405

Institute of Electrical and Electronics Engineers (IEEE)

445 Hoes Lane
Piscataway, NJ 08854

Motor Vehicle Manufacturers Association of the U.S.A. (MVMA)

320 New Center Building
Detroit, MI 48202

National Electrical Manufacturers Association (NEMA)

815 15th Street, Suite 438
Washington, DC 20005

U.S. Military Specifications

Naval Publications and Forms Center
5801 Tabor Avenue
Philadelphia, PA 19120

3. REGULATING AGENCIES

The following is a list of the agencies in the U.S. federal government that are concerned with noise. In the U.S. as in other countries, an environmental protection agency has been created but has not been given the entire authority for noise abatement. The transportation agencies continue to preserve at least some authority in their spheres of influence, particularly in aviation, and the labor or health authorities may have responsibility for occupational hearing protection. As a general rule it is sufficient to address a first enquiry to the "Department of the Environment" in most other governments.

Lower levels of government may also have some part in noise abatement. In the U.S., the states and cities for a time led the way in this field, but there is increasing federal preemption. In Canada and Australia, the provinces and states, respectively, have significant responsibility, and this has sometimes been delegated to municipalities. In Britain the central government assumes most responsibility; this is true also of France and Sweden.

International organizations are also taking their first steps toward unification of environmental policies. Both the Organization for Economic Co-operation and Development (OECD) and the European Economic Community (EEC) have sponsored noise studies.

Environmental Protection Agency (EPA)

Office of Noise Abatement and Control, 401 M Street SW, Washington DC 20460

The EPA also has regional offices at the following addresses:

JFK Building Room 2113, Boston, MA 02203.
26 Federal Plaza Room 907G, New York, NY 10007.
Curtis Building Room 225, 6th and Walnut Streets, Philadelphia, PA 19106.
1421 Peachtree St NE, Room 109, Atlanta, GA 30309.
230 S. Dearborn, Chicago, IL 60604.
1600 Patterson St., Room 1107, Dallas, TX 75201.
1735 Baltimore St., Kansas City, MO 64108.
1860 Lincoln St., Suite 900, Denver, CO 80203.
100 California St., San Francisco, CA 94111.
1200 Sixth Ave Room 11C, Seattle, WA 98101.

Department of Transportation (DOT)

Washington DC 20590

DOT agencies include the

Federal Highway Administration (FHWA)
Federal Railroad Administration (FRA)
Urban Mass Transportation Administration (UMTA)
400 7th Street SW
Washington DC 20590

and the

Federal Aviation Administration (FAA)
800 Independence Ave SW
Washington DC 20591

Department of Labor

Occupational Safety and Health Administration (OSHA)
Washington DC 20210

U.S. Government reports are sold by the National Technical Information Service (NTIS), U.S. Department of Commerce, 425 13th Street NW, Washington DC 20004, and by the Superintendent of Documents, U.S. Government Printing Office, Washington DC 20402. Federal legislation is published in the *Federal Register*.

4. USEFUL ACOUSTICS PERIODICALS

Journal of the Acoustical Society of America

American Institute of Physics
335 East 45th St
New York, NY 10017

Journal of Sound and Vibration

Academic Press Inc
111 Fifth Ave
New York, NY 10003
and also
24-28 Oval Rd
London NW1 7DX

Noise Control Engineering

9 Saddle Road
Cedar Knolls, NJ 07927

Noise Control Report

Business Publishers Inc
P.O. Box 1067
Blair Station
Silver Spring, MD 20910

Noise/News

Institute of Noise Control Engineering
P.O. Box 3206
Arlington Branch
Poughkeepsie, NY 12603

Noise Regulation Reporter

Bureau of National Affairs Inc
1231 25th NW
Washington DC 20037

Proceedings of Inter-Noise (annual conference)

Institute of Noise Control Engineering
P.O. Box 3206
Arlington Branch
Poughkeepsie, NY 12603

Proceedings of Noisexpo (annual conference)

Acoustical Publications Inc
27101 E. Oviatt Rd
Bay Village, OH 44140

Sound and Vibration

Acoustical Publications Inc
27101 E. Oviatt Rd
Bay Village, OH 44140

5. GLOSSARY OF ACOUSTICAL TERMS

This is a glossary of the acoustical terms most frequently used in this book. Explanations rather than definitions are used in order to avoid highly technical terminology which could confuse those without acoustics experience. Readers may obtain further explanations for these terms by using the index to find the various pages on which a given term appears. The text of the book may discuss these terms in contexts making for easier understanding.

In using this glossary to refer to terms beginning with either "noise" or "sound," bear in mind that these words are often interchangeable, so that a term not found under one of these words may be found under the other.

Absorption. A process, occurring generally at a solid surface but also in a fluid, resulting in sound energy being dissipated. When absorption occurs at a solid surface, an incident sound wave is not fully reflected.

AI. See *Articulation Index.*

Ambient Sound Level. The sound level in the absence of the sound under study.

Annoyance. In acoustics, the overall unwantedness of sound heard in a real-life (as opposed to laboratory) situation.

Articulation Index. An index which describes the suitability of a noise environment for distinguishing speech signals.

Attenuation. Reduction in sound level.

Audiogram. A graph showing, as a function of frequency, the amount in decibels by which a person's threshold of hearing differs from a standard.

Audiometry. The measurement of hearing.

Audiometer. An instrument for measuring the sensitivity of hearing.

A-Weighted Sound Level. A level of sound pressure in which the sound pressure levels of the various frequency bands have been weighted to accord roughly with human aural system frequency sensitivity. The A weighting is defined in standards such as IEC 179 (1973) and ANSI S1.4-1971. (Similarly for B, C, and D weightings.)

Background Noise. Noise other than that from the source being studied.

Bandwidth. The size of a frequency band.

CNEL. See *Community Noise Equivalent Level.*

CNR. See *Composite Noise Rating.*

Community Noise Equivalent Level. An index which describes the noise of aircraft flyovers, or community noise generally, over 24 hours.

Composite Noise Rating. An index which describes the noise in a period of aircraft flyovers.

Continuous Noise. Noise which continues without interruption over a period of time.

Criterion. A standard by which a noise is judged. For example, criteria for an acceptable noise are that it causes no annoyance, interferes with no form of behavior, and damages no being or object.

Day–Night Sound Level. A statistical descriptor of the sound over a 24-hour period taking account of the fact that sounds are more annoying at night than during the day. Calculated by determining the equivalent sound level over a 24-hour period after adding 10 dB(A) to the sound levels occurring in the period 10 pm to 7 am.

dB. See *Decibel.*

dB(A). The unit of A-weighted sound level. (Similarly, dB(B), dB(C), dB(D).)

dB(NP). The unit of Noise Pollution Level.

Deafness. Complete impairment of hearing.

Decibel. In acoustics, the unit of level for sound pressure squared, sound power, sound intensity, etc. Most commonly used for sound pressure level, when it is ten times the logarithm to the base ten of the ratio of rms sound pressure squared and a reference pressure squared.

Direct Wave. The sound wave arriving at a point without prior impingement on a solid object.

Directivity. The directional nature of the sound from a given source.

Drive-By Test. A test in which the source is driven in a prescribed manner past a fixed measuring position.

Effective Perceived Noise Level. A level describing the noisiness of a single aircraft flyover assessed at a number of points near an airport.

EPNdB. Unit of Effective Perceived Noise Level.

EPNL. See *Effective Perceived Noise Level.*

Equivalent Continuous Perceived Noise Level. An index which describes the noise in a period of aircraft flyovers.

Equivalent Sound Level. The level of a constant sound having the same sound energy as an actual time-varying sound over a given period. An energy-averaged sound level, usually but not always of the A-weighted energy.

Excess Attenuation. Attenuation of sound with distance other than that caused by the divergence of the sound wave.

Frequency. The time ratio of repetition of the sound pressure. The characteristic of a sound which influences our perception of it as high or low in pitch.

Frequency Band. A continuum of frequencies from a lower frequency to a higher.

Hertz (Hz). Unit of frequency.

Impact Noise. See *Impulse Noise.*

Impulse (Impulsive) Noise. Noise of short duration, typically less than a second. Also called transient or impact noise.

Infrasound. Sound of frequencies below the normal human audible range, often taken to begin at 16 Hz.

Intermittent Noise. Noise which falls below measurable levels one or more times over a given period.

L_{dn}. See *Day-Night Sound Level.*

L_{eq}. See *Equivalent Sound Level.*

Limits. Prohibitions, legal or otherwise, on the amount of sound. Amount may be expressed in terms of sound level, sound exposure, sound duration or any other characteristic of the sound output.

L_n. The sound level exceeded $n\%$ of the time over a given period. Thus L_1, L_{10}, L_{50}, L_{90}.

Loudness. The subjective magnitude of sound.

Masking. The process by which one sound has a reduced loudness when heard in the presence of another.

Misfeasance. The doing of a lawful act in a wrongful manner. In acoustics, the infliction of a sound which could be avoided, attenuated, confined to certain hours, etc., and therefore a product of the mindlessness of the noisemaker.

NC. See *Ranking Curves*.

NCA. See *Ranking Curves*.

NEF. See *Noise Exposure Forecast*.

NNI. See *Noise and Number Index*.

Noise. Unwanted sound. However, "sound" and "noise" are often used virtually synonymously.

Noise and Number Index. An index which describes the noise in a period of aircraft flyovers.

Noise Dose. See *Sound Exposure*.

Noise Exposure Forecast. An index which describes the noise in a period of aircraft flyovers.

Noisiness. The unwantedness of a sound heard in isolation from real-life situations. Also known as perceived noisiness.

Noise Reduction Coefficient. An index of the sound absorptivity of a material. Calculated by averaging the sound absorption coefficients at the frequencies 250, 500, 1000, and 2000 Hz.

Noise Pollution Level. An index describing the noise at a given point over a period of time, which takes account of the equivalent sound level and the standard deviation of the sound level over the period in question.

Notional Background Level. A background sound level established without measurement.

Noy. A unit of noisiness.

NPL. See *Noise Pollution Level*.

NR. See *Ranking Curves.*

Octave Band. A band of frequencies the highest frequency of which has twice the value of the lowest.

One-Third Octave Band. A band of frequencies the highest frequency of which is $(2)^{1/3}$ or $10^{0.1}$ greater than the lowest. There are three such bands in an octave band.

Phon. A unit of loudness.

PNC. See *Ranking Curves.*

Preferred Speech Interference Level. An index which describes the suitability of a noise environment for distinguishing speech signals. Calculated by averaging the sound pressure levels in the octave bands centered at 500, 1000, and 2000 Hz.

PSIL. See *Preferred Speech Interference Level.*

Psychoacoustics. The science of investigating acoustical matters from the standpoint of psychology.

Pure Tone. A sound of a single (i.e., discrete) frequency. Perceived as a "pure" note, e.g., whine, buzz, ring, squeal.

Ranking Curves. Curves which evaluate the suitability of a sound spectrum experienced inside a room or a vehicle. Include the NC, PNC, NR, and NCA curves.

Reverberation. The continuation of a sound in an enclosed space after the source has stopped; a result of reflections from the surfaces of the enclosure.

Reverberation Time. The time taken for the sound pressure level in a room to decay by 60 dB after the source has stopped.

SIL. See *Preferred Speech Interference Level.*

Sone. A unit of loudness.

Sonic Boom. The impulse sound from a supersonic flyover of an aircraft.

Sound. A fluctuation of particle displacement or pressure, particularly one resulting in a hearing sensation.

Sound Exposure. The cumulative acoustic stimulation at the ear of a person or persons over a period of time. Also known as Noise Dose when the exposure of one individual is described.

Sound Level. A frequency-weighted sound pressure level. Assumed to be the A-weighted sound level unless otherwise stated.

Sound Pressure Level. A level of sound pressure expressed in decibels.

Sound Wave. A disturbance consisting of vibrating particles moving through a medium.

Speech Interference Level. See *Preferred Speech Interference Level.*

SPL. See *Sound Pressure Level.*

Threshold of Hearing. The minimum sound pressure level that a person can hear in certain prescribed conditions.

TNI. See *Traffic Noise Index.*

Tone Correction. A number to be added to a scale of noisiness to account for the noisiness of pure tones.

Traffic Noise Index. An index which describes the noise of traffic over a period of time.

Transient Noise. See *Impulse Noise.*

Ultrasound. Sound of frequencies above the normal human audible range, which is often taken to end at 20 kHz.

Waveform. The graph of pressure vs. time at a point through which a wave passes.

Index

Index

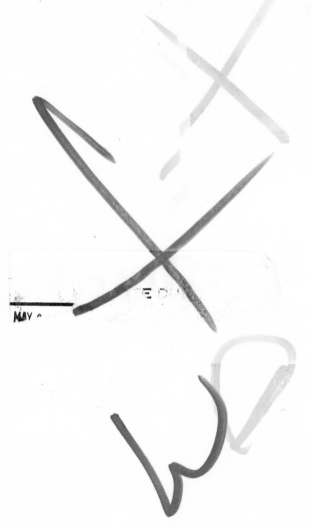

914.61